CAMBRIDGE LIBRARY COLLECTION

Books of enduring scholarly value

Life Sciences

Until the nineteenth century, the various subjects now known as the life sciences were regarded either as arcane studies which had little impact on ordinary daily life, or as a genteel hobby for the leisured classes. The increasing academic rigour and systematisation brought to the study of botany, zoology and other disciplines, and their adoption in university curricula, are reflected in the books reissued in this series.

Lake Superior

Written by Swiss-born geologist and explorer Louis Agassiz (1807–73), this 1850 publication was the first detailed scientific account of the natural phenomena of Lake Superior. Agassiz, who became a professor at Harvard and later founded the Harvard Museum of Comparative Zoology, was the first scientist to suggest that the earth had experienced an ice age. In the summer of 1848 he led an expedition of his students to Lake Superior, to examine the northern shores, which had previously received very little attention from scientists. The artist James Elliot Cabot (1821–1903), who was included in the party, wrote the 'narrative' of the tour to accompany the scientific report, and this makes up the first part of the work. The rest of the book describes the geological phenomena and zoological distribution in and around the lake, comparing it with similar regions of the world.

Cambridge University Press has long been a pioneer in the reissuing of out-of-print titles from its own backlist, producing digital reprints of books that are still sought after by scholars and students but could not be reprinted economically using traditional technology. The Cambridge Library Collection extends this activity to a wider range of books which are still of importance to researchers and professionals, either for the source material they contain, or as landmarks in the history of their academic discipline.

Drawing from the world-renowned collections in the Cambridge University Library, and guided by the advice of experts in each subject area, Cambridge University Press is using state-of-the-art scanning machines in its own Printing House to capture the content of each book selected for inclusion. The files are processed to give a consistently clear, crisp image, and the books finished to the high quality standard for which the Press is recognised around the world. The latest print-on-demand technology ensures that the books will remain available indefinitely, and that orders for single or multiple copies can quickly be supplied.

The Cambridge Library Collection will bring back to life books of enduring scholarly value (including out-of-copyright works originally issued by other publishers) across a wide range of disciplines in the humanities and social sciences and in science and technology.

Lake Superior

Its Physical Character, Vegetation, and Animals

Louis Agassiz

CAMBRIDGE UNIVERSITY PRESS

Cambridge, New York, Melbourne, Madrid, Cape Town,
Singapore, São Paolo, Delhi, Tokyo, Mexico City

Published in the United States of America by Cambridge University Press, New York

www.cambridge.org
Information on this title: www.cambridge.org/9781108032551

This edition first published 1850
This digitally printed version 2011

ISBN 978-1-108-03255-1 Paperback

G. Elliot Cabot from nat.

Lith of Sonrel

Löffler on stone.

LAKE TERRACES.

LAKE SUPERIOR:

ITS

PHYSICAL CHARACTER, VEGETATION, AND ANIMALS,

COMPARED WITH THOSE OF OTHER AND SIMILAR REGIONS.

BY

LOUIS AGASSIZ.

WITH A NARRATIVE OF THE TOUR,

BY

J. ELLIOT CABOT.

AND

CONTRIBUTIONS BY OTHER SCIENTIFIC GENTLEMEN.

ELEGANTLY ILLUSTRATED.

BOSTON:

GOULD, KENDALL AND LINCOLN,

59 WASHINGTON STREET.

1850.

PREFACE.

THE main object of the excursion, the results of which are given in the following pages, was a purely scientific one, viz.: the study of the Natural History of the northern shore of Lake Superior. Another end proposed by Professor Agassiz, was, to afford to those of the party who were unaccustomed to the practical investigation of natural phenomena, an opportunity of exercising themselves under his direction.

The party was composed of the following gentlemen: Prof. Agassiz and Dr. William Keller, instructors, and Messrs. George Belknap and Charles G. Kendall, students, of the Lawrence Scientific School; Messrs. James McC. Lea, George H. Timmins, and Freeman Tompkins, of the Dane Law School; Messrs. Eugene A. Hoffman, Charles G. Loring, Jonathan C. Stone, and Jefferson Wiley, of the senior class of Harvard College; Messrs. Joseph P. Gardner and J. Elliot Cabot, of Boston; Drs. John L. Le Conte and Arthur Stout, of New York; and M. Jules Marcou, of Paris.

Interspersed throughout the Narrative are reports, carefully made at the time, of the Professor's remarks on various points of Natural History, that seemed to him likely to interest a wider circle than those more particularly addressed in the second part of the book, which consists of papers on various points connected with the Natural History of the region, written, where not otherwise specified, by Prof. Agassiz. This portion of the work, however, does not aim at a mere detail of facts, but is intended to show the bearing of these facts upon general questions.

The Landscape Illustrations are taken from sketches made on the spot, by Mr. Cabot. Those of the Second Part were drawn and lithographed by Mr. Sonrel, a Swiss artist of much distinction in this branch, and formerly employed by Prof. Agassiz at Neuchatel, but now resident in this country.

Boston, March, 1850.

CONTENTS.

I. NARRATIVE.

CHAP. I.

BOSTON TO THE SAULT DE ST. MARIE.

CHAP. II.

THE SAULT TO FORT WILLIAM.

CHAP. III.

FORT WILLIAM BACK TO THE SAULT.

CHAP. IV.

FROM THE SAULT HOMEWARDS.

II. NATURAL HISTORY.

I.

THE NORTHERN VEGETATION COMPARED WITH THAT OF THE JURA AND THE ALPS.

II.

OBSERVATIONS ON THE VEGETATION OF THE NORTHERN SHORES OF LAKE SUPERIOR.

III.

CLASSIFICATION OF ANIMALS FROM EMBRYONIC AND PALÆOZOIC DATA.

IV.

GENERAL REMARKS UPON THE COLEOPTERA OF LAKE SUPERIOR.

BY DR. JOHN L. LECONTE.

V.

CATALOGUE OF SHELLS, WITH DESCRIPTIONS OF NEW SPECIES.

BY DR. A. A. GOULD.

VI.

FISHES OF LAKE SUPERIOR COMPARED WITH THOSE OF THE OTHER GREAT CANADIAN LAKES.

ILLUSTRATIONS.

I. LANDSCAPES.

II. NATURAL HISTORY.

ERRATA.

Page 10, Note, for *Tocelyn* read *Jocelyn.*

Page 31, for *Sault to Michipicotin* read *Sault to Fort William.*

Page 58, for *Juniperus Virginianus* read *J. virginiana.*

Page 384, first line of list, the specific names *cedrorum* and *cacalotl* should exchange places.

LAKE SUPERIOR.

NARRATIVE.

CHAPTER I.

BOSTON TO THE SAULT DE ST. MARIE.

We left Boston on the 15th of June, 1848, at 8 A.M., in the cars for Albany. The weather was warm, and we were well powdered with dust, when, at about 6 P.M., we arrived at the ferry on the Hudson. The Western appears to be more exposed to this nuisance of dust than the other railroads, probably from the many cuts through banks of crumbling clay and gravel. We were interested to hear that a contrivance for watering the track had been proposed and successfully experimented on.

At the hotel we found the New York members of our party, which now numbered eighteen. After tea we assembled in a large room up stairs, where Prof. Agassiz made the following remarks on the region over which we had passed :—

"The soil of this tract is of great variety, but everywhere presents this feature : that its surface is covered with loose materials, all *erratic*, (or belonging to rocks whose natural position is distant from the points where these fragments are found,) and all evidently transported at a very remote epoch. These erratics are of all sizes, from sand to large rocks ; the larger ones angular ; the smaller ones more or less rounded, scratched and polished, as are also the surfaces of the rocks on which they rest. These polished rocks have been noticeable to-day, especially to the westward of Worcester. These marks we shall find still more strongly shown as we proceed northward.

"We have nowhere seen *unaltered* rocks, but exclusively those of a granitic character, metamorphosed from originally stratified formations by the action of heat. Thus, for instance, the blackish mica slate, with veins of quartz,—which so frequently occurs on our route of to-day—is probably clay slate, altered by intense heat, which has produced several varieties of silicate of alumina. There is no clearly defined division between these slates; they pass without interruption from baked clay into chloritic slates. In one place in the Connecticut valley we saw red sandstone, generally in a horizontal position, except where disturbed by trap. Nearer Albany we passed through a region of highly metamorphic limestone, belonging to the oldest geological deposits. We have also seen indications of the Potsdam sandstone, one of the most ancient fossiliferous rocks.

"As to the *vegetation*, it is to be remarked in general, that the features of a country are given principally by its plants. These mark the variety of the soil, and its formation. The forests which we have seen to-day consist of a great variety of plants, mingled together. We have seen no forests composed of *one* species of tree. In the mountainous parts, indeed, certain species predominate, but elsewhere several are found in almost equal proportions. Among these are various pines; the white and pitch pines, the spruce, hemlock, red cedar, and a few larches. Then the Amentaceæ, viz., oaks, birches, chestnut, beech, poplar, and the platanus or button wood, (which is in a sickly condition, probably from injury done to the young wood by frosts,) hickories, elms, locust, ash, and maples, but the latter fewer in number. The hickories never form forests. About Niagara we shall find the beech abundant. Of shrubs, we have seen a great variety: e. g., sumachs of several species, (whereas in Europe there is but one,) elder, alder, cornus, viburnum, witch-hazel, willows, wild roses, and grapes. A remarkable feature of the vegetation of this country is, the number of species of grape, mostly useless for the manufacture of wine. Shrubs peculiar to America, are the Kalmias; viz., mountain-laurel and sheepsbane. In the meadows are various grassy plants, carices, and ferns; the latter in great variety. These spots exhibit probably a condition analogous to that of the Coal Period, in which the ferns, &c., prevailed. All the plants growing on the roadsides are exotics, as are also all the cultivated plants and grasses. Everywhere in the track of the white man we find European plants; the native weeds have disappeared before him like the Indian.* Even along the railroads we find few indigenous species. For example, on the railroad between Boston and Salem, although the ground is uncultivated, all the plants along the track and in the ditches are foreign. From this circum-

* Old Tocelyn says the Indians call the common plantain (*Plantago major*,) "the white man's foot."

stance, erroneous conclusions have been drawn as to the identity of species on the European and American continents.

"The *combination* of trees in forests is an important point in the physiognomy of a country. The forests of Europe are much more uniform in this respect than those of this country, from the greater variety of allied species here. Thus, in Central Europe, there are but two species of oak, and no walnut whatever; the so-called English walnut being a Persian tree. In the United States there are over forty species of oak; in Massachusetts there are eleven kinds of oak, and six of walnuts and hickories.

"Another important point is the distribution of water. We have crossed to-day three distinct basins, having no connection with each other, viz., that of the Atlantic coast, the Connecticut valley, and the valley of the Hudson. It would be interesting to examine how far each of these basins has a peculiar *fauna*."

June 16*th*.—At half-past seven this morning, after not a little worry, owing to the very defective arrangements at the railroad station, we set off in the cars for Buffalo. Weather hot, but as our course lay up the flat valley of the Mohawk, there were no more cuts, and the dust was not so troublesome as yesterday. We passed through level, well-cultivated fields, spotted in many places with the bright yellow flower of the mustard, just in blossom.

This rich alluvial plain very early attracted settlers. Part of it bears the name of the German Flats, from its first inhabitants, and the names of the towns along the route, such as Manheim, Palatine Bridge, &c., indicate an immigration from the Palatinate. The Dutch and German blood is still predominant here, as is shown by the names on the signs, the neat little red-painted houses, with open loggias and drive-ways, and the huge barns of this race of thrifty cultivators.

After an uncomfortable night in the cars, we found ourselves at daylight surrounded by the forest. Huge unbranching trunks, clear of undergrowth; occasional clearings, with log houses, and the corn or potatoes scattered among charred stumps. From Utica, westward, along this road, one is constantly reminded of the West. The land here, too, is much of it uncleared, cheap, and fertile; on the other hand, aguish. In short, the advantages and disadvantages are those of the West. From the abundance of pigs and children, and the untidy look of the cabins, one conjectures the settlers are mostly the former laborers on the railroad, or at least countrymen of theirs.

June 17th.—At 8 A. M. we arrived in Buffalo, after about thirty-six hours' actual travelling from Boston, a distance of 527 miles. We had previously ascertained that it would be advisable to wait until the 19th before embarking for Mackinaw, in order to give time for procuring stores, tents, &c., and had determined to spend the intervening time at Niagara. On our arrival we found that the morning train for Niagara was to start at 9; so leaving some of the party to make arrangements, the rest of us took the cars and arrived at the Falls about 11 o'clock.

The road thither presents a continuation of the same noble forest of "first growth," but often broken by clearings. Our European friends were much struck by the contrast with the region we had left only yesterday. A large proportion of the trees were elms, not the plume-like spreading elms of our avenues, but a straight, unbroken, scarcely tapering trunk of sixty feet height, then abruptly expanding with sturdy limbs at right angles into a round head.

In the afternoon we crossed to the Canada side. The museum here contains an interesting collection of the birds and fishes of the neighborhood. A camera-obscura, the field of which is some twenty feet in diameter, placed on the edge of the cliff, gives extensive views of the Falls. I was struck with the disproportionally high *tone* of the sky in the landscapes it presented. The effect was something like the glow that comes on after sunset.

In the evening we assembled in a hall leading to our lodgings at the Cataract Hotel, (in that part of the building which overlooks the Rapids,) and Prof. Agassiz, having displayed his portable blackboard, (consisting of a piece of painted linen on a roller,) gave us the following sketch of the region passed over since his last lecture:—

"East of Lake Ontario we have granitic formations, which were doubtless islands in the ancient time, on whose shores the later formations accumulated, by deposition from the water, in successive beds, the later covering the more ancient, except where these had in the meanwhile been elevated from the primeval ocean along the shores of the high land already dry. Thus the older deposits form strips around the granitic regions; the beds of sedimentary rock becoming continually narrower with the rise of the continent and the consequent contraction of the ocean. From this time there were three basins, viz., the coal basin of Pennsylvania, that of the West, and that of Michigan. It is evident that the north-east region was the earliest dry; to the westward all the formations are more recent.

"Wherever the water escaped towards the north-east, we have waterfalls over precipices; for instance, here at Niagara. Where depressions have been formed in soft rocks between harder ones, we have valleys, as that of the Mohawk.

"It is a remarkable fact, that the leading changes in the geological features of North America take place in a north and south direction. Thus the fissures forming the beds of the rivers, as those of the Connecticut, the Hudson, the Mississippi, and the rivers of Maine. In the Old World, on the contrary, most formations are parallel to the Equator, as the Alps, the Atlas, and the Himalayas. Only two mountain chains run north and south, the Ural, and the Scandinavian mountains, which are northern in their character. The longitudinal direction of fissures in this country is well shown by the New York State Survey. The lakes of Western New York lie north and south. So also Lake Huron and Lake Michigan. These longitudinal fissures are sometimes traversed by others at right angles, as in the instances of Lake Superior and Lake Erie. These fissures must have been formed by the upheaval of the continent, the layers of already solidified rock being lifted up or depressed. Rivers must have existed already in those early ages, as is shown for instance in the ancient channel of the Niagara, (above the Whirlpool,) which is filled with drift not found in the present channel.

"All the formations before spoken of are more ancient than the coal, yet many of them consist of soft clay. The hardness of rock is thus no proof or criterion of its age. These soft slates are nowhere more developed than in New York, and nowhere have they been more carefully examined and described. These details of facts are to be looked upon in the same light as a mere list of dates or occurrences in history. But geology aims at a full illustration of all these details.

"Passing to the vegetable kingdom:—As soon as we left the metamorphic rocks of Massachusetts, vegetation became much richer, because of the limestone and marl deposits. It is remarkable how limestone favors not only vegetable, but also animal life. In Switzerland, where the country is divided between the limestone and marl region of the Jura, the sandstone of the plain, and the granitic formations of the Alps, the cattle of the latter region are not more than one-third of the size of those of the former.

"Among the plants peculiar to this country, are many in whose analogues in Europe many interesting chemical products have been traced. Very little has been done here in organic chemistry, and it is a matter which might well occupy one's lifetime, to ascertain the chemical relations of analogous plants of the two countries, (for instance, *Angelica, walnut,* &c.) Tracing the forest vegetation, we have seen lately very few pines, but principally maples, elms, and ashes; and here at Niagara, almost exclusively elm, beech, hickory, ash, and arbor-vitæ, which is very rare in Massachusetts."

June 18th.—We met again this morning in the hall, where Prof. Agassiz had prepared diagrams illustrating the geology of Niagara, which he explained as follows:—

"The surface of the soil, both on the Canadian and on the American side, is covered with gravel, containing fossils in great numbers, and stones of all sizes, from that of a hen's egg to large bowlders. This stratum is now disunited by the action of the river, but was originally continuous, as is shown by the fossils, and by the fact that on the intermediate islands, where it has escaped the action of the water, it is still present. The fossils form a bed extending horizontally to the river bluffs, but not beyond; they occur in great numbers, covering the surface of the soil everywhere, and contributing to the great luxuriance of the vegetation. These fossil shells, doubtless, inhabited the river in former times, when its bed was the mass of gravel, &c., on which they now rest, the bluffs being at that time its banks. They are of species now living in the river, of the genera *Unio*, *Cyclas*, *Melania*, *Paludina*, and *Planorbis*. Hence we conclude that this bed was formed when the river filled the whole valley, at which time it had a breadth varying from one to seven miles, and averaging three or four. Probably at that time it resembled the present Rapids above Goat Island. Afterwards, from the acceleration of the current, owing probably to the opening of fissures which lowered the level of Lake Erie, the two present channels were cut down to the rock, and the river reduced to its present level."

Afterwards we went over to Goat Island, and blessed once more the good sense that has kept this place undisturbed. The decaying wood and fungi of the damp woods here afforded an abundance of specimens to our entomologists. The variety of trees and shrubs on these islands is remarkable. On the little islet (only a few feet in extent,) connected by a foot-bridge with the toll-house, Prof.

Agassiz pointed out seven different kinds of trees, viz., arbor vitæ, red cedar, hemlock, bass-wood, chestnut-oak, white pine, and maple. The Professor also pointed out the shell-bed of which he had spoken. The shells are very numerous, as may be readily seen in the crumbling bank on the outer side of the island. At the upper end of the island, vast numbers of delicate ephemera-like insects, with long filaments, were fluttering about, particularly under the trees.

Some of us had never seen the Falls, and none of us at this season of the year, when the mass of water is greatest. Coming at length in sight of them, we were struck with the thickness of the sheet at the pitch of the English Fall, particularly in that part of it between the apex of the Horseshoe and the middle of the cataract on the Canadian side.* It bends over in a polished, unbroken mass, as of green glass over white. Some one said the average depth of water at that point was about fourteen feet. Other remarkable features are, the distance to which the water is projected, the rocket-like bursts of spray from the falling sheet, and the sudden spouting up of the mist at intervals from below, as if shot from a cannon. These sheets of mist rise high above the Fall, and move slowly down the river in perpendicular columns, like a procession of ghosts. On the whole, the difference of season is in favor of that when the river is lowest, the features of the scene, particularly the Rapids outside of Goat Island, being rather obscured than improved by a greater depth of water.

After tea, the following remarks on what we had seen were made by Prof. Agassiz:—

"If we follow the chasm cut by the Niagara River, down to Lake Ontario, we have a succession of strata coming to the surface, of various character and formation. These strata dip S.W. or towards the Falls, so that in their progress to their present position, the Falls have had a bed of very various consistency. Some of these strata, as the shales and the Medina sandstone, are very soft, and when they formed the edge of the Fall, it probably had the character of rapids. But wherever it comes to an edge of hard rock, with softer beds below, the softer beds, crumbling away, leave a

* The "Horseshoe" at present is a triangle, but it has been a nearly regular semicircle within the recollection of persons now living.

shelf projecting above, and then the fall is perpendicular. Such is the case at present; the hard Niagara limestone overhangs in *tables* the soft shales underneath, which at last are worn away to such an extent as to undermine the superincumbent rocks. Such was also the case at Queenston, where the Clinton group formed the edge, with the Medina sandstone below. This process has continued from the time when the Niagara fell directly into Lake Ontario, to the present time, and will continue so long as there are soft beds underneath hard ones. But from the inclination of the strata, this will not always be the case. A time will come when the rock below will also be hard. Then, probably, the Falls will be nearly stationary, and may lose much of their beauty, from the wearing away of the edge, rendering it an inclined plane. I do not think the waters of Lake Erie will ever fall into Lake Ontario without any intermediate cascade. The Niagara shales are so extensive that possibly at some future time the river below the cascade may be enlarged into a lake, and thus the force of the falling water diminished. But the whole process is so slow, that no accurate calculations can be made. The Falls were probably larger and stationary for a longer time, at the "Whirlpool" than anywhere else. At that point there was no division of the cataract, but at the "Devil's Hole" there are indications of a lateral fall, probably similar to what is now called the American Fall. At the Whirlpool, the rocks are still united beneath the water, showing that they were once continuous above its surface also."*

Afterwards, some of us went to bathe by moonlight in the "Hermit's Fall," a little cascade eight or ten feet in height, between Goat Island and the islet at its upper end. It is so called from a crazy Englishman who lived for some time in a hut on the other side of the island, and was finally drowned in bathing at this place. There is, however, little danger, as the water is shallow, and just below the pool a large log extends across the stream, which is only some twenty feet wide. The "Hermit" was probably tired of his own society at last, as he had been already of other people's, and took this method of getting rid of it. The place, indeed, one could conceive might be dangerously attractive to one tired of life. It is so shaded and shut off by the overhanging trees of the island, that one might fancy it a mountain stream a hundred miles from any

* The data on which these and the previous remarks on the geology of the Falls are founded, are derived from Prof. James Hall's investigations in the New York State Survey. A.

human habitation. The little cascade, near at hand, drowns the roar of the great one, and though by day it cannot boast of any great privacy, yet at night very few even of the most romantic moonlight strollers get so far as this.

The power of the water was greater than I expected, and difficult to bear up against, even in a sitting posture. It was not a simple pressure, but a muscular force, like a kneading or shampooing by huge hands. We crawled in at the side of the Fall, and found a hollow underneath the shelving edge, large enough for several to sit at once, quite free from the water, which shoots over like a miniature of the great cascade below. With some difficulty, from the pounding of the falling water, we penetrated through the sheet in front, and came out into the pool, the bottom of which is smooth rock. Close to the surface there was a strong current of air down the stream, not perceptible at the height of two feet.

Afterwards, in walking round the island, we saw on the cloud of mist over the English Fall, a lunar rainbow, glimmering with a pale, phosphorescent, unearthly light, and showing prismatic colors, but not quite joined at the top. Some of the party afterwards saw it complete.

June 19*th.*—Took an accommodation car on the Lockport Railroad as far as the Suspension Bridge, (about a mile below the Falls,) of which the piers were finished and a rope stretched across, bearing suspended a basket, in which some adventure-loving person was being hauled across. From the bridge we walked along the bank through the woods to the Whirlpool. The river, when thus seen from above, is of such a dark and solid green, that it is difficult to persuade one's self that it is not occasioned by some colored matter suspended in the water. At intervals we got glimpses of the Fall, between the high perpendicular banks enclosing it as in a frame. The slow, heavy plunge of the water was distinguishable to the eye even at this distance, but the roar was hardly audible.

Rattlesnakes are found among the rocks about these cliffs, and one had been taken alive the day before, in the path leading down to the Whirlpool. There is said to be a mound of their bones in the neighborhood, erected in token of full revenge by some Indians whose chief had been killed by a rattlesnake's bite.

Returning to the Suspension Bridge, we went on board the little steamer, "Maid of the Mist," which runs up to the foot of the Falls. I confess I was doubtful as to the advantages to be gained by any one who had crossed the ferry so often as I had, but I was old traveller enough to know that one oftener repents of not going than of going, and went accordingly, instead of returning by the cars with the more skeptical of the party. The result showed the soundness of the principle. Many things are to be learned by such close proximity, (for the boat, true to her name, runs actually into the mist at the foot of the Fall,) and may be studied more conveniently in the steamer, with a chance to dodge any extraordinary shower of spray, than in an open skiff. I saw plainly here, what I had not been able to satisfy myself of before, that the *catenary curves* in high waterfalls, insisted upon by the "Oxford Graduate,"* are fully exemplified in the greatest cascade of the world.

At half-past two P. M. we took the cars for Buffalo, and as the steamer was not to start until seven, we had some time on our hands after our arrival there, which we spent in making some last purchases, and in seeing the place.

The number of Germans here is a prominent feature. At the Post-office there is a separate delivery for "Deutsche Briefe." Another feature, striking to a New Englander, though common to all the towns in New York, (which justify themselves probably by the example of their great city,) is the number of *pigs* running at large in the streets. When at length we went on board the "Globe," we found everything in confusion. Bye and bye, however, the confusion subsided; even the escape-pipe abated its vehemence by degrees, and at last became silent, and still there seemed to be no movement towards starting. But in proportion as the boat became quiet, the passengers became noisy for departure, and at last, after much expostulation, and finally the threat of leaving altogether, at half past ten we got under weigh.

June 20th.—Weather pleasant, wind S.S.W., strong. The water green, but less so than at Niagara. This forenoon we took possession of a little cabin in the after part of the vessel, to listen to the

* Modern Painters, (Am. Ed.,) I., 363.

following account from the Professor, of the forest trees about Niagara, illustrated by specimens gathered the day before on the spot:—

"1. *Coniferæ*, (pine family,) remarkable for the apparently whorled arrangement of their branches, and for their evergreen leaves; in most cases they form hard *cones*, but one has soft, berry-like fruit. The seeds are naked, winged, resting on the scales. The leaves are peculiar, the nerves not being spread, but often gathered into compact bundles. The *Coniferæ* existed at a very early geological epoch. This was the first family that became numerous after the ferns. Their remains are easily recognized under the microscope by the circular disks on their wood-cells.

"2. Sterile flowers grouped together, in spike-like branches, forming catkins; fertile flowers surrounded by a cup. They all belong to temperate climates. Gen. QUERCUS (oak,) characterized by their fruit, and by the fact that the female flowers are scattered, and the stameniferous flowers form bunches. There are more than forty species in the United States. Gen. CASTANEA, (chestnut,) allied to the oaks, but the fruit surrounded entirely by the cup (burr). There are two species in the United States. Gen. OSTRYA, (hop-hornbeam,) only one species. Gen. CARPINUS, (hornbeam,) fruit supported by flat leaf. May be distinguished from OSTRYA by the more prominent ribs, and less deeply marked serratures of the leaves.

"3. *Amentaceæ;* both kinds of flowers in catkins. Gen. BETULA, (birch,) distinguished by the shape of its catkins, which are long and cylindrical, and its winged fruit. Gen. POPULUS, (poplar,) seeds in a pod, very minute, and surrounded by down. P. *tremuloides* (American aspen,) like the other species, has the leaf-stalk very much compressed, hence the tremulous motion of the leaf.

"4. *Juglandeæ*, fruit with an external soft husk, the nut separating into two halves. There are two genera of this family in the United States: JUGLANS. All have compound leaves, that is, each leaf is divided into leaflets. Two species, black walnut and butternut, the latter distinguished by the silkiness and whitish color of the underside of the leaf. CARYA, the nut does not divide so well as in JUGLANS, but the husk is divided and falls off in pieces, which is not the case in Juglans. At Oeningen, in Switzerland, are found fossil hickories. The trees of the tertiary epoch of Europe correspond to the species existing at present in this country.

"5. *Oleaceæ*, (the ash family,) leaves like those of hickory, but the large lateral nerves do not run to the points of the serratures, as in the hickories. Fruit in bunches, with dry capsules. Flower in the ash, without corolla.

"6. *Hamamelidæ*, (witch hazel,) named probably from its flowering in the fall. Fruit in four little nuts. No species of this family in Europe.

"7. *Tiliaceæ*, leaves unsymmetrical. Tilia *americana*, (bass-wood,) leaves smooth below.

"8. *Acerineæ*, Gen. Acer, (maple,) leaves in three main lobes, subdivided into five.

"9. *Ampelidæ*, (the grape family,) petals dividing below sooner than at the apex. Great variety of species in America, but not suitable for making wine. Three species on Goat Island."

The south shore of Lake Erie is flat and monotonous; red, crumbling banks, surmounted by a forest broken only by an occasional log-house. At one time high land visible on the horizon, being a spur of the Alleghanies.

In spite of all glorification on the score of the "Great Lakes," it must be confessed that the Lower Lakes at least are only geographically or economically great. Any one accustomed to the sight of the ocean has to keep in mind the square miles of extent, to preserve his respect for them. Their waves, though dangerous enough to navigators, have not sufficient swing to carve out a rocky shore for themselves, or to tumble any rollers along the beach, and thus the line where land and water meet, in which, as has been well said, the interest of a sea-view centres, is as tame as the edge of a duck-pond. Much of this character is doubtless owing to the flat prairie country by which they are mostly surrounded.

In the afternoon heavy clouds rolled up from the N.W., and a squall was evidently approaching. At this time we saw a steamer in the distance outside of us, with her flag union down. On reaching her we found she had broken her crank. After some clumsy manœuvring we got alongside, and her captain persuaded the owner of our boat, who was on board, to "accommodate" him by towing him into Cleveland. This kind turn would delay us many hours, and was by no means necessary for the safety of the boat, since there were other ports under the lee. Nevertheless, our owner (although, as we learned, he was to be paid nothing for the trouble,) agreed, and took them in tow. But shortly after, the squall coming on, it was found that our machinery would not stand the additional strain, and she was accordingly cast off to shift for herself. We arrived at

Cleveland at half past ten P. M., and spent there some hours. It is a thriving town, and a regular stopping place for steamers, but like almost all the towns on this lake, is without a natural harbor, the only shelter to vessels being a long pier stretching into the Lake.

June 21st.—Weather fine and warm, with smooth water. Arrived at Detroit at half past eleven, and left at three P. M. Near the entrance of Lake St. Clair we were surrounded by numbers of black terns, (*Sterna nigra,*) which, at a moderate distance, were distinguishable from the swallows by which they were accompanied, only by their superior size. Numbers of slender gauze-winged insects, (*Ephemera, Phryganea,*) with long antennæ, and some with two long filaments projecting behind like the tail feathers of the Tropic bird about the boat, and on the water. In the St. Clair straits there were a few ducks, even at this season, though nothing like the vast flocks to be seen here a little later in the season.

We were sounding constantly through these straits, having on an average about three feet below the keel in the channel, our boat drawing seven feet. The shores are low, marshy and aguish, with woods at a distance, and scattered log-houses. This remarkable extent of mud-flats, (some twenty miles across,) is covered with only a foot or two of water in most parts, and even the channel is so shallow that the larger boats have to discharge a part of their cargo into lighters while passing it, and are often delayed here many hours. Even our boat continually touched, as was evident from the clouds of mud she stirred up. To make and maintain a proper channel for such a distance, is an undertaking much called for, but not to be expected of single States, nor is there any one State principally interested in it. One would hope, therefore, that the General Government may before long do something about it.

The water over these flats is still as green as that of Lake Erie, and not more turbid. About 10 P. M. we put in to wood, and remained until 7 A. M., taking in sixty-four cords of wood.

June 22d.—We entered Lake Huron about breakfast time; the weather calm, and what the sailors call "greasy," the water darker than in Lake Erie, partly owing, no doubt, to the greater depth of water, and partly to the cloudy sky. The dark sullen water, and the unbroken line of forest, retreating on either hand as we issued from the straits, gave a kind of grim majesty to this lake, by contrast to those we had left. Many sea-gulls about. Land in sight on the left all day, except in crossing Saginaw Bay.

On entering Lake Huron, we began to feel that we were getting into another region. Canoes of Indians about; the weather cool morning and evening, and the vegetation northerly, the pine family having a decided preponderance in the landscape. We might be said to have left the summer behind at the St. Clair, for thenceforth there was hardly a day during some part of which a fire was not necessary for comfort.

Just before sunset, when the sun was three or four degrees high, we noticed in the opposite quarter of the heavens, rays of light converging towards a point apparently as much below the horizon, as the sun was above. It had the appearance of a cloudy sunrise. We afterwards saw the same thing in the St. Mary's River; and it may be remarked, in both cases before rain.*

June 23d.—Arrived at Mackinaw early in the morning, and landed on the wharf in a shower. We had been about eighty hours on the way from Buffalo, a distance of 663 miles, and we were vexed to hear that the weekly steamer for the Sault had left the evening before, and that if we had taken the other boat, which started punctually a couple of hours before us, we should have been in time.

We landed on the little wooden wharf in face of a row of shabby cabins and stores, with "Indian curiosities" posted up in large letters to attract the steamboat passengers during the brief stop for fish. Over their roofs appeared the whitewashed buildings of the Fort stretching along the ridge. The inhabitants of the place, looking down upon us from all sides, as from the lower benches of a theatre, soon perceived that we had not departed with the steamer, and we were soon plied with invitations to the two principal lodging-houses. From previous experience, I advised the "Mission House," and thither we went.

On the beach some Indians were leisurely hauling up their canoes, or engaged upon their nets, regardless of the rain. The Professor was soon in the midst of them, and bought white-fish and large pike,

* See a notice of a similar phenomenon by Bory St. Vincent, in Goethe's Farbenlehre: [Entoptische Farben, cap. XXXI.]

which had been taken with nets or lines set the night before. An excellent breakfast (at which white-fish figured,) and comfortable rooms, showed that the character of the "Mission House" was still kept up.

It continued to shower at intervals during the day, but this did not prevent us from seeing the Natural Bridge, with its regular arch, ninety feet high, rising on the border of the island, the huge conical rock called the "Sugar Loaf," the Fort, &c. I do not know whether any of the party visited the cave where Alexander Henry was concealed by his Indian friend during the massacre of the English—as I did on a former occasion, when, bye the bye, I found a fragment of a human skull among the rubbish on the floor of the cave, attesting the correctness of that part of Henry's narrative.

The wet weather was not unfavorable to vegetation, which is luxuriant on the island, though the trees, (maple and beech,) are of small size, this latitude being nearly the northernmost limit of the latter. The flowers were beautiful; the twin-flower, (*Linnœa borealis,*) so fine that I thought it must be another new species; then the beautiful yellow ladies' slipper, Lonicera, and Cynoglossum.

The island is of a roundish form, two or three miles in diameter. On the N.E. the crumbly lime-cliff rises abruptly from the water to the height of a hundred feet or more; but on the south there is a sloping curve of varying width between the bluff and the beach.

The village lies on this slope, a single street of straggling log-cabins and ill-conditioned frame houses, parallel with the beach, and some of a better class standing back among gardens at the foot of the bluff. On the edge of the bluff, which rises abruptly from the slope at the distance of some three hundred yards from the Lake, stands the Fort, a miniature Ehrenbreitstein, with a covered way leading down the face of the bluff.

We were disappointed at finding only three or four lodges of Indians here. In August and September (the time for distributing the "presents,") there are generally several hundreds of them on the island.

Notwithstanding the rain, the Professor, intent on his favorite science, occupied the morning with a fishing excursion, in which he was accompanied by several of the party, most of them protected by water-proof garments, while he, regardless of wet and cold, sat soaking in the canoe, enraptured by the variety of the scaly tribe, described and undescribed, hauled in by their combined efforts. Not content with this, he as usual interested and engaged various inhabitants of the place to supply him with a complete set of the fishes found here.

With a view of indoctrinating those of us who were altogether new to ichthyology with some general views on the subject, he commenced in the afternoon, scalpel in hand, and a board well covered with fishes little and big before him, a discussion of their classification:

"These fishes present examples of all the four great divisions of the class. This pike, (*Lucioperca americana,*) belongs to those having rough scales and spinous fins. The rays of the first dorsal, and the anterior ones of the ventrals and the anal are simple and spinous; the other rays are divided at the extremity, and softer. The scales are rough and remarkably serrate. These are the CTENOIDS. They have five sorts of fins, viz: the dorsal, caudal and anal, which are placed vertically in the median line, and can be raised or depressed, and the ventral and anal, which are in pairs. In the Ctenoids the ventrals are placed immediately below the pectorals, though fishes having this arrangement of fins do not all belong to this division. There are but two families of Ctenoids found in fresh water: the *Percoids* and the *Cottoids;* the former are characterized by having teeth on the palatal and intermaxillary bones, but none on the maxillary. Also by a serrate preoperculum and by the spines on the operculum. Of this family are the genera *Perca, Labrax, Pomotis, Centrarchus,* &c. The fish before us belongs to the genus *Lucioperca.* They have a wide mouth and large conical teeth, like the pickerels, and two dorsals. There are two species in Europe and two in the United States. This is *L. americana;* its color is a greenish brown above, with whitish below, and golden stripes on the sides. On opening the fish we find the heart very far in front, between the gills, and consisting of a triangular ventricle, a loose hanging auricle, and a bulbous expansion of the aorta. All the Percoids have three cœcal appendices from the pyloric extremity of the stomach. These probably take the place of a pancreas. Below is the air-bladder, which is a rudimentary lung. Above this are the ovaries, which extend from one extremity of the abdomen to the other. Behind is the kidney, extending along the spine.

This trout belongs to the CYCLOIDS. In this division there are only two families which have spinous rays in their fins, (the tautog and the mackerel.) We have before us specimens of two families of Cycloids.

1. *Salmonidæ.* Distinguished by having the intermaxillary and upper maxillary in one row, which seems to me to indicate the highest rank in the class of fishes. They all have a second dorsal, of an adipose structure. The anterior dorsal and the ventrals are in the middle of the body. Genus *Salmo:* characterized by teeth on every bone of the mouth and on the tongue. There is but one genus in the class of fishes that has teeth on more bones than the salmon. In no genus are the species more difficult to distinguish. Sixteen species have been described as belonging to Europe, which I have been obliged to reduce to seven. The same species presents great variety of appearance, owing to difference of sex, of season, food, color of the water in which they live, &c. In this country I have examined two species, the brook trout, (*S. fontinalis,*) the spawning male of which has been improperly separated as *S. erythrogaster;* and the present species, the Mackinaw trout, *S. amethystus* of Mitchill. Dekay has described a variety of this species, as *S. affinis.* In this species the *appendices pylorici* before spoken of are very numerous. The small intestine arises from the lower extremity of the stomach, and curves only twice throughout its length. The gall-bladder is very large: the liver forms one flat mass; the ovaries and kidney extend along the whole spine. All this family spawn in the autumn.

"(2.) *Cyprinidæ.* Like the salmons they have the ventral and dorsal fins in the middle of the body, but no adipose dorsal. Branchiostegal rays, three. Upper maxillary forming another arch behind the intermaxillary. Teeth only on the pharyngeal bone behind the gills, at the entrance of the œsophagus. No pyloric appendices. Intestine long and thin, as in all herbivorous fishes. Air-bladder transversely divided into two lobes, communicating by a tube with the intestinal canal.

"This family is the most difficult one among all fishes. As yet there is no satisfactory principle of classification for them. I have studied them so attentively that I can distinguish the European species by a single scale; but this not from any definite character, but rather by a kind of instinct. Prof. Valenciennes, a most learned ichthyologist, has lately published a volume on this family, in which he distinguishes so many species, and on such minute characters, that I think it now almost impossible to determine the species, until all are well figured.

"Here are specimens of two genera: (*a*) *Leuciscus*, with thin lips; only one species here, an undescribed one characterized by a brownish stripe above the lateral line. (*b*) *Catostomus*, with very thick lips and prominent snout."

3

June 24th.—Rather than wait here a week for the next steamer, we engaged a Mackinaw boat and some Canadians to take us to the Sault. These boats are a cross between a dory and a mud-scow, having something of the shape of the former and something of the clumsiness of the latter. Our craft was to be ready early in the morning, but it was only by dint of scolding that we finally got off at 10 o'clock. A very light breeze from the southward made sufficient excuse to our four lazy oarsmen and lazy skipper for spreading a great square sail and sprit-sail, and lying on their oars. Unless it was dead calm, not a stroke would they row.

At about 1 o'clock, Mackinaw still plainly visible at a very moderate distance to the southward, we stopped to lunch at Goose Island, a narrow ridge of rough, angular pebbles, about half a mile long, covered with thick bushes and stunted trees, among which the principal were arbor-vitæ and various species of cornus. It passed through my mind whether this could be the *Ile aux Outardes*, where Henry parted with his Indian friend. It is difficult to say what bird of this region could have reminded the French colonists of a *bustard.*

Getting off again we continued at rather a better rate (the wind being now fortunately ahead) until twilight, when our steersman said it was time to look out for a camp, and proposed landing us on a little island near the western shore of the strait. The more ardent naturalists of the party, however, seeing a sand-beach, (capital hunting-ground for Coleoptera,) backed by a grassy bank among the trees, were anxious to land there, but this was promptly opposed by the whole of our native ship's-company, who urged that we should be devoured by "*les mouches.*" This suggestion seeming reasonable, it was arranged that those who wished it should be landed on the beach, while the rest proceeded to encamp and get supper ready on the island. This was done; but hardly had we disembarked and lighted a fire, when cries were heard from the main land, and on looking round we saw our friends, some with their heads bound up in handkerchiefs, others beating the air with branches of trees; all vociferating to us to "Send the boat!" and on the whole, manifesting the most unmistakable symptoms of musquitoes, which were abundantly confirmed when they joined us.

Our island was a mass of large irregular stones, about a quarter of a mile long, with a narrow ridge covered with long grass and arbor-vitæs, many of them dead, and (particularly on the west,) hung over with pendant lichen (*Usnea*). Here, (after some trouble from not having brought tent-poles, which had now to be cut,) we pitched four tents, for only two of which was there any room on the grass, the others looking out for the smallest stones. However supper and three blazing fires soon settled all down into a comfortable state, and before long the white tents and the ghost-like trees with their hoary drapery were the only upright objects to reflect the light of the fires, and the long melancholy notes of some neighboring loons (a sign of bad weather, they say,) the only sounds to be heard. As my lot was cast upon the stones, I took the precaution of thatching them with some armfuls of usnea, which with a couple of blankets made an excellent bed.

June 25th.—Our island was only about thirty miles from Mackinaw, and so, as it behoved us, we were off by half past four o'clock this morning, with the wind aft, to try to make up for lost time. Our course lay along the American shore of the strait, amid innumerable islands and islets, generally low and wooded with venerable lichenous arbor-vitæs. The shore also was uniformly low, and covered with a forest which reminded me of the lower summits of the White Mountains.

We stopped to breakfast just beyond the light-house at the Détour, at the log-house of some lime-burners, a tavern moreover, rejoicing in the name of "the saloon," where we experimented upon tea with maple-sugar, and bread of the place, somewhat like sweetened plaster-of-Paris. Drummond Island, interesting from its fossils, we were obliged to pass without stopping.

By noon the wind had got so high that we thought prudent to make a lee under a point on St. Joseph's Island. As we landed, a rather rough-looking, unshaven personage in shirt-sleeves walked up and invited us to his house, which was close at hand. We found his walls lined with books; Shakspeare, Scott, Hemans, &c., caught my eye as I passed near the shelves, forming a puzzling contrast with the rude appearance of the dwelling. A very few moments sufficed to show a similar contrast in our host himself. He knew Prof. Agassiz by reputation, had read the reports of his lectures in the newspapers, and evinced a warm interest in the objects of our excursion. When he found out who the Professor was, he produced a specimen in spirits of the rare gar-pike of Lake Huron, and insisted upon his accepting it, and afterwards sent him various valuable specimens. His conversation, eager and discursive, running over Politics, Science and Literature, was that of an intelligent and well-read man, who kept up, by books and newspapers, an acquaintance with the leading topics of the day, but seldom had an opportunity of discussing them with persons similarly interested. He turned out to be an ex-Major in the British army, and he showed us a portrait of himself in full regimentals, remarking with a smile that he had once been noted as the best-dressed man of his regiment. Whilst in the service he had travelled over Europe, seen what was best worth seeing, and become acquainted with the principal modern languages, particularly Italian, which he read here in the wilderness with delight. In company with a friend he had purchased the entire island of St. Joseph's and devoted himself to farming, bringing up his children to support themselves by the sweat of their brow. He said it would be time enough to give them a literary or professional education when they manifested a disposition for it, for he did not approve of the indiscriminate training of all for what comparatively few have any real talent for. He was preparing them, he said, to be American citizens, for he thought the Canadas would form a part of the United States within three years at farthest; and though he for his part was a loyal subject of her Majesty, and would fight to protect her dominions if it came to that—yet he had no objections to his children being republicans.

After chatting several hours with the Major, and discussing an excellent white-fish which he placed before us, the wind having meantime moderated, we continued our course. St. Joseph's, according to the Major, forms a triangle, of which the two longest sides measure twelve and twenty miles. The climate he described as temperate, being influenced probably by the great mass of flowing water by which the island is surrounded. His custom was to work throughout the winter in his shirt sleeves; he did not remember to have seen the thermometer lower than—10° Fah., and that only for very short periods. The soil excellent, except near the shores.

Passing the end of the island we saw two solitary chimneys, the remains of the fort that formerly stood here. Our course lay among small islands, reminding one of the little wooded islets of Lake George, with a brilliant background of sunset sky. We noticed the same appearance in the east, spoken of June 22nd. The twilight continuing late, we pushed on until about ten o'clock, when our men proposed to land on a small rocky island, but they being alarmed at a discovery (probably imaginary) of snakes among the rocks, and we for our part not finding room enough among the stones to pitch a tent, we continued our course to another island which bears the name of " Campement des matelots." Here it was voted too late to pitch tents, so we rolled ourselves in our blankets, some on shore and some in the boat, taking care to include our heads, for the musquitoes had roused themselves and were making active preparations to receive us.

June 26th.—The musquitoes of the night before must have been merely those who occupied the spots where we lay down, for when in the morning, being awakened by sundry energetic exclamations in my neighborhood, I extricated my head from the blanket and looked about me, my first impression was wonder, at the swarms that surrounded the heads of my companions. Having fortunately a musquito-veil in my pocket I was soon a disinterested spectator of their torments. It was with some difficulty that the necessary arrangements for embarking (with no thought of breakfast) were completed, and it was more than an hour after we left the place before with all our exertions we could get the boat rid of them.

Soon afterwards it began to rain. Our course lay up the boat-channel, (twelve miles shorter than the main passage,) over mud-flats covered with only a few feet of water, the banks on either side flat and covered with a monotonous forest which in one place was burnt, and for miles a tedious succession of blackened trunks. We crowded together in the middle of the boat and covered ourselves as well as we could with tarpaulins and India rubber cloaks, the importance of which rose considerably in the general estimation. This muddy expanse of the river or strait, goes by the name of Mud Lake. It resembles Lake St. Clair on a smaller scale, being eight or ten miles wide. Here, as we were afterwards told, is found a great abundance and variety of fishes, and also the salamander which the Indians call the "walking fish" (*Menobranchus*), and which even to them is a great curiosity. At last we reached the Lower Rapids, where with all the exertion of our men we for some time made little progress. Soon a cabin or two made its appearance; then we saw the palisades of Fort Brady, and at noon arrived at the wharf, where even the rain did not prevent a considerable concourse of the idle population. Carts drove down into the water for our luggage, and at length our drenched state was relieved by the comfortable accommodations of the "St. Mary's Hotel."

CHAPTER II.

THE SAULT TO MICHIPICOTIN.

June 27th.—The Sault de St. Marie, on the American side, is a long straggling village, extending in all some two or three miles, if we reckon from the outposts of scattered log-huts. The main part of it, however, is concentrated on a street running from the Fort (which stands on a slight eminence over the river,) about a quarter of a mile along the water, with some back lanes leading up the gradual slope, rising perhaps half a mile from the river. Behind this again is an evergreen swamp, from which a rocky wooded bluff rises somewhat abruptly to the height of a hundred feet or thereabouts.

The population is so floating in its character that it is difficult to estimate; some stated it at about three hundred on the average, consisting of half-breed voyageurs, miners waiting for employment, traders, and a few Indians. The chaplain at the Fort, however, estimated the number of inhabitants on both sides of the river at one thousand, of whom the majority belong to the American side.

The most striking feature of the place is the number of dram-shops and bowling-alleys. Standing in front of one of the hotels I counted seven buildings where liquor was sold, besides the larger "stores," where this was only one article among others. The roar of bowling alleys and the click of billiard balls are heard from morning until late at night. The whole aspect is that of a western village on a fourth of July afternoon. Nobody seems to be at home, but all out on a spree, or going a fishing or bowling. There are no symptoms of agriculture or manufactures; traders enough, but they are chatting at their doors or walking about from one shop to another. The wide platforms in front of the two large taverns are occupied by leisurely people, with their chairs tilted back, and cigars in their mouths. Nobody is busy but the barkeepers, and no one seems to know what he is going to do next.

The cause, probably, may be in part the facilities for smuggling brandy from the Canadian side of the river, where it is cheaper than on ours. But the mischief lies chiefly in the unsettled state of things, the irregularity of employment and wages of labor. Money is not earned and spent from day to day, at home, but comes in lumps, and seasons of labor are followed by intervals of idleness. In short, the life of most of the inhabitants is essentially that of sailors, and brings accordingly the reckless character and the vices of that class.

Something also is due to the admixture of Indian blood, which has a fatal proneness to liquor. Whilst we were here a number of Indians arrived with the son of a chief, from Fort William, and after parading about the town with an American flag, speechifying and offering the pipe at all the grog-shops to beg for liquor, they dispersed and devoted themselves to drinking and playing at bowls. In the evening, two of us passing one of the bowling-alleys, saw in front of it, lying on a heap of shavings, a dark object which proved to be the chief's son, extended at full length, dead drunk, with several Indians endeavoring to get him home. The only sign of life he gave was a feeble muttering in Indian, copiously interspersed with *the* English curse; another instance of the naturalization of John Bull's national imprecation in a foreign tongue. It is said the Indians have no oath in their own language. Finding it impossible to make him walk, they squatted around him on their haunches and remained still for some time, apparently considering what to do. They were all perfectly sober and evidently greatly troubled at the state of their leader. At length, seeing us watching them, they came up and stood staring with their faces close to ours, but without speaking. We did not know exactly what they were at, but my companion by signs explained to them that they should take up the drunken man by the legs and arms and carry him home. The idea struck them as a good one, for they immediately "how, howed," set about it, and bore him off, one to each leg and arm.

The river opposite the village is about a mile wide. Just above are the Upper Rapids, which give the name to the place, nearly three-fourths of a mile in length. There is no very great vertical descent,* but the stream is much compressed and moreover very shallow, whence the great rapidity of the current at this spot. On the opposite bank is a thin, straggling village, and a large building belonging to the Hudson's Bay Company.

Our explorations of the neighborhood showed a great abundance of birds for the season. Prof. Agassiz as usual had got all the fishes of the neighborhood about him; among others several specimens of the gar-pike of Lake Huron, dried or in spirits, were presented to him by the various coadjutors whom he had interested in his favor. One of the most zealous of these was a fisherman whom he had captivated by a distinction (at first stoutly and confidently combatted) between two closely-resembling species. In the evening he unrolled his blackboard and gave us the following account of them:

"The gar-pike is the only living representative of a family of fishes which were the only ones existing during the deposition of the coal and other ancient deposits. At present it occurs only in the United States. The species of South Carolina was described by Linnæus as *Esox osseus*, from a specimen sent to him by Dr. Garden. But it is not an Esox, though it has the peculiar backward dorsal of that genus. It differs in the arrangement of the teeth, which in Esox are seated on the palatal bones and the vomer, but in this genus, Lepidosteus, on the maxillary and all other bones which form the roof of the mouth. Moreover, the snout of the latter is much longer, the upper jaw bones being divided into ten or twelve distinct pieces. The intermaxillary is a small bone pierced with two holes for the admission of the two anterior projecting teeth of the lower jaw. In Esox the scales are rounded and composed of layers of horny substance, and overlap each other. In Lepidosteus the scales are square and overlap only very slightly. Each scale is composed of two substances; first, a lower layer of bone, forming that part of the scale which is covered by the next; second, enamel, like that of teeth. The scales are also hooked together; a groove in each, with a hook from the next fitting into it. Nothing of this kind occurs in other fishes of the present day. From these peculiarities I have named this family the Ganoids. Their vertebræ are not articulated together as those

* According to Bayfield the total descent is twenty-two and one-half feet, but this probably includes both the Upper and Lower Rapids, as the whole difference of level between Lake Superior and Lake Huron, in a distance of forty miles, is only thirty-two feet.—*Bouchette's British Dom. in N. America*, I., 128.

of other fishes, but unite by a ball-and-socket joint, as in reptiles. The scales also resemble in some particulars those of the Crocodilean reptiles, which immediately succeeded the fossil Ganoids, during whose epoch no reptiles existed. The embryology of the gar-pike, of which nothing as yet is known, would be an exceedingly interesting subject of investigation, since it is a general law that the embryo of the animals now living resembles the most ancient representatives of the same family. As probably connected with the preservation of this ancient family of fishes in this country, may be mentioned the fact that there was an extensive continent formed in North America at a time when all the rest of the earth was under water. Thus physical conditions have been more unaltered here than elsewhere.

"The white-fish, (*Coregonus albus,*) has all the characters of the salmons, but no teeth. Among those I obtained to-day, is a new species, characterized by a smaller mouth and more rounded jaw. To the same family belongs the lake "herring," which is no herring at all. This species has a projecting lower jaw and is undescribed. Here is a little fish which on hasty examination would seem to belong to the salmons, but has a projecting upper jaw, and teeth on the intermaxillary, the upper maxillary forming another arch behind, without teeth. It has pectinated scales, like the perch. It is a new genus, allied to the family of Characini of Müller. Fossil fishes of this family occur in great numbers in the cretaceous period; they are the first of the osseous fishes. This again is an instance similar to that of the Lepidosteus. The fish before us presents a curious combination of the characters of the Cycloids and Ctenoids. Here is a fish belonging to the *Cyprinidæ*, but characterized by thick lips and a projecting upper jaw, whence I propose to call it *Rhinichthys marmoratus*.

"This fish, one familiar with the fishes of Massachusetts would suppose to be a yellow perch, but it differs in wanting the tubercles on the head and operculum. It is *Perca acuta* Cuv. In the tertiary beds are found Percoids, with thirteen rays in the anterior dorsal; this is also the case in the North American species. Again the variety of minnows found in this country has a parallel in the tertiary epoch."

June 28th.—To-day we made our first acquaintance with the genuine *black fly*, a little insect resembling the common house-fly, but darker on the back, with white spots on the legs, and two-thirds as large, being about two lines in length. They are much quicker in their motions, and much more persevering in their attacks, than the musquito, forcing their way into any crevice, for instance between the glove and the coat-sleeve. On the other hand, they are easily killed, as they stick to their prey like bull-dogs.

June 29th.—Among the birds here, the most abundant is the white-throated sparrow, (*Fringilla pennsylvanica,*) evidently breeding in great numbers in the swamp, for from the top of nearly every dead tree a male bird of this species was pouring forth his loud, striking note, something like the opening notes of the European nightingale. The females were not to be seen, and were doubtless sitting. I found the nest and new-laid eggs of the song-sparrow, but could not discover those of the *pennsylvanica.* In the evening the Professor made the following remarks on the classification of birds:

"Animals have usually been classed merely according to the characters of the adult. In some instances, however, the importance of an examination of the embryonic state also has already been acknowledged by naturalists. For example, the barnacle, though in fact a crustacean, has in the adult state so much the appearance of a mollusk, that its true relation could hardly be recognized without the investigation of the embryo, which has all the aspect of the ordinary crustaceans. Hitherto embryology has been applied principally to the study of functions and organs, and not of classification, but I think it of the highest importance to the right understanding of the affinities of all animals.

"Birds are at present classed according to the form of the feet and bill. They form a very distinct group in the animal kingdom, all having wings, naked bills, and the same general form of feet. Yet no class has puzzled naturalists more.

"Great weight has been given to the form of the toes. In one great group, (*Palmipedes,*) at least three of the toes are united by a web (four in the pelican and gannet,) throughout their whole length. In all other birds the toes are free, though in some the upper joints are united.

"The form of the claws has also been considered of great importance. In *birds of prey* an agreement in the form of the claws is accompanied by a resemblance in the shape of the bill. In others, however, this is not the case; thus the parrots, with crooked bills, and the woodpeckers with straight bills, have been united as climbers. Again, the passerines, classed together from the shape of the bill, agree very well in other respects; but in the water-birds, species of very various characters have been brought together.

"Taking all these things together, ornithologists have very generally agreed on four or five great divisions, though with some differences. Thus the waders, or those birds having the tarsus and a space above it naked, are put in one group by some, and by others made into two. The arrangement of the water birds now most generally admitted is: *Palmipedes:* with the feet united, except in one group, (the grebes, &c.) This division, I incline to think, is made on an insufficient consideration of their true affinities. *Grallatores:* with three toes before, and one behind. The *gallinaceous* birds form a very natural group, having the upper jaw arched, and feet like those of the grallatores, but with short and curved claws. The *climbers* have two toes before and two behind, of which one may generally be moved in either direction. Sometimes there is only a trace of this arrangement, in a closer union of two of the toes with each other than with the rest. The *passerines* have curved claws, or sometimes the hind-claw is straight; three toes before and one behind. Some make three groups of them, bringing together those with flattened bills, (Insectivora;) those with conical bills, (Granivora,) and those with the upper mandible much stronger than the lower, (Omnivora.) Some again separate from these the swallows, pigeons, &c.

"The toes in all birds have the same number of joints. The hind toe always consists of a single joint, the inner toe of two, the middle of three, and the outer of four. This arrangement is important in distinguishing the fossil tracks of birds from those of other animals, it being peculiar to them.

"In examining birds within the egg, I have recently found some characters to be less important than has been supposed. Thus the foot of the embryo robin is webbed, like that of the adult duck; so also in the sparrow, swallow, summer-yellow-bird, and others, in all of which the adult has divided toes. The bill also is crooked and the point of the upper mandible projecting, as in the adult form of birds of prey. These latter, then, it would seem, should be brought down from the high place assigned to them on account of their voracious and rapacious habits, as if these would entitle an animal to a higher rank. For the resemblance of an adult animal to the embryo of another species, indicates a lower rank in the former.* Probably the true classification of birds would include various series, each embracing representatives of all the various types now admitted as distinct."

Mr. Ballenden, of the Hudson's Bay Co., to whom the Professor had letters, paid him a visit to-day, and showed the most obliging

*For further details see Prof. Agassiz's Lectures on Comparative Embryology, delivered at the Lowell Institute, January, 1849; published in the *Daily Evening Traveller*, and afterwards in a pamphlet form by the same publishers.

readiness to forward his plans, giving him letters to the gentlemen in charge of the various posts on the lake, which were highly serviceable to us.

Dr. C. T. Jackson and the gentlemen engaged with him in the geological survey of the copper region of the south shore of Lake Superior, also arrived to-day, and his assistant, Mr. Foster, gave the Prof. some valuable information, particularly concerning Neepigon Bay, which he had visited.

Mr. McLeod, of the Sault, lent to the Professor Bayfield's large map of the Lake, (which we had not been able to procure,) enriched with manuscript notes, and gave him the results of various geological excursions on the lake.

June 30th.—Rainy. Nevertheless, our preparations being made, we decided to start. It was necessary to convey our multifarious luggage to the upper end of the portage, above the rapids, a distance of about two-thirds of a mile. Walking thither in the rain, over a road made across the swamp, the surface of which is strewed with bowlders of various sizes, we found a collection of warehouses and a few log-cabins, just at the commencement of the rapids. Here our boats were moored at a wharf at the extremity of which was a huge crane for unloading copper ore. Here also lay at anchor several schooners, and a propeller that runs along the south shore, and occasionally crosses to Fort William.

Our boats were three in number; one large Mackinaw boat and two canoes of about four fathoms' length. One of these canoes was kindly lent to us by Prof. James Hall, of Albany, the other we hired; the boat we had been obliged to buy, giving eighty dollars for it. It proved a considerable hindrance to speed, being always behind, except when the wind was aft and fresh. Our luggage, however, with the collections of specimens and the apparatus for collecting, could not be carried in canoes without uncomfortably loading them. From my own subsequent experience I should say that what is called a "five-man-boat," is the craft best adapted for such an occasion as ours, and this opinion was confirmed by a gentleman at the Sault who had tried the experiment. The canoes were precisely what one sees from Maine to Michigan, birch-bark stretched by two layers of thin, flat, wooden ribs, one transverse, the other longitudinal, placed close together, with a strip of wood round the gunnel, and the whole sewed with pine-roots. It is said that after the materials are cut out and fitted, two men to put them together, with six women to sew, can make two seven-fathom canoes in two days. While on the lake the canoes are not usually paddled, but rowed, the same number of men exerting greater force with oars than with paddles. By doubling the number of men, putting two on a seat, more of course can be accomplished with paddles. The gunnel of a canoe is too slight to allow of the cutting of rowlocks, or the insertion of thole-pins: so a flat strip from a tree, with a branch projecting at right angles, is nailed to the gunnel, and a loop of raw hide attached, through which the oar is passed.

Our boats were stowed as follows: On the bottom were laid setting-poles and a spare paddle or two, (to prevent the inexperienced from putting their bóot-heels through the birch-bark,) and over these, in the after part, a tent was folded. This formed the quarter-deck for the *bourgeois*, (as they called us,) and across it was laid the bedding, which had previously been made up into bolster-like packages, covered with buffalo-robes, or with the matting of the country, a very neat fabric of some fine reed which the Indians call *paquah.* These bolsters served for our seats, and around them were disposed other articles of a soft nature, to form backs or even pillows to our sitting couches. The rest of the luggage was skilfully distributed in other parts of the canoe, leaving room for the oarsmen to sit, on boards suspended by cords from the gunnel, and a place in the stern for the steersman. The cooking utensils were usually disposed in the bow, with a box of gum for mending the canoe and a roll or two of bark by way of ship-timber. Our canoe was distinguished by a frying-pan rising erect over the prow as figure-head, an importance very justly conferred on the culinary art in this wilderness, where nature provides nothing that can be eaten raw except blueberries.

The voyageurs (some ten or twelve in number,) were mostly half-breeds, with a few Canadian French and one or two Indians. All except the Indians spoke French, and most of them more or less English, but there were only two who spoke English as well as they did French. The half-breeds were in general not much if at all lighter in complexion than the Indians, but their features were more or less Caucasian, and the hair inclining sometimes to brown. They were rather under medium height, but well made, particularly the chest and neck well-developed. The Indians were Ojibwas (ŏjíb-wah), and had the physical peculiarities of their tribe, viz.: a straighter nose, rather greater fulness of the face, and less projecting cheek-bones, than the Western Indians. But I was most struck with the *Irish* appearance of the Canadians, and though I ascertained that they had no Irish blood in their veins, yet the notion often recurred during the trip, and I found myself several times surprised at missing the brogue. They were blue-eyed, with flaxen hair, a rather low and square head, and high-pitched voice. This resemblance, which also struck others of the party, is interesting as showing perhaps the persistance of blood and race. It was not until afterwards that I was informed that the French of Canada are Bretons and Normands by origin; thus coming from that part of France in which, whether as most remote from invaders, or from having been recruited from the British Isles, the Celtic blood is best preserved. I do not know whether the Celtic features are so noticeable at this day in that part of France, but no one would have ever taken these men for Frenchmen.

Our preparations occupied some time; finally, just as we were about to start, it was suggested and on short consultation decided that we must have an additional canoe; those provided proving insufficient to hold us all comfortably. Two of the party accordingly remained behind to attend to this matter, and we got under weigh.

We had but three in the canoe besides the boatmen, which gave us an advantage over the others, so that we immediately took the lead, and soon ran the other boats out of sight. The rain ceased, but the weather was still unsettled, and the wind, strong down the river, much retarding our progress. Our men had a hard pull of it, yet they kept up an unceasing chatter in Ojibwa, (which sounded occasionally much like Platt-Deutsch,) interspersed with peals of laughter. About five o'clock we reached the Pointe-aux-Pins, about six miles from the Sault, and as the wind had become very strong, and the other boats were far behind, we decided to wait for them.

The Point is a mass of sand and gravel, mingled with large stones; towards the main land are a few pitch-pines and willows; the ground covered with moss and low bushes, and a few strawberries. Some flocks of pigeons were whirling about, at times dashing down to the ground, and then rising high in the air; a couple of these were shot, as well as a young creek-sheldrake, (*Mergus cucullatus*,) from a small flock in a creek emptying into the river. On returning to the neighborhood of the boat, we found a fire lighted and preparations making, under the superintendence of Henry, the steersman, for getting a supper from a ham and some flour which had been providently stowed in our canoe. The process of frying the ham, and roasting the birds on a spit stuck in the ground, was neither new nor interesting to me otherwise than as conducive to supper. But the process of making bread with mere flour, water, salt, and a frying-pan, excited my curiosity. Nothing to my knowledge was put in to make the bread *rise*, neither had anything been provided by us for that purpose, yet the dough, after having been kneaded for a long time, pressed down into the frying-pan and toasted before the fire, turned out excellent bread, perfectly light and well-tasted. By what mystery the fermentation was accomplished or gotten over, I leave to the initiated to make out. Perhaps the vigorous and long-continued kneading may have supplied the place of yeast; at all events, some of the party, whose cooks were more sparing of their labor than ours, used to have heavy bread, a misfortune that never befell us.

Shortly before dark the other canoe arrived, and we learned that the bateau had been driven back by the force of the wind, and had put in for the Canada shore.

We were now established for the night. There was nothing very cheery about the aspect of the Pointe-aux-Pins; — a desolate mass of sand, with the tent standing out against the bleak sky, backed by a few stunted willows, the river a couple of hundred yards in front, and a horizon of forest beyond.

A bleak, desert situation, so exposed to the wind that we had to carry a guy far to windward, attached to the peak of the tent, to prevent it from being blown over. No vestige of human habitation in sight, and no living thing, except the little squads of pigeons scudding before the wind to their roosting place across the river. Yet I felt as I stood before the camp-fire, an unusual and unaccountable exhilaration, an outburst, perhaps, of that Indian nature that delights in exposure, in novel modes of life, and in going where nobody else goes. We slept comfortably on the sand, which makes a good bed, easily adapting itself to the shape of the body, with the drawback however of getting into one's hair and blankets.

July 1st.—Early this morning our companions in the bateau joined us. They had run some danger of swamping, the day before, and had been forced to put in on the Canada side, not much above the Sault, where they found good quarters on board a steamboat that had been seized for smuggling and laid up in ordinary by the Canadian government. After breakfast we started in company and got up to Gros-Cap, about fifteen miles, where we halted, there being no good camping-ground for some distance beyond.

From the Pointe-aux-Pins to the mouth of the river, some four or five miles, the width of the stream varies from one to two miles. Here it enlarges rather suddenly, so that Gros-Cap and Point-Iroquois, the Pillars of Hercules of Lake Superior, as some one calls them, are six or seven miles apart. This is the true entrance of the lake. The shore continues low and marshy for some distance beyond; then the high land of the Cape comes in sight, stretching across at right angles with the course of the river, and soon the scenery in the immediate neighborhood also assumes the proper character of the lake. I was struck with the similarity to some portions of our sea-coast, for instance, in the neighborhood of Gloucester in Massachusetts, or Cape Elizabeth, near Portland. Rocky points, covered with vegetation, rising abruptly from deep water, alternate with pebble beaches; back of this, the land slopes gradually upward, densely covered with white pine, canoe-birch and aspen, to the foot of the cliff, which rises steeply to the height of seven hundred feet, showing vertical faces of bare rock, and crowned on the top with evergreens.

We encamped early in the day in a narrow cove, formed by a point of low rocks, running almost parallel to the shore. Here we encamped among large aspens, and thickets of the beautiful white-flowering raspberry of the lakes, (*Rubus Nutkanus.*) Our friends joined us from the Sault with a large seven-fathom canoe pulling three oars, which was christened the "Dancing Feather."

After dinner, two of us set off for the top of the cliff. The slope forming the border of the lake in this spot seems to be merely the *débris* fallen from the face of the cliff, which rises so abruptly that we were obliged to skirt along its base for some distance before we found a practicable ascent in a gully in the face of the rock, and here even only by help of the trees. Climbing along the ledges and from one trunk to another, we at length reached the top, a mass of rock, intermingled with spruce trees. The wind blew fresh and we were in hopes to be free from the flies and musquitoes, which were rather troublesome below. The result showed that we had reasoned correctly as to the musquitoes, but not at all as to the flies, who, as we now learned for the first time, by actual experience, affect high and dry places. They surrounded us in such swarms that it was impossible to remain quiet for a moment; brushing them away with branches was of no use, and even a musquito veil proved no protection. The meshes being rather larger than their bodies, they alighted for a moment upon it, and then deliberately walked through. When the wind blew very hard they would make a lee for an instant, and then reappear in clouds. On arriving at the camp, we were speckled with blood, particularly about the forehead and back of the ears. Our faces looked as if charges of dust shot had been fired into them, each sting leaving a bloody spot.

It was discovered this evening that some things had been left behind, and our short experience had already taught the need of some others, so two of the party volunteered to go back in a light canoe to fetch them from the Sault.

July 2d.—It was thick and rainy to-day, so we did not leave our camp. In our immediate neighborhood were several lodges of Indians; "*gens du Lac*," as our men called them, from whom we bought trout. They had the general features of the Ojibwas, but ragged and dirty. They subsist by fishing, and seem to bear out the remark that among savage nations, the fishing tribes are the most degraded. Their lodges were composed of a dome-shaped framework of poles, over which were laid pieces of birch bark. We often afterwards met with these frames at our encampments, but without the bark covering, which they probably carry off with them. They are perpetually shifting their quarters, for no reason but mere restlessness, often leaving a prosperous fishery to go off to some other place where the prospects are entirely uncertain.

During our stay at this place, finding it inconvenient to eat our meals all together, we separated into four messes, each having its boat and its tent, and making its separate camp-fire and *cuisine*.

This arrangement is indeed on many accounts an advisable one. Otherwise there is a great deal of squabbling among the men, for each is willing to look out for his own canoe and *bourgeois*, but not for the rest, and they try to shift the labor from one to the other. Except that we usually encamped in the same neighborhood at night, and were sometimes within hail of each other during the day, we might henceforward be considered as four separate parties.

In our canoe everything settled down after this into a very methodical routine, which I may as well describe here. We were provided in all respects with an independent equipment, embracing provisions for a day or two, viz., salt pork, ham, potatoes, peas, beans, flour, hard bread, rice, sugar, butter, coffee, tea, pickles and condiments. When we landed in the evening, as soon as the canoe was unladen and hauled up, two of the men proceeded to pitch the tent, while the other collected wood, made a fire, put on the tea-kettle, and brought up the mess-chest, which contained tin plates, knives and forks, &c., and also in bottles and tin cases those of our stores that would be injured by moisture. Then they devoted themselves to preparing supper. One kneaded dough in a large tin pan; another fried or roasted the fish, if we had any, or the pork or ham, if fish was wanting. A large camp-kettle, suspended by a withe from a tripod of sticks, over the fire, contained a piece of pork, and dumplings, which the men preferred for themselves, or occasionally a rice pudding for us. When all was ready, an India-rubber cloth (which served to protect the luggage, and on occasion for a sail,) was spread on the ground, and the dishes arrayed upon it. Around this we reclined in the classical fashion, and Henry stood by to serve coffee and fetch anything that might be wanted. As to provisions, if I were consulted about the outfit of such a party as ours, I should recommend a full supply of rice and sugar. Maple sugar (which can usually be had in these regions,) is as good as any, for one's taste becomes unsophisticated in the woods; the rice, I may observe, must be boiled in a bag, and not loose in the camp-kettle, as the Professor's man did it one day, when it came out in the shape of mutton broth without the mutton. Salt pork is very well where one goes a-foot, or paddles his own canoe, but in a life of so little exertion as ours, the system cannot dispose of so much carbon, and rejects it accordingly. For the same reason, perhaps, I found that I not only did not miss the milk in the coffee, but could not drink it when it was sent to us at the trading posts. Potatoes would no doubt be a good thing, but our men did not know how to cook them. Before we started, the question being raised as to the relative quantities of tea and coffee to be bought, the most thought they should drink very little coffee, but depend upon tea. On the contrary, however, I believe there was hardly a cup of tea drank on our whole tour, (except by the men,) when coffee could be had. The truth is, that tea is very refreshing after a hard day's work, and it was prized accordingly by the men, but we did not take exercise enough to care for it.

After we had done our meal, the men took theirs. At dark Henry brought us a candle, and then he and the other men turned in, all lying close together, sometimes entirely in the open air, sometimes with their heads under the canoe, or if it rained they made a kind of tent with the India-rubber cloth. They had each a very comfortable supply of blankets, &c., and somewhat to my surprise each was provided with a pillow. Our own bedding consisted, in my case, for instance, of a buffalo robe by way of mattress, and two very heavy Mackinaw blankets, which I had brought from Boston, as they are dearer and of inferior quality at the Sault. Others had the same, or an equivalent. I have heard of travellers who brought blow-up mattresses of India-rubber, and if these things are manageable, I should recommend their being taken, as we were often inconvenienced by the large angular stones of the beaches on which it is usually necessary to encamp. At all events I should decidedly take a pillow of this description, for we soon found the voyageurs were wiser in this matter than we. In the morning we started about sunrise, and usually made ten or twelve miles before breakfast, giving the men a rest of about an hour at breakfast time. At noon we stopped to lunch, making no fire. Our usual time for encamping for the night was seven o'clock, but this depended somewhat upon our reaching a good camping-ground. Once an hour or so during the day the men would lie upon their oars, and one of them would light a short clay pipe, filled with *kinni-kinnik*.* After a

* A mixture of dried bear-berry leaves (*Arctostaphyllus uva-ursi*) and plug-tobacco, rubbed together between the thumb and fingers. Their tinder was a fragment of a tough, yellowish fungus that grows on the maple and birch.

puff or two he would pass it to the next, and when each had had his turn, it was put away and they took to their oars again.

While detained in our tent by the rain to-day, we employed ourselves in manufacturing a musquito net out of some muslin we had brought for the purpose. This being provided with cords, was stretched at night from one tent-pole to the other, (the tents being roof-shaped, with flat gables and a tent-pole at each end,) and pegged down to the ground at the sides, thus forming a tent within the tent; an arrangement quite essential to a comfortable night's rest in these regions.

The point forming the breakwater of our harbor, and to which the bateau was moored, presented the first example we had seen of drift scratches and grooves. Some of the grooves were several feet in length, the surface a curve of eighteen inches radius, and as smooth and even as if cut with a gouge. These marks were almost entirely confined to the inner side of the point, where some of the scratches could be traced as far below the surface of the water as we could distinctly see, that is, some five or six feet; the lake side presented rough points of rock, occasioned, as Prof. A. explained, by the decomposition of the surface on that side, from its greater exposure to the wind and waves. In the afternoon, the rain having ceased, we assembled to hear the Professor's remarks on the specimens of various rocks collected in the neighborhood.

"Geology," he said, "investigates the great masses of the rocks; mineralogy the forms and composition of their materials. Geologists are apt to neglect the study of mineralogy, and thus to overlook the differences, in different countries, of rocks bearing the same name.

"If geology had been studied first in this country, the text-books of the science would read very differently. For example, there is no rock in this region answering the description of true granite. We have granitic rocks enough, but none of an amorphic structure. All are more or less stratified. At the beginning of the century, each of the two great schools in geology maintained that all rocks had but one origin, disagreeing, however, as to what this origin was. The reason was, each had examined only the rocks in its neighborhood. About Edinburgh the rocks are trap; Hutton, therefore, referred everything to the action of fire. Near Freiberg there is nothing but sedimentary rock; Werner, therefore, would admit no influence but that of water.

"Most of the rocks in this region are Plutonic, that is, they manifest the action of fire. The only sedimentary or aqueous rock found here is sandstone, the age of which is uncertain, as no fossils have as yet been found in it.* Probably it belongs to the Potsdam sandstone. It passes frequently into quartz and quartzose rock. If quartz were broken up, mixed with clay and lime, and subjected to the action of heat, the forms of metamorphic rocks would be produced which we see here. Some varieties, however, are quite peculiar, as, for instance, a red felspar porphyry, with numerous specks of dark epidot."

The canoe from the Sault arrived this afternoon.

July 3d.—The air was very chilly this morning, when at about half past five our canoes issued from the little cove into the open lake. But the prospect before us was sufficient to divert our thoughts from any discomfort. On our right was the deep bight of Goulais Bay, terminated by Goulais Point, a high promontory of the character of Gros Cap. Directly ahead rose the fine headland of Mamainse, ("*little sturgeon*,") distant about thirty miles. We were yet in the shadow of Gros-Cap, and all the shore in sight seemed to have the same mountainous character. Ridge over ridge, distinct at last only by the cutting line against the sky, it had the freedom and play of outline, which, rather than size, distinguishes a mountain from a hill. So different was the scene from anything on the Lower Lakes, that although I knew in general that the shore of Lake Superior was much bolder and more rocky than that of the others, yet it took me by surprise, and I was disposed to think this part of it an exception, until assured, by one who had been here before, that the grandeur of the scenery constantly increased to the northward.

Opposite Mamainse stands White-Fish Point on the south shore, and the two approach each other somewhat, repeating on a large scale the feature of Gros-Cap and Point-Iroquois, which is again repeated on a gigantic scale by Point Keewaiwenaw and the land of which Otter Head forms the outer extremity. White-Fish Point has the outline of a raven's head, with a projecting sand spit for the bill: the high land above was just visible. We passed this morning Isle Parisien and the Sandy Islands, low, flat islands covered with

* Remains of chambered shells have been since found in this rock, on the southern shore of the lake.

trees, like all those in this part of the lake. Several loons flew by to-day, and whenever one appeared, the men all began to shout "*oory, oory*," which seems to be the Indian "hurrah,"* whereupon the bird would usually fly in circles round the boat. This was regularly repeated whenever a loon came in sight; the experiment was tried on gulls and sheldrake, but not with the same success.

The sun and wind rose together, so that by eleven o'clock it was very warm, and at the same time so windy that we were obliged to make for Maple Island, a low, sandy island, densely covered with trees. On the lake side the trees were covered with long lichens, (*Usnea*,) and presented a weather-beaten aspect, much in contrast with the side towards the land. The shore here was evidently wearing away, and the roots of many of the trees were exposed. The beach was covered with large fragments of red porphyry, and slabs of dark red sandstone, often ripple-marked.

When the bateau arrived we found they had caught some fine trout on their way hither. This excited the emulation of the other boats, and hooks, &c., were forthwith prepared. The tackle consists of small cod-line, with a hook (or often two,) with a large sinker of lead melted round it. The bait is a piece of pork, or better, a trout's stomach, drawn over the hook and tied at the shank. A simple plate of brass, with a couple of hooks on the lower edge, is said to be very effective without any other bait, and I have heard of a pewter spoon being used with success. This is allowed to trail a dozen fathoms astern of the canoe, and kept in constant motion by jerking the line. After the first excitement, as the fish did not bite oftener than half a dozen times a day, and sometimes not at all, the lines were handed over to the steersmen, who made them fast round their paddles, and thus kept up the requisite motion without any trouble. The fish we caught were the lake trout, (*Salmo amethystus*,) and Siscowet, (*Salmo Siscowet Ag.*, see Plate I.); their average weight five or six pounds. The latter fish is so exceedingly fat that we found it uneatable. It is said to be much improved by pickling. White-fish and lake-herring are taken only in nets, and the other fishes only in the streams. The wind did not allow us to get off to-day.

* See Kip's Early Jesuit Missions, pp. 60, 140.

July 4th.—Thermometer one would guess about 40° Fah. this morning. Goulais Point is separated from Mamainse by Batcheewauung Bay, by far the most considerable inlet on the E. and N. E. part of the lake, (being about ten miles deep, by five across the mouth,) unless we count as such Michipicotin Harbor, which is rather the commencement of a new direction of the shore, than an indentation in it. The general outlines of the lake are simple, and though cut into innumerable narrow coves, yet bays of any considerable size are rare.

Not long after starting we encountered several canoes of Indians, (*gens du Lac,*) on their way to the Manitoulin, to receive their annual "present" from the British Government. Among them was a chief, who stood up and addressed our men in his own tongue, which, as we were informed by Henry, was a separate dialect of the Ojibwa, but intelligible enough to them. In an unwritten language, dialects soon spring up. A lifetime, the men said, was sufficient to make a noticeable change in their language, though where large numbers are collected together and any kind of schooling exists, the bibles and catechisms must do much to arrest the process. We stopped for breakfast at ten o'clock, at a point under Mamainse, much resembling Maple Island in its general features. Charred logs and beds of matted leaves on the beach, showed it had been recently visited.

From Mamainse onward the character of the shore changes. Instead of the low sandy islets, we now passed among isolated rocks of greenstone, rising abruptly from deep water, generally bare, but sometimes crowned with a tuft of trees at the top. The rock, which about Gros-Cap is sandstone, often unaltered, now becomes more highly metamorphic. But the larger islands and the edge under the cliffs, continue of sandstone, and are flat and low for some distance to the northward. The line of cliffs is continuous, rising at a distance of a quarter of a mile at most from the water, with an average elevation of two to three hundred feet. The whole surface, down to the very beach, was covered with trees: indeed I may say once for all, that with the exception of some ancient terraces of fine sand and gravel to be described hereafter, and a few summits of bare rock, the entire shore of Lake Superior, as far as we went, is continuously covered with forest. The trees continued the same, except that the white pines and maples had disappeared. The number of species is small; black and white spruce, balsam fir, canoe birch and aspen, with arbor vitæ in the moist places, and here and there a few larches and red pines, with an occasional yellow birch; the spruces prevailing on the high land, and the birch and aspen near the water, yet everywhere a certain proportion of each. From the great similarity of the evergreens on the one hand, and the white-stemmed aspens and birches on the other, at the distance of a couple of hundred yards the forest seemed to be composed of only two kinds of trees. The trees are not large, usually not exceeding thirty or forty feet in height. Yet the whole effect is rich and picturesque. Here, as in all the features of the lake, the impression is a grand uniformity, never monotonous, but expressive of its unique character.

The resemblance to the sea-shore often recurred to my mind. According to Dr. Leconte, several insects found here are identical with species belonging to the sea-shore, and others corresponding or similar. The beach-pea, *Lathyrus maritimus*, and *Polygonum maritimum*, both of them sea-shore plants, are abundant in this neighborhood; the former, indeed, throughout the north shore of the lake.

Although so cold this morning, yet by noon the heat was intense. The weather, indeed, during the whole time we were on the lake, was such as we sometimes have in Massachusetts in September; cool morning and night, and warm in the middle of the day. The sun has great power, and blisters the hands and face unless well guarded, but the air is cooled by the vast expanse of water, (which contains ice during the largest part of the year, and even on the surface is rarely above 40° Fah. at any season,) so that it was never warm in the shade, or when the sun was below the horizon. We in our canoe being induced to land by a white pebble beach which at a very short distance had the appearance of sand, and thus promised an entomological harvest, indemnified ourselves by a bath in the icy, crystal water. Here was another resemblance to the sea; we could dive from the rocks into thirty feet of water, which, moreover, was of about the ordinary temperature of the ocean at Nahant. Above the beach and parallel to it was a terrace of sand about fifteen or eighteen feet in height. Others of the same kind but of various heights we traced during the day, sometimes only by the terracing of the forest on the different levels.

The cliff, which rose a few hundred yards from the beach, was cloven to the base, presenting a wide chasm of bare, splintered rock, several hundred feet deep, nearly parallel to the shore. The surrounding woods had been burnt, leaving the black stems, some standing and some lying crossed at various angles, like jack-straws. The ground was already covered with the fire-weed, (*Epilobium angustifolium,*) striving to conceal the ruin with its showy blossoms. Black flies very numerous and troublesome. They appear to have a fondness for the burnt woods, in which we always found them abundant.

In the course of the day we passed a deserted mining "location," marked by ruinous log-huts; and in another place we saw on the rocks the wreck of one of their bateaux. At about five o'clock we came in together at the Pointe-aux-Mines, or Mica-Bay, as they call it now. This establishment belongs to the Quebec Mining Company, who have already commenced operations here. It is a deep cove, protected on either side by ranges of rocks, with a broad beach at the bottom, and above this a steep bank, on which, at the height of thirty or forty feet above the water, stands the very neat wooden cottage of Capt. Matthews, the superintendent, and about it the storehouse, the lodgings of the workmen, &c. We were very hospitably received by Capt. and Mrs. Matthews, and enjoyed in their house the luxury of a civilized *tea*, before which, however, we visited the mine, which is about half a mile from the house, by a Brocken-like wood-path, nearly all the way up hill.

Capt. M., avoiding the errors of his predecessors on both sides of the lake, spent eighteen months in making his preparations, securing a thorough system of drainage, ventilation, &c., before attempting to get out any ore. The work seemed to be carried on with great method and thoroughness, and to be in very successful operation. The present state of the concern he represented as most promising.

July 5th.—The Professor before starting showed us a rock at the south entrance of the bay, which he considered a proof positive of the correctness of the glacial theory. Its surface was a couple of hundred yards in extent, sloping regularly north to the water's edge. The whole was polished and scratched, except where disintegrated. The scratches had two directions, the prevailing one north 10° to 30° west, the other north, 55° west. The scratches on the outer or lake side seemed to have a rather more westerly direction than the rest. Great numbers of these striæ could be traced below the water's edge, from which they ascended in some places at an angle of 30° with the surface, showing, as the Professor remarked, that they could not have been produced by a floating body. The rock is granitic, with an astonishing number of veins and injections of epidotic felspar, granite, and trap, often crossing each other so as to form a complicated net-work. Wherever exposed, it was ground down to an even surface.

The day was calm and very warm. About noon we stopped at Montreal River, (one of several of this name on the lake.) This river, forty yards wide at the mouth, empties through a kind of delta, partly overgrown with large trees. The water is deep and clear, but of a rich umber color, such as we often see in the small streams in New England. This is the case with all the rivers we met with on the lake; the color was there attributed to the presence of pitch, an explanation the Prof. thought likely to be correct. At its entrance into the lake is a broad beach, which on the south forms a point somewhat jutting across the mouth.

On the northern side, at a short distance from the water, the beach, which was of small pebbles, had a slope of 30° that is, nearly as steep as it could stand. We frequently met with such steep beaches, often of a considerable height. Outside there is a bar which extends entirely across, six feet below the surface. The stream issues from the hills through a chasm sixty or eighty feet deep and a few yards wide, with straight walls of rock, somewhat overhanging on one side. From this gorge the river issues with great force. Higher up there was a cascade some forty feet in height, falling from a dark, still lakelet, and above this again a succession of rapids. This is the general manner in which the streams on this side of the lake make their way down from the table-land through the barrier of rock. On the delta below were several of the largest red pines (*P. resinosa*,)

I ever saw. I regret that I did not take the girth of one of them, which must have been five feet in diameter. But the black flies and musquitoes were so annoying as to absorb much of one's attention; the only refuge was the beach, where we had made fires to drive them off. The heat of the day made a bath very agreeable ; we found the current of the river at the mouth so strong as to make some difficulty in swimming even this short distance across.

One of the men killed here a squirrel of the kind that takes the place of our "Chipmunk" in these regions, the *Tamias quadrivittatus*. It resembles our animal, except that it is a little smaller, has a longer tail, and four black stripes instead of three, on its back. We found it afterwards much more abundant than any other species, particularly on hill-sides among broken rocks, attracting the attention by its loud, peculiar cry.

On the bank was the skeleton of an Indian lodge, and a well-worn trail ran up along the stream. The Indians here as everywhere love the neighborhood of rivers, where we always found traces of their camps. As we left the river we saw some of their handiwork on a rock over the beach. It was the picture of a schooner under sail, scratched out from the black lichens so as to show the lighter surface of the rock.

The Professor pointed out here the difference of water action from that of ice. The former, he said, leaves the harder parts prominent, although the whole is smoothed, as was the case in this instance, but the latter grinds all down to a uniform surface, scratching it at the same time in straight lines.

This afternoon, the water being smooth, we tried an experiment as to its transparency, by lowering a tin cup at the end of a fishing-line. It went out of sight at forty-two feet. It is said that when the water is entirely unruffled and the sky clear, a white object may be seen at the depth of one hundred and twenty feet.

Passing Montreal Island, a large, low island covered with trees, some three or four miles from the shore, we threaded our way through a group of rocky islets and came out into a wide bay, which we *traversed*, i. e., took the direct line across, instead of following the curve of the shore. The voyageurs are in general unwilling to keep out more than a quarter of a mile or so, and usually coast along the rocks. But

G. Elliot Cabot from nat. Lith. of A. Sonrel Löffler on stone.

RIVER TERRACES. (TOAD RIVER.)

this time the weather being so calm, they ventured on a course which brought us at one time about two miles from the shore. Their caution seemed to some of us, accustomed to a bolder style of navigation, somewhat exaggerated. But if the rocky character of the shore, the suddenness with which both wind and sea rise here, and the frailness of the vessels be taken into consideration, perhaps it is not so unnecessary as it would seem at first. Moreover it is to be remembered that although a swim of a mile might under ordinary circumstances be no very desperate undertaking, yet in this icy water, a person swamped at that distance from the shore would in all probability be disabled long before reaching it. And even if the shore were reached, the prospect of having to make one's way on foot through this rugged, gameless, fly-possessed region to the nearest trading-post or mining location, would be dismal in the extreme. Deprived of salt pork and biscuit, one's subsistence would depend on the chance of snaring a hare or two, with *tripe de roche* as the sole alternative.

As we pushed out into the bay a weather-beaten veteran in the Professor's boat struck up a song, the others in the canoe and those of the "Dancing Feather" joining in the chorus and repeating each verse as he got through with it. Their singing had nothing very artistic about it, being in fact only a kind of modified recital, in a quavering and rather monotonous voice, coming, with little modulation, from the mouth only, but they kept time well, and it had a heartiness and spirit that rendered it agreeable. Their songs were all French; according to the Professor, the wanton *chansons* of the *ancien régime*, which the ancestors of these men had no doubt heard sung by gay young officers, in remembrance of distant beloved Paris. A strange contrast, as he said, between these productions of the hot-bed civilization of a splendid and luxurious court, and the wilderness where alone they now survive! The tunes, I fancy, are indigenous; at least, their singing had a certain *naïveté* and sometimes sadness about it quite at variance with the words. Neither the Canadians of the bateau, nor the Indians (of whom we had one, with a couple of half breeds in whom the Indian blood decidedly predominated, in our canoe) joined at all in the singing, either now or afterwards, though the Indians had a low monotonous chant which they occasionally grumbled to themselves.

We were looking for a stream called Flea River, where there were said to be falls of 90 feet, but not finding it, we decided to encamp on a sandy beach at the bottom of the bay, where we heard the noise of rapids. This was the Rivière aux Crapauds, or Toad River. There seems to be about this continent some pervading obstacle to the giving of reasonable names to places. In this region, indeed, one is not troubled with the classicality of New York, for instance, but, as in the case of those just mentioned, there is nothing very happy in the choice; and as for repetition, it is fully as bad as anywhere. There seems to be no end to Black Rivers and White Rivers and Montreal Rivers, occasionally varied into Little Black and Large Black, and so on.

As we neared the shore several canoes of Indians came out to sell fish. Their appearance as they squatted in their canoes, wrapped in their blankets, brought to mind the pictures of the South Sea Islanders. Their faces were round, full and rather flat, with no great projection of the cheek bones, the mouth very wide, with thickish lips, and gaping like a negro's. The hair brownish, and not so straight and coarse as that of the Indians in general. They were very filthy, and their clothing in general ragged. They seemed, however, good natured and happy, and grinned widely as they accosted us with the customary salutation of "Boojou, boojou!" (*Bonjour*, *bonjour*). Their canoes are very small, generally not more than nine to twelve feet in length, yet each usually contains a whole family; the man in the stern, the squaw in the bow, and the intermediate space filled up with two or three children of various ages, and generally at least one dog. In exchange for their fish they prefer flour or tobacco to money, of which they do not know the value very well. Indeed in any case they seem to regulate their demands rather by what the buyer offers than according to any notion of relative values. Thus when we offered in exchange for some fish a quantity of flour that would have overpaid it at the Sault, they thought it too little. On the other hand, a fifteen-pound trout was bought for a small fish-hook. We were afterwards told at Michipicotin (*Mishi-picótn*) that an Indian came there once from a distance to buy supplies, and produced a bundle, in which, after taking off wrapper after wrapper, there appeared enclosed — a ninepence! He had taken it in exchange for a number of valuable skins.

Pulling in for the beach we soon encountered the brown water of the river, but its mouth was not to be seen, the sand-beach extending apparently unbroken across the cove. When close in, however, we discovered an opening in the corner, whence issued a rapid current, and crossing a bar, we entered the mouth of the river, which is thus shut off by a spit of sand extending from the south or left bank of the river, northward across the stream, leaving only a narrow outlet. Inside, the river has a breadth of forty or fifty yards, flowing through a wide expanse of sand. This sand-beach is terraced, showing different heights of the river, and above the beach a succession of terraces was marked in the forest. On the south side the sand spit is cut away by the current, forming a vertical bank, in which is seen the horizontal stratification of the sand and gravel. The same general features were noticed subsequently at other rivers, and seem to depend on a general law.

On landing I walked towards the rapids, about a quarter of a mile up the stream. The flies and musquitoes made their appearance as soon as I entered the woods, and jumping down into the bed of the stream with the intention of sketching the mass of water that was foaming down over the rocks, I was instantly surrounded by such swarms that there was no getting on without a smudge. Even standing in the midst of the smoke, so many still clung to me that my paper was sprinkled with the dead bodies of those killed as I involuntarily brushed my hand across my face. We took refuge on the sand, at a distance from the woods, and here were comparatively free from them. But here their place was supplied by sand flies, the *brûlots* or "no-see-ems," an insect so minute as to be hardly noticeable, but yet more annoying where they are found than the black flies or musquitoes, for their minuteness renders musquito nets of no avail, and they bite all night in warm weather, whereas the black fly disappears at dark. Such is their eagerness in biting that they tilt their bodies up vertically and seem to bury their heads in the flesh. We found, however, that an anointment of camphorated oil was a complete protection, making a coating too thick for them to penetrate, and entangling their tiny wings and limbs.

July 6th.—Weather calm and overcast. Stopped to breakfast at the mouth of a river much like the last. Hearing the noise of rapids, some of us made our way up the stream until we came in sight of the fall, but the musquitoes were so unendurable that we hastened back.

As the day advanced the wind rose, and gave the bateau an opportunity to use her sails, but only for a short time, speedily coming ahead. The prospect in front of us was a noble one, lofty headlands rising one beyond the other until fading away in the distance. The shore, which had continued to present an uninterrupted ridge three or four hundred feet in height, becomes more abrupt and broken about Cape Gargantua, with deep chasms from decomposed dikes. The aspect of the coast here is exceedingly picturesque, steep broken points and rocky islands and islets generally sloping towards the north, and often worn smooth, grooved and scratched on the north side. We passed inside of one cliff, that showed a vertical face of at least two hundred feet in height, dyed with an infinite variety of colors by the weather and by the lichens, whose brilliancy was increased by the moist atmosphere. One orange-colored lichen in particular, was conspicuous in large patches. Here and there a tuft of birch aided, by the contrast of its bright green, the delicate gradation of tints on the gray rock. On a little strip of beach at the foot of a cliff in a cove called Agate Bay, we picked up an abundance of very pretty agates and other interesting minerals. At lunch-time we stopped at a curious rock, part of which seems as if cut away nearly to the level of the water, while the rest rises steeply to the height of thirty or forty feet. One of the common Indian legends about the deluge and the creation of the earth attaches to this rock, and the Indians still regard it with veneration. According to one of the men, "the Evil Spirit," (N. B. The gods of the aborigines here as elsewhere are to their Christianized descendants nothing but the devil, the *elder* spirit of all mythologies.) after making the world, changed himself and his two dogs into stone at this place, and the Indians never pass without "preaching a sermon" and leaving some tobacco. Even our half-breeds, though they laughed very freely about it, yet I believe left some tobacco on the top. This rock is remarkable in a mineralogical point of view. It is an amygdaloid porphyry containing asbestos and quartz, with thin layers of chlorite, and injections of granite. Numbers of martins and barn-swallows (*H. viridis and americana*) frequent these cliffs, and often a pair of screaming sparrow-hawks. Farther on, the hills were burnt over for a great distance, showing rounded summits of white scorched rock, the lichens and earth mostly washed off from them, but the blackened tree-stems still upright.

At Cape Choyye, where we encamped, the cliff comes boldly down upon the lake, the rocks rising from the water to the height of three hundred feet, with narrow chasms, sometimes vertical, sometimes slightly inclined, and strewed all the way up with stones, like the "slides" at the White Mountains. Beyond this it falls away into a vast basin of green sloping hills, curving inland and then sweeping out to rocky points beyond. The cliff, wherever the slope allows any soil to rest, is covered with birches to its base, leaving room for a wide slope of débris, and a beach that rises in five terraces, the lower one falling steeply to the water some twenty feet, showing that it alone can be connected with the present level of the lake, and that the rest must belong to former epochs.

At the water's edge were several unconnected masses of dark red sandstone in place. One mass, which John, our "middleman,"* christened "fire-boat" (i. e. steamboat) we waded out to, in order to avoid the flies while we bathed. Further on was a broad sheet of the same rock, sloping gradually from below the water up to the beach, full of "pot holes," worn into the rock by the action of the waves on stones lodged in its crevices. One of these stones, which was nearly round, might have weighed fifty pounds. Some of the holes were three or four feet deep, and as many in diameter. One was in the shape of a cloven foot; others formed steps, the stone having worn down at one side of the hole for a certain distance, worked on horizontally awhile, and then downwards again. The outer part of the rock, over which the water still washed at ordinary times, was covered with winding channels, of only a few inches' depth, running off into the lake, formed apparently by the grating back and forth of sand and small pebbles.

July 7th. — We were off by four this morning, but the wind

* The bowman and steersman of a canoe are called the "*bouts*" and are usually picked men, receiving higher pay than the "*milieux.*"

was up before us; and when we started, we foresaw that we should have head wind to contend with to-day.

At sunrise, the bay north of Cape Choyye presented a noble landscape. On all sides but one, an unbroken extent of rounded hills, so evenly wooded, that as the sun touched the curves at the top, it looked like a bank of grass. At one spot, far in the bottom of the bay, a white streak down the hill, and a faint roar at intervals, betokened the cascade of a stream that enters here.

The cove where we breakfasted, narrow and rocky at its mouth, and expanding inside, had something so liveable and civilized about it, that one might almost look for a cottage or two on some of the beautiful points of abrupt birch-clad rock.

On the rocks here, we found the purple flower of the wild onion, and the pretty Potentilla fruticosa: also brilliant lilies, reminding one of home. I was quite puzzled at finding our common red cedar, (*Juniperus Virginianus*,) which we had not seen hitherto, creeping on the rocks; not forming a tuft like the creeping savin, but a wide-meshed net-work of long straight shoots.

The shore on the northern side of the bay becomes yet bolder and higher, attaining, according to Bayfield's chart, the height of 700 feet. Between Cape Choyye and Michipicotin, a distance of about twenty miles, I did not notice but one beach, and that of only a few yards' extent. The rocks rise from the water, often vertically, several hundred feet, scored with deep rents and chasms, from decomposed trap-dykes, and striped down with black lichens. In some places, huge basalt-like parallelograms of rock stood out like pulpits. Along the top of the ridge, stretched the never-ending spruce forest, and wherever a gully or break varied the perpendicular face, a few birches crept downward from crevice to crevice.

On turning the point of Michipicotin harbor, we encountered the full force of the wind, now fresh from the west; and what was worse for us, something of a sea. Our course was such as to bring the wind abeam, and afford little shelter from the shore. We edged along from point to point, so close to the rocks that often the oars almost touched, and we were hardly lifted on the crest of a wave, before it broke against the cliff, and rushed up into the chasms at its foot. This was much closer proximity to a lee-shore than one would think prudent under the circumstances, yet our men dipped confidently on, and never ceased their chatter or their laugh for a moment, even when the bow man occasionally got a wet jacket from a wave that broke too soon. In truth, they had such perfect command of the canoe, that their course was no doubt the safest, for not only did we thus get some partial shelter from an occasional rock or point, but also the force of the wind was deadened by the nearness of the cliff.

At the little beach before spoken of, we stopped to rest. Here was an abundance of Labrador tea in blossom, Pinguicula, and Potentilla fruticosa. A rapid stream came in at the centre of the beach, about the mouth of which were multitudes of brook trout; some were caught, being the first that we had seen since leaving the Sault, although they were said to be numerous in all the streams. Beyond this, we found the rocks along the water much grooved and polished; one groove, about six inches deep, I traced for some twenty feet.

A sudden exclamation from the men, as we passed a deep narrow cleft, called our attention, but too late to see what they maintained they saw, namely, a quantity of *snow* at the bottom of the chasm. This seemed at first impossible in this burning July weather, with the thermometer about 80° at noon; but on reflection, this chasm, open to the N. W., must doubtless be filled with some hundred feet of snow in the winter, and the sun can never penetrate into it for a moment, so that the process of melting in the short summer must be slow. And then the summer was after all but just set in; Gov. Simpson, if I remember rightly, found the lake full of ice about the first of June.

We came in sight of the bottom of the bay, a wide and high sand-beach about a mile in length, but seeing nothing of the river, we approached a dark object on the beach, (which we had ascertained to be an Indian squatting on the sand) to make inquiries, but he retreated rapidly, and we had to coast for some distance, before we discovered the entrance.

Michipicotin River, a rapid stream of clear dark brown water, some two hundred yards wide, here cuts through the beach at right angles, leaving a somewhat projecting sand spit on the south. The name Michipicotin was declared by some of the men to signify "Big

Sandy Bay," certainly quite descriptive of the place, but they were not unanimous, some of them maintaining that nobody could say what it meant. It was a pretty hard pull to the factory, half a mile up on the left bank. Our approach had been already announced, probably by the Indian whom we saw on the beach, and we found Mr. Swanston, the gentleman in charge, at the landing when we arrived. He received us kindly, and showed us where to pitch our tents, in an open sandy space behind the factory, surrounded by whitewashed cabins, and the birch-bark lodges of the Indians. A large seine was suspended from a series of poles, and, near the water, a platform for dressing and packing fish.

This open space was bounded on the west by a steep ridge of stratified sand and gravel, some sixty feet high, cut through by the present channel of the river, and also by an ancient, now deserted channel further south. The river just above the factory takes a sharp turn to the north, doubling back in a direction nearly parallel to its course below. The interval between the factory and the lake, is thus a peninsula, the base of which is cut across by the former channel. It is evidently a range of sand-dunes, thrown up by the winds and waves, so as to divert the stream from a direct passage to the lake, to a course for some distance nearly parallel with it. From its mouth, to the Falls, it is a series of abrupt windings, though its general direction is straight; indicating, the Professor said, a bay repeatedly closed by sand-bars, one outside of the other, and successively cut through by the river. It evinced, he said, a contest between the river and the lake, beginning at a time when the level of the water was somewhat higher than at present.

Michipicotin is the principal post of the Hudson's Bay Co. in this district. From it, the other posts are supplied, and the line of communication with Hudson's Bay passes through here. It is sixteen days' journey up Michipicotin and Moose Rivers to James' Bay.

The agent's house is a little one-story cottage, uncarpeted, unpainted, and if my memory serves me aright, even unplastered, with panelling and projecting beams of pine, colored only by age; yet by no means uncomfortable in its aspect. The casings of darkened wood, the heavy beams of the ceiling and cornice, the ancient

G. Elliot Cabot from nat. Lith. of A. Sonrel, Cambridge. Löffler on stone.

CAMP AT MICHIPICOTON.

unpainted settle, and the wide niche for the capacious stove, now stowed away for the summer, had all a cosy and liveable look. And Mr. Swanston, although he had inhabited this wild country in the service of the H. B. C., at one or another of their posts, over twenty years, yet for anything in his manner or appearance (unless it were that he wore moccasins instead of slippers) might have left the pavement of Fenchurch Street only yesterday.

The life at these posts is a very quiet, and, doubtless, monotonous one; busy during the seasons when the hunters come for their supplies, or to bring in their furs; at other times, with only the fish to be seen to when the nets are drawn in the morning, some to be cleaned and salted, if there is a good haul, and perhaps put into barrels to be sent to the Sault. An arrival from some other post, a straggling party of explorers for copper, and above all, an occasional packet of newspapers from below, — these are the great events. In such a life, a man changes slowly, but gathers moss in another sense than that of the proverb.

A few hundred yards above the factory are very pretty falls, on the Magpie River,* which here empties into the main stream. Two miles up there was said to be a fine cascade, and a still more remarkable one fifteen miles up, which could be reached by a short cut of six miles by land.

Neither the love of the picturesque however, nor the interests of science, could tempt us into the woods, so terrible were the black flies. This pest of flies, which all the way hither had confined our ramblings on shore pretty closely to the rocks and the beach, and had been growing constantly worse and worse, here reached its climax. Although detained nearly two days, in order to supply the place of the Professor's canoe, (too small for his accommodation, and moreover rotten and unserviceable,) with a larger and fresh one, which had first to be put in order,—yet we could only sit with folded hands, or employ ourselves in arranging specimens, and such other occupations as could be pursued in camp, and under the protection of a

* The magpie of these regions, bye the bye, is no magpie at all, but a jay (*Garrulus Canadensis*), the "moose-bird" or "carrion-bird" of our lumberers; a confusion that might lead to error as to the range of the American magpie.

"smudge."* One, whom scientific ardor tempted a little way up the river in a canoe, after water-plants, came back a frightful spectacle, with blood-red rings round his eyes, his face bloody, and covered with punctures. The next morning his head and neck were swollen as if from an attack of erysipelas. Mr. S. said he had never seen the flies so thick. Year before last there were hardly any; last year they increased very much, and this season went beyond all his experience in this region. He consoled us, however, by the information, that it was nothing to what they have further north. On Mackenzie's River, the brigades are sometimes stopped by the musquitoes, and very often are able to advance only by having fires in the canoe.

The little plain on which we were thus collected, presented a stirring scene, with the buildings of the factory, the lodges, the white tents, the figures crossing from one fire to another, the half-starved Indian dogs prowling about to pick up anything loose, and the Indian women and children staring at the unwonted spectacle. The dogs were small, and fox-like in their appearance, and perhaps take rather after the foxes, since they bark, (contrary to what is said of Indian dogs in general,) and like them in a high key. Even the crying of the children had a wild, animal sound, resembling the barking of the dogs. A bull and some cows, (N. B. Mr. Swanston sent us fresh butter and milk, for tea,) and a robin hopping along the ground with an occasional chirrup, gave it by comparison quite a home look.

The hunters were most of them in the woods making canoes, and preparing for the winter campaign. In August they come for supplies of ammunition, &c., and are gone until the weather becomes too severe to be endured abroad. This is usually in January, but sometimes they do not come in until March.

According to Mr. S. they generally remain attached to the post of the district where they are born, obtaining their supplies on credit and paying for them in skins. It is said that they are very scrupulous about discharging their debts, and although they sometimes have credit for over £100 currency, yet these wild fellows, whose notions of morality seem in most points so loose, and in the

* Readers familiar with the Maine or New Hampshire woods, will know that a *smudge* means a smoke made to drive away the flies. Green evergreen boughs, or damp lichen thrown on the fire will make a good smudge.

midst of the wilderness, beyond the reach of all compulsion,—rarely or never neglect to pay every farthing. Their sense of honor among themselves, too, seems, in some points at least, acute. We were told that if an Indian finds a beaver-lodge, he cautiously traps a beaver or two, and then leaves them alone for the season, since otherwise the animals would forsake the place altogether. This he does year after year in perfect security that no one will meddle with them after he has proclaimed his discovery, and it is said that a beaver-lodge sometimes descends thus from father to son.

July 8th.—Being in Mr. S.'s room this morning, a hunter came in from the woods to get a supply of tobacco, which, with ammunition and apparatus for making fire, are the hunter's indispensables, and are never refused them. His first words (in Indian, for he understood no English,) were an exclamation at the astonishing quantity of flies.

Happening to be in want of a tobacco-bag, I made a proposal through Mr. S. for a rather ornamental one, (of broadcloth of various colors, with hanging tassels, and worked with beads,) which the Indian wore at his girdle. He signified his acquiescence, and handed me the pouch; but when in return I gave him a five franc piece, he eyed it curiously, and bursting into a giggle, asked Mr. S. what he should do with it? Mr. S. satisfied him on this point by telling him how much cloth it would buy, whereat he seemed satisfied, and requested to have the things out of his pouch. These consisted of a quantity of *kinni-kinik*, and fire apparatus, being a small cylinder of wood, hollow at one end, round which was an edge of steel. A quantity of the fibrous inner bark of the arbor-vitæ being placed in the hollow, is ignited by striking a stone across the mouth.

So large a number of Indians are collected here, (I think Mr. S. said about 150,) that it would seem to be a good opportunity for doing something towards civilizing them. There is certainly room enough for improvement. They have no church, no schools, no marriage ceremony, unless it be in the Indian style, every man having as many squaws as he can support. They do not attempt any agriculture, but depend on hunting, and when that fails, on the charity of the traders; they build no houses but the birch-bark lodges of their ancestors.

Speaking of agriculture, there is an extensive potato patch attached to the factory, some of the produce of which we carried with us when we left. The potatoes, however, are small, and other vegetables are said not to ripen here, on account of the shortness of the summer. Yet the winters are not very severe, the quicksilver, Mr. S. said, never sinking below—20° Fahrenheit.

The fur trade, he said, was very much on the decline, which he ascribed to the use of various substitutes for beaver in making hats. The principal furs at this post are lynx, martin, otter and beaver. The lynx and the martin are never abundant together. If the lynxes are plenty, there are few martins, and vice versa. Probably as their prey is similar, the lynx, being the stronger, drives off its rival.

Great quantities of fish are seined here; white-fish, lake-herring, trout, &c., not only enough for the use of this and other posts, but also some are sent down to the Sault for sale. The number of white-fish annually put up on the whole lake, Mr. Swanston estimated at three thousand barrels, worth on an average $5 a barrel. Of these, about one thousand barrels are sent away for sale. At Fort William, about five hundred barrels are taken. Out of some fifty thousand specimens that he had seen at Fort William, there were two with red flesh, like salmon.

July 9th.—This forenoon the canoe was finished; the sewing of *wattap* being renewed throughout, and a fresh coat of gum applied. This *wattap* is usually said to be spruce roots, but as well as I could make out, on this occasion the roots of the ground-hemlock (*Taxus canadensis,*) were used.

We had now got thoroughly used to our men, and they to us. Our steersman, Henry, whose culinary skill (a prominent qualification of a voyageur,) has been already celebrated, was careful and obliging, but rather slow both in wits and senses in comparison with John, who, though *milieu*, was decidedly the genius of the crew. This man was wholly or mostly of Indian blood, and his real name an unpronounceable jumble of letters that would take up half a line. No hawk's eye was ever keener than his; nothing escaped it;

nothing was too distant for it to make out. A wiry, sinewy fellow, of astonishing strength and endurance, and always on the watch for dangers above and below the water, but his chatter and his merriment were unceasing; he laughed more than all the rest, and made all the jokes beside. Henry spoke English in a very deliberate and rather inarticulate tone, having probably a diplomatic dread of committing himself by blunders in grammar. John understood no English nor French, but he knew instantly what you wanted, and did not often need even the assistance of pantomime.

They were all thoroughly practised in their craft; not only as to the navigation of the canoe, but also in doing and contriving every thing needful to our comfort. When we landed they waded into the water to carry us ashore on their backs, (for except where a rock projected favorably, the canoe could never be brought near enough to step ashore dry-shod,) then carefully lifted the canoe on to the beach, and after taking out its contents, turned it bottom up. Next, a good spot being selected, the tent was pitched, and drift-wood (of which there is generally an abundance at hand,) collected in good supply. This occasioned sometimes a good deal of good-natured rivalry among the various crews, the men of each boat considering their interests identified with those of their *bourgeois*, and accordingly making haste to pounce upon the best logs and the softest camping-ground. This was generally at the top of the beach, to secure level ground, and moss where there was any. Then they brought up from the water whatever things they observed we liked to have in the tent, to one his gun, to another his insect-net, and carpet-bags and bedding for all. In the morning, unless we were up of our own accord, we were aroused by their "*embarquez, embarquez*," and wo to him who lingered many minutes after this warning, for he was sure to find the tent tumbling about his ears without further preface, and his loose effects transported to the canoe by these inexorable fellows.

For this is remarkable about these men, that obliging and respectful as they are in general, there are certain things for which they stand out, and will have their way. John, for instance, though the best fellow in the world, would never allow the due sweep of his oar to be obstructed even by an inch, and any one whose back or head came in the way, was reminded of the impropriety by a dig from the end of it at every stroke, until he withdrew within his proper limits. About these matters, (which, however, were confined entirely to the management of the boat, &c., and respected exclusively the public interests,) they never argued nor attended to arguments, but quietly persisted in doing as they thought proper.

The immediate shore on our course this afternoon, was lower than we had had it since leaving Gros-Cap; rounded, gradual slopes of rock down to the water, bare in some places, and the rest covered with a scanty growth of trees. At some distance back, rounded hills rose to a greater height.

We were struck here and elsewhere by the regular succession of coves and points, owing apparently to the trap-dykes, which, instead of being more easily decomposed than the surrounding rock, and thus forming chasms, as on the other side of the bay, were here harder, and so stood out from the rest.*

At several places we observed terraces, and carried two of them, at various heights, but preserving their relative positions, about two miles, to the Riv. a la Chienne, where they turned up the valley and extended along its left bank as far as we could see, having an elevation of about two hundred feet. Here, according to intention, we encamped at sunset, fifteen miles from our starting place. This river is deep, and about ten fathoms wide, umber-colored as usual, with a broad expansion inside, which, with the wideness of the valley and the scanty growth on the terraces (doubtless of sand) forming its left bank, permitted an extensive view up the stream into an amphitheatre of high rounded hills, behind which the sun was setting. There are rapids and a fall of about ten feet a quarter of a mile up. We pitched our tents on a spit of sand, broad at the base, and running out in a point across the mouth of the stream to within a few yards of the steep rock of the right bank. Just inside the point, the bottom sunk sheer down twenty feet. Outside there is a bar, having only a few feet of water on it.

One of the men collecting firewood on the bank found a bear's

* This contrast between the different dykes induced the Professor to examine into their relative ages, and thus led to the views set forth in the paper on the Outlines of the Lake.

skull, with two shoulder-blades and some vertebræ, stuck in the crotch of a tree. The jaws were very neatly bound together with *wattap*, and the bones painted with broad stripes of black and vermillion. Inside of the skull was some tobacco, plugged in with birch bark. This is said to be a common token of an Indian grave, marking the dead as a brave hunter. On the bank above were remains of an Indian lodge.

July 10*th*.—Very cool this morning. The rocks on our course uniformly sloping south-west to the water, in consequence, the Professor said, of glacial action. He explained that in order to form satisfactory evidence of the action of ice, it was necessary that the slopes and the rounding and scratching of the surface should have a direction different from the stratification of the rock.

We passed this morning several mining "locations," indicated by poles set up on the rocks. At "Les Ecrits" were rude pictures of canoes, caribou, horses, snakes, &c., cut out of the black lichens, on a perpendicular face of rock. We stopped to lunch at a rocky point forming a shelf nearly level with the water, which was thirty feet deep alongside. To this the canoes were moored by a mountain-ash sapling at head and stern, the small end tied to the canoe, and the large end loaded with large stones. One of the men shot a spruce partridge, (*Tetrao canadensis*,) the first we had seen, though they are said to be abundant here.

I climbed up the point, and on the top entered a thick growth of shrubs, Labrador tea, and various species of Vaccinium. The whole surface of the ground was covered with rich green moss (*Sphagnum*), spreading over the loose rocks a uniform velvet carpet, into which I several times sunk to my middle. Larches began to appear. The woods much like those of northern New England, except the prominence of the lichens and mosses here, and the smaller size of the trees. Contrary to my expectation, and to what had been told me of the country, the forests are not remarkably dense, and there is rarely any difficulty in penetrating, except in the cedar swamps. The ground is generally rough, since it is, in fact, the broken slope of the lake shore. We never penetrated far into the interior, which is said to be in general thinly wooded. The most striking feature of these woods is their stillness and loneliness, though as to this the season must be taken into account. Even in Massachusetts, in July and August, there are comparatively few birds to be seen or heard, and travellers, among others Prince Max of Neuwied, (who is a naturalist to boot,) have founded on this fact very false conclusions as to the scarcity of birds in the United States. The truth is that owing perhaps to the absence of marked climatic divisions, the birds of this country extend their migration very far, so that any such comparison should be made in spring or fall. Then much allowance must be made for the change wrought by civilization. Birds and animals (except the carnivorous ones,) always increase about settlements; a well-known fact which our experience confirmed, for about the posts, and at the Sault, both were always more numerous than elsewhere. In Chicago, a few years ago, a gentleman told me that the grouse and quails had increased in that neighborhood eight-fold within his recollection; I myself saw numbers of quails in the main street and on the houses, and was assured that they sometimes entered the shops. The cause is simply the increase of food. Even deer continue to increase for some time about settlements.

The shore now became higher and more precipitous, until at Les Ecourts, marked on Bayfield's chart, "no landing for boats," the cliffs of sienite rose to the height of eight hundred feet above the lake. Here were swarms of swallows, and a pair of sparrow-hawks, the invariable inhabitants of these cliffs. Michipicotin Island was now plainly visible to the south, distant about ten miles. We had intended to take it on our way, but decided to put this off until our return.

The sunset was beautiful, but autumnal; the clouds in large well-defined masses, tinged with a suffused roseate hue. Afterwards the air became cool. It was nine o'clock when we encamped, on a beach just inside of Otter Head. The bateau, which had detained us much during the day, remained behind at dark. The "Dancing Feather," on the other hand, had the start of our two canoes, and went round the Head.

The beach where we landed rose some twenty feet from a narrow margin on the water, at an angle of twenty to thirty degrees. The little semi-circular plateau above seemed by the dim light to be surrounded on all sides by a dense forest. In stumbling about after drift-wood, we made the discovery that the upper part of the beach was strewn with lichens, in large tufts or clods, often eight to ten inches deep by eighteen inches to two feet across; a few armfuls of this made a very comfortable bed. After the sunset faded, the moon shone out brilliantly, and we sat on the edge of the slope talking of many things, long after our men were snoring comfortably under the shelter of the canoes below.

July 11*th*.—Daylight showed us that our plateau was a niche cut in the rock, which rose steeply and with great regularity from all sides, fringed and covered with trees. We rounded the point of Otter Head, so called from an upright parallelogram of rock, (having, however, so far as I could see, no particular resemblance to the head of an otter,) resting on the top of the point, and, joining the "Dancing Feather" at breakfast time, we put ashore and decided to wait for the bateau. On the way a solitary Indian, excessively dirty and ragged, came off in his canoe to sell us fish, and turned out to be the *brother-in-law* of one of our men, a very decent-looking Canadian Frenchman.

The woods here also carpeted with moss, and sprinkled with Linnæa and bunch-berry; here also we found very few flies, and began to give some credence to the assertion of some of the men, that they disappear towards the end of this month. Perhaps the change of temperature may render them sluggish, for we had now crossed the 48th degree of latitude, and the greatest heat of summer, in these northern regions coinciding more nearly with the solstice, was now past.

One of my companions and myself making the circuit of a muddy pond, formed by the damming up of a small stream by the lake beach, incautiously attempted to return through a patch of burnt arbor vitæs. It is difficult to persuade one's self at a short distance that these burnt places are so impracticable as they really are, even though one may have had full experience of them before. You can see through the trees every where, and the ground is plainly visible among the stumps. But when fairly engaged, you find the fallen trunks are piled together in such wild confusion that you seldom touch the ground at all, but are obliged to get along squirrel fashion (only not so quickly and easily), by climbing and jumping from one log to another. Moreover the effect of the fire is not at all uniform; some of the wood, without much change of the outside, is converted into mere punk, so that if you step on it you are precipitated among the charred logs, and in your passage made feelingly aware that many of the small branches and ends have been merely sharpened and hardened by it into spikes. So slow and laborious was our progress that, having with great difficulty made my way to the edge of the pond, I waded along, with the water up to my middle, in several inches of mud, as far as the fallen trees would allow, rather than take to the bank. We were about twenty minutes in making less than a quarter of a mile, and my companion assured me that once on the south shore of the lake it took him a whole day of hard work to get over seven miles of this ground.

The shore now became very varied and broken; not very abrupt, but rounded hills and points of considerable size coming successively in sight, and on the water-side numerous picturesque wooded islets of granite, with abrupt faces towards the south, and polished and rounded slopes northward. Wide trap-dykes in the reddish sienite rock all ground down to an even surface. The wind blew in puffs from the N. W., alternating with dead calms. The fluctuation of temperature was astonishing. So long as it was calm, the unclouded sun beat down upon us with all the fervor of our own July, but the moment the wind sprung up it was October.

Evening coming on, the bateau and the "Dancing Feather" encamped, but we in the other two canoes decided to keep on to the Pic (Peek), which was only ten miles off. Not that we were particularly anxious to get on, but having hitherto taken the journey rather leisurely we thought the men seemed inclined to take advantage of our good nature. So after tea we started again, the moon shining brightly and the sunset just fading away.

The Northern Lights, visible to some extent almost nightly, were unusually beautiful this evening, forming three concentric bows in the north, the upper one about thirty degrees from the horizon. From this bow as a base sprang up long flickering streamers quite to the zenith, where there was a flecky appearance, as if of light clouds, which, however, were stationary. Hence radiated tremulous flashes of light toward every point of the compass.

We reached the Pic about one o'clock, the moon down, and no

objects discernible except some Indians and their dogs, and the indistinct forms of their lodges on the beach.

July 12th.—Before we were stirring this morning, our friends of the "Dancing Feather" made their appearance, and we learned to our surprise that they had been encamped for some time and had already finished their breakfast. The fact was their voyageurs were a little piqued at our having pushed on ahead of them, and were resolved we should not gain any advantage by it. So getting up very early they came up with all speed, and silently passing the spot where we were encamped, pitched their tent at some distance beyond, and made haste to get breakfast before we were up.

The Pic is a post of the Hudson's Bay Company; the smallest of the three on the lake*; the name is derived not as we at first supposed, from the pointed hills across the river, but from an Indian word, *Peek* or *Neepeek*, signifying, I believe, "dirty water." The same word occurs in Neepeegon. It is situated near the mouth of a rather sluggish stream of turbid, brown water, about two hundred and fifty yards broad, flowing through a valley, wide near its mouth and narrowing higher up, apparently a delta of the river. There are considerable falls at some distance up the river. A sand-bar, on which there are six feet of water, extends across its mouth, and particularly on the northern side there is a very broad beach of white sand, like that of the sea-shore, drifted into hills, and at the top of the beach into a high ridge or dune, like that at Michipicotin, but smaller, whence there is a steep descent into the pitch-pine woods behind the post. Near the beach is a remarkable dyke of pitchstone.

The establishment consists of a number of whitewashed red-trimmed buildings of one story, like the fishermen's cottages of our coast, ranged round a hollow square and surrounded by a high palisade. The

* The following lists of the furs obtained for the two last years, as given by Mr. Beggs to one of the gentlemen who remained behind here, may be of some value as an indication of the relative abundance of the different species:—1847,—bears, 21, beavers, 125, lynxes, 237, fishers, 83, cross foxes, 6, red do., 18, silver do., 3, martins, 710, minks, 297, musk-rats, 2,450, otters, 137, wolverine, 1, ermines, 32.—1848,—bears, 20, beavers, 126, lynxes, 61, fishers, 66, red foxes, 6, white foxes, 6, martins, 1,167, minks, 402, musk-rats, 1,999, otters, 179, ermines, 118. The inverse proportions of lynxes and martins confirm what Mr. Swanston said. It is to be observed that the number of hunters is much smaller here than at either of the other posts.

ground inside of this courtyard is covered with plank, and a plank road, also enclosed by a palisade, leads up the slope from the river to the gate-way, which is surmounted by a sort of barbican.

July 13th.—There was a dense mist and an easterly wind this morning, much like one of our chilly sea-fogs. This was the first instance of fog after sunrise we had met with on the lake, though it was often foggy early in the morning. The air was never colder than the water, so that condensation could take place only when the saturated atmosphere was cooled by the lake, unresisted by the action of the sun, that is, before sunrise. That the air was full of moisture seemed to be shown by the fact that we could often see our breath when the air was by no means cold, the atmosphere being so charged with moisture as to raise the dew point, or degree of temperature at which the vapor becomes visible, unusually high.

The pitch-pine woods behind the post had been burnt over, and the trees, though yet standing, were mostly dead, affording food for myriads of wood-beetles, (*Monohamus scutellaris,*) whose creaking resounded on all sides. These in their turn were fed upon by the Canada jays, and by two rare species of woodpeckers, (*P. arcticus, and P. hirsutus.*) The *arcticus* in particular was very abundant and noisy, having a shrill, startling cry.

The Professor got a number of fishes, among others a brilliant green pickerel, a new species. A sturgeon was caught in the river opposite our tent, in a net belonging to one of the Indians, who dispatched him after some contest, with a fish-spear. Prof. Agassiz requested me to make a sketch of this fish, which was some four or five feet long. This took some time, and meanwhile we observed that all the inhabitants of the lodge to which it belonged were assembled and crouching in a row in front of us. We supposed this to be mere curiosity, but one of our men happening to come up, discovered that the whole family had been without food all day, and were waiting to eat the fish as soon as we were done with it. We were shocked at having committed such a breach of propriety, but the sketch not being finished, we proposed to them to lunch meanwhile on some of our pork and biscuit, to which they readily agreed.

July 14th.—Started this morning with a strong head wind. We were obliged to leave behind one of our number, who had been ailing with a feverish attack ever since Mica Bay, and was now pronounced by the medical men too ill to proceed. Fortunately we were able to leave him in good hands. One of the party volunteered to stay with him, and Mr. and Mrs. Beggs gave him the best accommodation the post afforded.

This was the only case of sickness during our excursion, although the mode of life was quite new to most of us, and some degree of hardship was anticipated. But speaking for myself, the only serious inconvenience was the scorching heat of the sun, which severely blistered the skin wherever exposed.

Our course this forenoon fortunately lay through a labyrinth of islands, by which we avoided the force of the wind somewhat. Just after leaving the Pic we passed through a river-like channel, about fifteen feet wide, the steep sides of which were deeply scored in a direction diagonal to the chasm, showing, the Prof. said, that the body by which the marks were made, had a momentum sufficient to disregard the shape of the ground over which it passed. The striæ hereabouts were inclined at an angle of 39° with the surface of the water.

We stopped for lunch on a point covered with *Vaccinium uliginosum*, and similar shrubs. The slimy water-plants floating along this point were filled with astonishing numbers of drowned insects, and many fine specimens were obtained. From here it was necessary to make a traverse of some three or four miles with quite as much wind as we could stand up to. This brought us into a cluster of islets abreast of Pic Island, a fine bold peak seven or eight hundred feet high, stretching off into a rocky ridge. The whole skeleton and structure of the peak were distinctly visible, from the effects of a fire that had streamed up the side of the mountain from a cove on the north, where there is a camping-ground. The Indians and voyageurs in their carelessness and wantonness allow their camp-fires to extend into the woods, which on these rocky slopes are dry and inflammable. The consequence is that the foliage of the trees being destroyed and their roots killed, they no longer hold together the soil, and it is accordingly swept off by the next rains, leaving a clean surface of white, calcined rock for Nature to cover again in the course of ages, by the slow succession of lichens, shrubs and trees.

While passing this island, two canoes came in sight from the opposite direction, evidently making a wide traverse for the Pic. They passed rapidly along under sail too far off to be spoken, but we had no doubt that it was Gov. Simpson of the Hudson's Bay Company, who was expected at the Pic on his annual tour. We afterwards learned that this conjecture was correct, and that he arrived about eight o'clock that evening, thus making in three hours (for it was about five when we passed them,) what we had taken all day for.

The Governor is much noted for his rapid travelling. On one occasion he is said to have dined one day at the Sault, and breakfasted the next at Michipicotin, a distance of one hundred and twenty miles. We encamped this evening on a most picturesque rocky islet near the shore, where we slept on natural beds of solid moss and huckleberry bushes, a foot deep.

July 15*th*.—Rain early this morning, but cleared away cold, with an autumnal sky and high wind. We passed the Slate Islands, high and blue, at the distance of seven or eight miles, and ran into a cove, at the bottom of which opened what seemed to be a well-ordered lawn, with balsam firs and larches judiciously disposed at intervals. In landing, the rich green grass turned out to be bear-berry, and the soil mere sand, which the bear-berry loves, but which accounted for the scantiness of trees.

The woods were crossed and recrossed in every direction by rabbit (or rather hare) paths, and we saw some trails that some of us fancied might be caribous', with many tracks of a dog or wolf. Caribous are found all through this region, but not in great abundance. An Indian who passed last winter on Isle St. Ignace, killed twenty-five caribous in the course of the winter, and was thought to have done very well. We saw here, for the first time, *Parus hudsonicus*, in company with a number of its cousins, the chickadees, from which it was to be distinguished only by its brown head, its slenderer and higher note, and a slight difference in habit, fluttering more about the ends of the twigs.

We made a long stay here, and some of the men amused themselves with lighting a fire, which unfortunately ran along the ridge of the beach, and, in spite of their utmost exertions, marched with a broad front into the woods. It was an exciting spectacle, the eagerness of the flames to seize upon each fresh tree, winding round it like serpents, crackling and rushing furiously through its branches to the top, until every fragment of dry bark, lichen, &c., was consumed. The fire seems too dainty to take the more solid parts, and so, for instance, the bunch of upright cones at the top of the balsams, remains distinguishable in the forest as a blackened tuft. Our beautiful bear-berry lawn looked now more like a peat-bog. When we left, the fire was in full progress, and was probably stayed only by a swamp beyond.

Nature, however, generally provides that no land that can be of much value to man shall be subject to this fate, for the heavily-timbered (and thus fertile) land of these latitudes is mostly too wet to burn, except the solitary birches, which if you set a torch to them, go off like rockets, but do not set fire to the other trees.

We passed terraces several times to-day, and in one place in particular, on a grand scale at the bottom of a bay, forming a series of vast unbroken arcs of about a mile chord, ascending one above the other to the height of several hundred feet, and, from the scantiness of the vegetation, evidently composed of sand.

Camped on a beach of coarse, dark sand, under a high abrupt promontory, enclosing it with precipitous walls. Among the rocks in our neighborhood were discovered veins of copper, suggesting to the Professor some remarks, which he illustrated on his black canvas, pinned against the side of his tent:

"Veins are formed sometimes by the cracking of igneous rocks as they cool; sometimes also by the subsidence of strata; cracks being formed, are filled from the melted mass below, pressed upon by sinking strata and thus forced upwards, or thrown up by other causes. The injected mass, even though originally the same as that into which it penetrates, may yet produce a vein of a different character, from the difference of cooling. Where the injected mass is very great it alters the surrounding rock, more or less in proportion to its vicinity to the melted substance. In these *metamorphic rocks*, as they are called, such as we have seen in great abundance throughout our passage along the lake shore, there is accordingly the greatest variety of character, and one species of rock passes into another by so many intermediate forms that it is often difficult to say what name should be given to it, the rock, originally sandstone, perhaps, with various admixtures, being changed into sienite or porphyry, or into rock partaking in various degrees of the characters of both, by the influence of large veins of melted materials. Metallic veins are sometimes formed in the same way, by injection, and they also in the same manner modify the surrounding rock, as in the instance before us. Sometimes, also, they are formed by sublimation into crevices, or by electro-magnetic action, causing an interchange of particles between various parts of the rock."

July 16*th*.—Early this forenoon the Island of St. Ignace appeared looming up in the distance. We passed the "Petits Ecrits," a rock ornamented with representations of various animals, canoes full of men, &c., together with various fabulous monsters, such as snakes with wings, and the like, cut out of the lichens; the work of the Indians, or perhaps of stray miners or searchers for copper, who, as appeared by dates and initials, have adopted from them this mode of attracting the attention of the passer-by. These pictures were of various dates, as was shown by the various degrees of distinctness, as the rock was either quite laid bare, or the black lichens had more or less completely recovered possession of it. We now entered the vast archipelago of islands occupying the whole N. W. corner of the lake, as far as Pigeon River, a distance of about two and a half degrees of longitude, viz.: from 87° 30′ to 90° W. It is difficult to convey any notion of the vast number of islets and rocks in this part of the lake. Capt. Bayfield in his (unpublished) chart of Lake Huron, is said to have laid down thirty-six thousand islands, on twenty thousand of which he has landed; the number in Lake Superior cannot, I should suppose, fall much short of this. In both lakes the islands lie almost exclusively along the northern and eastern shores. In Lake Superior, with the exception of the group called the Apostle's Islands, there are very few islands on the south shore, or on the north-west shore beyond Pigeon River. In Lake Huron there is scarcely an island outside the Georgian Bay, and in the lower lakes islands are almost entirely wanting.

As we were passing under an overhanging cliff where nests of the barn-swallow were niched into the rock within reach of the hand, an Indian in his canoe with his squaw and child suddenly glided alongside from some cove, and offered fish in exchange for tobacco.

He was a huge fellow, with a great head, covered with dishevelled hair, yet not ill-shapen, and having something of the picturesqueness of a bowlder of granite. The woman had on a sort of cloak of white hare-skins, with a hood attached, which was drawn up over her head. Somebody gave the man a cigar, and showed him which end to put into his mouth and how to light it, which he did, and smoked away very cleverly. Signs were made to him to give the woman a puff, but she unluckily put the lighted end into her mouth, and after that good-naturedly but firmly declined to have anything to do with these new-fangled pipes.

The wind meantime had risen, and coming out from the lee of the islands into an open bay, we found the head wind and sea too strong to be contended with, and so put back into a cove, the entrance of which we had just passed. Passing through a narrow strait we came into a quiet bay that seemed like a land-locked lagoon, but was in fact separated from the lake only by a couple of islands. The sides of the cove rose steeply from the water's edge with only a narrow circlet of sand between the water and the trees, in some places hardly leaving room to pass outside. Thus protected, the little bay, with its fringe of birches and arbor-vitæs, as unruffled as some inland pool of a still September afternoon, presented a strong contrast with the turbulence of the weather without. I climbed up the steep bank, which was everywhere covered with deep beds of moss, and penetrated with some difficulty to the outside of the island, for an island it was, and the reader must understand that at the "Petits Ecrits" we quitted the shore, which here trends to the northward, and pursued a westerly course among the almost continuous islands, intending to pass outside of St. Ignace.

The spruce woods here were very dense, and encumbered with fallen birch trunks, as if the spruces had usurped the place of a birch forest. Part way a sort of path was broken, and fresh tracks of some large animal, sinking a foot deep into the moss; — probably a lynx, as they abound here. Hare tracks in all directions. Snares were set in the evening, and two hares caught. The method of setting these snares, which is extensively practised by the Indians, is this. A well-frequented hare-path being selected, is blocked up by a fence of sticks, leaving only a narrow passage over which a running-noose is stretched; the animal in jumping through gets caught by the neck. It is said that they can hardly be made to leave the path, and they are thus very easily caught. The Indians rely much upon them for support, particularly in winter.

On the outside of the island were rough beaches of large stones, and rocky points against which the waves were beating furiously.

This evening as we were arranging the musquito-bar in our tent (a nice job and one requiring abundance of light), our candle proved to be missing, and we supplied its place by piling on the fire a large quantity of usnea, which streamed from all the trees. This is not an unimportant article in the economy of these regions. There is no better material for the packing of specimens; it makes capital bedding, and it is so inflammable that a tree covered with it makes the best possible beacon or signal-torch. The Indian women use this as well as moss for stuffing the bottom of their portable cradles.

The wind fell in the course of the night, and there was rain before morning.

July 17*th.*—Cloudy and warm. Made a traverse at sunrise of three or four miles, and then began again to thread our way through endless woody islands of greenstone, often showing vertical sides. The main shore was now several miles distant and constantly receding in high domed summits. St. Ignace, high in front, black to the top with spruce forests; and a dim, majestic outline in the far distance, seeming only to divide one part of the sky from the other, our voyageurs declared to be Thunder Cape, seventy or eighty miles off. The ends of all distant points were turned up by the effects of the *mirage*, a very common phenomenon here, owing to the contrast in temperatures between the air and the water.

We ran into a narrow bay on the east end of St. Ignace, the bottom of which approached a peak marked on Bayfield's chart as thirteen hundred feet above the lake. This bay is a quiet little nook, hedged around with larches and other trees, over whose tops appeared the peak. A small clearing had been made here, it being a mining "location," and on a board fixed to one of the trees was an inscription signifying that the spot had been "taken possession of by the Montreal Mining Company, June 5, 1846." They had even gone so far as to put up a log-house, yet standing in tolerable repair,

G. Elliot Cabot from nat. Lith. of A. Sonrel Löffler on stone.

ISLAND OF S^t. IGNACE.

with a crib for sleeping inside, and "Douglass' Hotel" written on a board by the door. This was one of the many places (there are several on this island), where works were commenced without any proper exploration of the ground, the only indication of ore being some veins of calc-spar, which by a too hasty induction was supposed to be a sure sign of copper. Small quantities of native copper were found, but not sufficient to pay for the trouble of getting it.

After breakfast, the weather being favorable it was decided to make the ascent, and we started accordingly, taking a narrow gorge that one of the men, who acted as guide, said led to the peak; but stopping behind for a moment, I lost the party, and could not distinguish the trail amid the multitude of hare-tracks through the woods. I shouted, and was answered repeatedly, but the voices were so echoed back and forth in the narrow valley, that I could not make out their direction, and went back to the camp.

In the afternoon they returned, reporting a very fatiguing climb, the barometer broken, and the flies very troublesome. The black fly is fond of high and dry situations, and is always found in greater numbers about the top of a hill than at the foot. They had ascended the peak, however, and christened it Mount Cambridge, in case it had not already been named. The summit was steep and rocky, the rocks polished and scratched to the top. Contrary to expectation they found no change whatever in the vegetation.

The woods here were filled with Linnæa, and several species of Pyrola. We left at five o'clock, passing outside of the island.

St. Ignace seems to be a collection of peaks, and in the middle a long interrupted ridge, that seemed still higher than Mt. Cambridge. We encamped this evening on a long narrow island lying north and south, consisting of two beaches meeting in a ridge in the middle, and composed of large angular fragments of porphyry with only the corners worn off. Each side of the island was ploughed from one end to the other with furrows a foot or more in depth, parallel to the water. The stones were covered with great clods of lichen, and a few mountain-ashes and spruces grew along the dividing ridge.

July 18*th*.—Started at sunrise with our India-rubber cloth for a sail, the wind being for once favorable. In rounding the end of the island we found furrows like those above described, but at right angles with them, running across the end of the island. Our course lay through long river-like channels, formed by parallel series of rocks and islets. Near evening we passed a number of Indian lodges clustered on an island, with the usual number of barking dogs and squalid children, and hoped to get fish from them, but they had none except dried, which is tough and tasteless, in texture and appearance somewhat resembling parchment.

In the night it blew hard from the westward, and we waked up in some anxiety lest our tent should be capsized, but John was already on hand and secured it.

July 19*th.*—Detained here by the violence of the wind (*dégradé*, the voyageurs call it,) until about three P.M., when we pushed on past Point Porphyry, and encamped in a deep narrow bay to the northward, stopping on the way to examine an interesting locality where altered red sandstone and trap were seen in close contact.

In the sandstone were ripple-marks and cracks, such as one sees in a dry mud-flat. The surface in many places had an oily smoothness, and in looking down upon it one might easily have taken it for a bed of red mud just left dry.

This cove was evidently a favorite camping-ground, from the marks of recent fires, and the large number of lodge-poles on the bank. Near the water's edge was a quantity of spruce bark, saddled in sheets one over the other on a horizontal stick, like the roof of a house. We at first took it for a grave, but it afterwards appeared that it was only the bark-covering for the lodges, thus disposed in order to keep it sound. It rained hard in the night, with thunder for the first time on the lake.

July 20*th.*—Calm and cloudy. At a distance to the northward were two twin hills, called "*les mammelons*," by the voyageurs, and by the Indians, much more aptly, "the Knees." One could easily fancy the rest of the gigantic body lying at ease on the plateau, with the head to the north, and the knees drawn up in quiet contemplation of the sky ; perhaps Nanaboujou, or the First Man.

We soon came in full sight of Thunder Cape, a magnificent ridge, 1,350 feet high, according to Bayfield, running out into the lake directly across our path. It is composed of metamorphosed sandstone, the horizontal stratification plainly visible, from a distance, on

G Elliot Cabot from nat. Lith. of Sonrel Löffler on stone.

THUNDER CAPE.

the face of the vertical wall of basalt-like columns rising out of the forest that clings about its base and sides. Near at hand, the horizontal lines disappear, being in fact rather suggested than clearly made out, and only the vertical chasms are seen. As we passed the end of the cape we found the ridge narrow and precipitous on both sides, forming a wall across the mouth of Thunder Bay. Another fragment of this wall we had in the southern ridge of Pie Island, on our left. It is continued by the high, narrow islands beyond, and repeated in the parallel ridges of Isle Royale.

We stopped to lunch at Hare Island, a little bit of gravel with few stunted spruces, but covered with grass and an abundance of flowers. We now had before us a traverse of about fourteen miles to Fort William, the white buildings of which were visible amid the dark swamp across the bay.

The wind was rising, but we set off, and the boats were soon far apart. Our canoe and the Professor's made for the southernmost entrance of the river on which the post stands, as the nearest, and were glad to escape into quiet water from the rough waves of the bay, several of which found their way into our boat in spite of all Henry's care and skill. The entrance of the river is wide and shallow, enclosing a large delta, cut through the middle by the stream, so that the river has in fact three mouths, the northern and southern ones some two or three miles apart. Some distance outside the mouth the water became very shoal, and islands were forming, on which a few willows had already taken root.

The river-water is of the usual dark brown, and tolerably clear. The banks swampy, densely wooded, and lined with water-plants, among others the elegant heads of the sagittaria, also nuphar, equisetum, bull-rushes, &c. Such was the luxuriance of the vegetation, that it reminded one of a swamp in the tropics, rather than of a northern river.

The name of Fort, applied to this post of the Hudson's Bay Company, dates from the old days of the Northwest Company, (to whom it formerly belonged,) and their quarrels with the Hudson's Bay. At that time the place was strong enough to induce Lord Selkirk, who came up with hostile intent, to take the trouble to bring with him a field-piece, which he planted on the opposite bank of the river, to make them open their doors. In those days a grand annual council of the company was held here, and we hear traditions of banquets, and crowds of clerks, and armies of hangers-on of all kinds. But all this has now disappeared. The trade has fallen off, the gross receipts being now, they say, only about £600 per annum; and moreover the Northwest is merged in its old rival, and all those troubles at an end, so that although the court-yard is surrounded with a palisade, and there is a barbican gate-way, as at the Pic, yet these fortifications are not very formidable at present; the old blockhouse behind is falling to pieces, and the banqueting hall has probably been burnt up for firewood, at least, we saw nothing there that looked like it. Even the little flower-garden opening out of the stone-paved courtyard was overgrown with weeds.

The general arrangement here is much the same as at the other posts, only the soil (a yellowish sandy loam) being better, and the climate less severe, the cultivated ground is more extensive, and they have a herd of some thirty cows. Sheep also are kept here, and several of the dogs were in disgrace, with heavy clogs fastened to their necks, for sheep-stealing. As the pasturage on the other side of the river is much better than about the Fort, these cows *swim across* regularly every morning and back in the evening, a distance of two or three hundred yards. I was much surprised, the morning after our arrival, when the cattle were let out of the yard, to see a cow walk down and deliberately take to the water, of her own accord, the whole drove following her, swimming with only their noses, horns and tails showing above water. An evolution so out of the usual habits of the animal, that I could account for it only by supposing it to be an ancient custom, established with difficulty at first, on the strong compulsion of necessity, and subsequently yielded to from force of example by each cow that successively entered the herd.

The land about the post is low and flat, mostly a larch swamp; a wide gap being broken in the rocky rim of the lake by the valley of the Kaministiquia. To the northward the hills retreat to the distance of eight or ten miles. Southerly the line is resumed by McKay's Mountain, a ridge of greenstone gradually ascending towards the north-west, to the height of one thousand feet, and there broken into an abrupt precipice.

The post is still an important one, as being the portal to the Red River country, Lake Winnipeg, and the north-west, and furnishes various supplies to other posts, among other things, of canoes, of which some seventy or eighty were lying here in store. It stands on the left bank of the northern mouth of the river Kaministiquia, about half a mile from the lake. Outside, close to the water, are the log-cabins of the Canadians attached to the post, and on the plain across the river the birch-bark lodges of the Indian hunters.

Mr. Mackenzie, the gentleman in charge, received us very kindly, and handed to us a number of letters and newspapers that had been forwarded hither from the Sault, by the propeller, which had come up the south shore and touched at Prince's Location, about twenty miles west of this.

July 21*st.*— Spent the day here. Wild pigeons, cross-bills, and ravens about the fort, and partridges in the swamp. Bathed in the river; the bottom muddy, and the water warm. Mr. M. says that before a gale from the northward the river falls sometimes eighteen inches in twenty-four hours. This they supposed to be owing to a heaping up of the water on the southern shore (where these gales usually commence,) by the wind, causing a corresponding depression on this side. The fact, more accurately described perhaps as a difference of atmospheric pressure on the two sides of the lake, was afterwards confirmed by several persons. We decided to ascend the river as far as the Kakabeka (Kah-káhbeka) Falls, twenty-five miles, to-morrow. Mr. Mackênzie kindly offered to go with us, and furnished us with whatever was necessary for the excursion.

This evening our men, with some of the employés of the post, had a dance in a cabin near the Fort. The music consisted of a squeaking fiddle, and none of the fair sex honored the assemblage with their presence, yet they stamped away half the night with the greatest jollity.

July 22*d.*—We started this morning accordingly, in three canoes, Mr. M. following after in a little cockleshell about a dozen feet long. The men in the two large canoes were placed two on a seat and furnished with paddles instead of oars, and there was a good deal of rivalry between them for the first few miles, the paddles dipping with wonderful rapidity, so that they looked like a row of tailors sewing against time. I did not time their stroke, but the rate must have been upwards of sixty dips per minute, for their common oar-stroke was forty-five per minute, and this seemed twice as quick.

A mile or two up, the river is narrow and the forest closes again upon its banks, which are somewhat higher; the trees larger than any we had seen on the lake; at first mostly aspens, afterwards spruce and elm. Five or six miles up, the banks are often thirty or forty feet high, and in some places broken away, showing horizontal layers of yellow, sandy loam, occasionally interrupted by sand and by narrow beds of clay. The margin of the river filled with sagittaria and other water-plants. Mr. M. says ducks and geese are very abundant here in spring and fall. At present there were only a few creek-sheldrakes.

The course of the river is very winding, and our men cut off half a mile or more in one place, by making a portage through the woods from one bend to another. They carried a surprising weight of luggage, suspended on the back by a *portage strap*, a broad thong of leather passed across the forehead.

For the distance of eleven miles the current is very sluggish. Then we came to rapids, where it was thought advisable to get out and make our way by land, leaving the men to pole the canoes up. We disembarked on a piece of marshy bottom-land, covered with a fine growth of elms. After proceeding some distance through rank grass and undergrowth, we came to the bluff, which was a very stiff fifteen minutes' climb. This brought us on to a table-land covered principally with scrub-pine (*P. Banksiana*,) much like our common pitch-pine, but more pyramidal in shape, with shorter leaves and curious contorted cones. This table-land was dry, sandy, and thinly covered with wood, with wide openings covered only by scanty, withered grass. The fire had been through in several places, and here woodpeckers and black flies abounded. This seems, from what we heard, to be the general character of the interior, except on the water-courses.

A fast walk of two hours and a half brought us to the river, where we waited about an hour before the boats made their appearance. All of them had touched repeatedly, and received some scratches; one had been obliged to put in to gum up a leak. We reëmbarked, but the current was still rapid; in some places we estimated it at six miles per hour. At the Décharge des Paresseux we again landed, and walked up some hundred yards while the men pushed the boats up with poles, which they grasped by the middle, using the ends alternately on each side.

We encamped at sunset, climbing up a steep clay bluff to an open spot above, for we could find no landing on a level with the water. Very cold in the evening, silencing the swarms of musquitoes that greeted us on our first arrival.

July 23d.— Very cold this morning also, and the dew heavy. Even inside of the tent some of the blades of grass were hung with dew-drops, and outside every thing was as wet as if from a smart shower. Without breakfasting we walked through the dripping woods to the Falls. On the way I noticed an old martin-trap, made like the *culheag* of our woods, viz. the butt of a sapling arranged to fall like a portcullis across the mouth of a hole in which the bait is placed. We came out first in an open space, bounded by a broken cliff of slate-rock, whence we could hear, but not see the cataract. The river here flows between high perpendicular walls of rock, and here commences the Portage de la Montagne. Following up the portage path about a quarter of a mile, we struck off through the thick arbor-vitæ woods, guided by the roar of the fall, until we came out on an open grassy bank in front of it, and so near that we were drenched by the spray.

From where we stood we could look up a long reach of the river, down which the stream comes foaming over a shallow bed, thrown up in jets of spray, like the rapids at Niagara. At the brink the stream is compressed, and tumbles over in two horseshoe-shaped falls, divided in the middle by a perpendicular chimney-like mass of rock some feet square, the upper part of which has been partly turned round on its base. The entire height of the fall is about one hundred and thirty feet, but somewhat filled up by fragments from above. Its breadth is about a hundred and fifty yards.

The rock is clay-slate, the strata dipping two or three degrees southward, that is, from the fall. Just above the pitch, the slate is broken into very regular steps, and the same structure is visible in the face of the cascade itself, particularly on the right, from the broken water where they project. On the other side, where the descending sheet is less broken, the rich umber color of the stream tinges the foam half-way down.

The name Kakabeka was explained by some of the men to mean "straight down:" i. e., falls *par excellence*, it being the most considerable waterfall in this region.

In the afternoon our friends of the "Dancing Feather," who had determined to return to the Sault by way of the south shore, made haste to depart, as we had appointed the 15th of August to meet at the Sault, and they had much the longer way to go. Mr. Mackenzie left us at the same time.

The Professor this afternoon invited some of us to make the attempt with him to push up the stream as far as a small island at the foot of the Falls, in order to see them from below. For a short distance we got along very well, taking advantage of a counter-current near the opposite bank. Soon, however, this assistance failed us, and we were exposed to the full strength of the stream. For a moment or so with all the men could do we could only hold our own, and then began to go astern, but Jean Ba'tiste caught the branch of a tree and checked the boat, and then jumping into the water actually dragged her along, the rest straining their utmost with the setting poles. The stream here was shallow, and hurried along with great force, eddying and spouting into the air over the stones with which the bottom is covered. For a moment or two it was a fair struggle between muscle and the force of gravitation; then we got under the lee of the island, and without farther difficulty landed on the lower end. The island consists merely of a heap of large angular stones, with a tuft of bushes in the middle.

At the upper end we sat down on the rocks, with the falling hill of water directly in front of us, its outline against the sky. Our position was a favorable one for feeling the full force of the mass of water, but did not command the whole of the fall, each side being partially hidden by the projecting cliff. Indeed there is no position from which the whole can be taken in at once.

The distinguishing feature of these falls is *variety*. In the first place each of the two side-falls has worn out for itself a deep semicircular chasm, which, with the foot of the cliff projecting from below,

G. Elliot Cabot from nat. Lith. of A. Sonrel, Cambridge. Löffler on stone.

KAKABEKA FALLS.

gives the appearance of two horseshoes joining in the middle, as if two separate streams had happened to come together here. This peculiar conformation throws the masses of water together in the middle, whence they are thrown up again by the resulting force, as if shot out of a cannon. The turmoil is farther increased by projecting rocks, (perhaps piles of fragments from above,) which, on the right particularly, shoot the water inwards towards the centre, at right angles with the course of the river. Then the sharp projecting shelves which project, especially on the right side, through the falling sheet, cause a succession of little falls in the face of the great one.

All these peculiarities are due no doubt to the nature of the rock, which, dipping slightly from the fall, and not being underlain by softer strata, as at Niagara, its recession is not regular, but depends on the accidental dislodgment of blocks on the edge, by frost, collision of ice, &c., and the blocks again, when fallen are not so readily decomposed or removed. Hence, also, the shallowness of the channel below. Some of our friends who meanwhile had been exploring above the Falls, reported a small fall, ten or fifteen feet in height, about half a mile above, where the slate was replaced by sienite.

We had some thought of proceeding up the river to Dog Lake, two days' journey to the north. But our men grumbled very much at the thought of the portages, (one of which from its destructiveness to shoes is called Knife, *or* Devil Portage ;) then our canoes were too large for the undertaking, and might possibly be knocked to pieces ; so we concluded to give that up.

July 24*th*.—Last night was warm and rainy, and we started down the river this morning in a drizzle. We stopped at the clay-bank, above which we had encamped before, to get some clay-stones, which occur here in abundance at the water's edge. These are nodules of clay, some soft, others of the hardness of chalk or harder, often in shapes requiring little aid of the knife to transform them into fantastic images. Capt. Bayfield says the bottom of Lake Superior is of clay, which readily indurates on exposure to the air.*

Kaministiquia, according to our native authorities, signifies " the river that goes far about," which this river certainly does, though in

* Bouchette's British Dominions in North America, I., 127.

the course of its windings it presents such a variety of beautiful scenes of overshadowing forest, that we did not grudge the delay. Two or three miles down, long after we had lost the roar of the Falls, it suddenly came to us again, quite distinctly and unmistakably, probably owing to some shift of wind.

This valley is the only spot we saw on the lake that seemed at all to invite cultivation; indeed, if we except the posts, almost the only place where cultivation seemed possible. The better quality of the soil was abundantly manifest in the size of the forest trees. The crumbling banks of loam and sand furnished abodes to large numbers of sand-martins and kingfishers. We were seven hours in reaching the Fort, and found our companions had left two hours before.

CHAPTER III.

FORT WILLIAM BACK TO THE SAULT.

July 25th.—We proposed to visit the copper-mine at Prince's Location, on the shore of the lake about twenty miles to the westward, and thence to cross to Isle Royale. In order to travel more rapidly we sent the bateau back to Point Porphyry to await our return, and proceeded with the two canoes only.

Starting at about ten o'clock, we found the wind strong ahead and encamped early in a bay about fourteen miles from the Fort. On the way we passed Pie Island, a large mountainous island, so called from an isolated peak on the west, which bears a strong resemblance, not at all to a pie, but to a French pâté, or pasty, with high sides; and this is its true name. A porcupine was killed on the beach as we landed, and proved very good meat.

In the evening the Professor made the following remarks on the distribution of animals and plants:

"There is no animal, and no plant, which in its natural state is found in every part of the world, but each has assigned to it a situation corresponding with its organization and character. The cod, the trout, and the sturgeon are found only in the north, and have no antarctic representatives. The cactus is found only in America, and almost exclusively in the tropical parts. Humboldt, to whom the earliest investigations on this subject are due, extends the principle not only to the distribution of plants according to latitude, but also according to vertical elevation above the surface of the earth in the same latitudes. Thus an elevation of fourteen thousand feet under the tropics corresponds to 53° north latitude in America, and 68° in Europe. The vegetation on the summit of Mt. Etna would correspond with that of Mt. Washington, and this again with the summits of the Andes, and the level of the sea in the Arctic regions. In the ascent of a high mountain, we have, as it were, a vertical section of the strata of vegetation which 'crop out' or successively appear as we advance towards the north over a wide extent of country.

"But in dwelling on the resemblances between the plants of high latitudes and those of high mountains, we must not lose sight of their not less constant differences. In the northern regions in general, we find the number of species comparatively small. Thus in the region through which we have passed, and which has already a northern character, we find vegetation characterized by great vigor; the whole country covered with trees and shrubs, and lichens and mosses in great profusion, but the species few, and the proportion of handsome flowering shrubs small. In the Alps, on the other hand, vegetation is characterized by great beauty and variety, and the number of brilliantly flowering plants, of Gentianaceæ, Primulaceæ and Compositæ, is very great. The plants, however, are dwarfish, and vegetation comparatively scanty; the lichens and mosses much less abundant. There is, then, not an identity, but an analogy only, and an imperfect though very interesting one, between Alpine and Arctic vegetation."

July 26th.—We pursued our way this morning under the shadow of magnificent walls of basaltic rock, with Pie Island rising in the distance outside of us like a Gibraltar. We reached the Location early in the forenoon, and were most kindly received by Mr. Robinson, the agent of the Montreal Mining Co., who have begun operations here.

A high rocky promontory, running S.W., (parallel to Thunder Cape and the other high ridges hereabouts,) is here cut across by a sort of fault or interval, leaving a strip of land rising gently from the lake on either side, to a ridge in the middle, backed on the north-east by cliffs seven hundred feet in height. The slope from the little curved beach where we landed was shaded by scattered trees left from the forest. Under these the workmen were busy in putting up cabins for a number of miners who had just come up with Mr. Robinson, and who, for the present, were living in tents on the beach. Back of these, was a row of cabins, and the little one-story house of the agent. Mr. R. showed us a large number of minerals collected hereabouts, and kindly offered us whatever of them we chose to take. Among them were very brilliant specimens of calc-spar associated with cobalt, manganese, and blue and green sulphurets of copper.

Afterwards he carried us by a path running back of the house past the opening of the shaft, through a clearing planted with potatoes, and a young orchard of cherry, apple and pear trees, down to the cove on the other side of the point, whence we sailed across the strait to Spar Island.

This island receives its name from a vein of calc-spar, some twenty feet wide, quite pure and white, except where brilliantly colored by metallic salts, running across the island and down into the lake on the other side, visible with a phosphorescent light for a considerable distance under water. This is the locality of most of the specimens we had seen at the office; splendid masses of white translucent spar, tinged with brilliant blue and green by the associated minerals. We noticed drift-scratches on the outer side of the island, having a direction nearly E. and W.

The day was showery, with driving thundery clouds and mist, through which we got a fine view of Pie Island, dim and majestic in the distance. We were driven for shelter into an unfinished building of squared logs, which the company are erecting with a view to continuing the mining operations which have of late been suspended on the island. Such a building (about forty feet square and of two stories,) they say can be put up in four or five days. On our way back the weather improved, and we had a good view westward of hills over hills towards Pigeon River, the boundary between the United States and Canada, distant about twenty miles.

When we got back towards evening, we found the miners amusing themselves after their day's work, by pitching, or "putting" stones, and I was surprised to find the puny Canadians had rather the advantage of the burly Cornish men. Mr. Robinson invited us to supper, and I believe none of us experienced any of the difficulty of the traveller, who, after a trip over the prairies, found himself, on his return to civilized life, constantly tempted to draw his feet up into his chair. In our case the benches were felt to be a decided improvement.

After supper Mr. R. carried us into a shaft they are sinking at the foot of the cliff. Here we got fine specimens of Iceland spar. No ore had as yet been sent to market from this mine, but the prospects seemed favorable, and the whole establishment had a thriving look.

July 27th.—We had intended to cross to Isle Royale, which lay like a blue cloud along the horizon, twelve or fourteen miles off, and vanishing into the distance eastward. Having got outside of the chain of islands, however, we found the wind so strong as to render the traverse dangerous, and we accordingly landed on one of the Victoria Islands, west of Spar Island, to wait for some change of weather.

The beach where we landed was a mere niche cut into the side of the cliff, which rose steeply on all sides, thickly wooded. The ground everywhere covered with moss. Among the trees on the bank was the skeleton of a lodge, and a birch canoe apparently in good condition. Some playthings of the Indian children were lying about, among others a little boat scooped out of a chip of wood, with mast and bowsprit, precisely such as the boys make with us, and not at all resembling the Indian canoes. The frequency of these traces of Indian encampments, with the small number of Indians living on this part of the lake, shows their restless, wandering disposition.

While we were detained here, the Professor made some remarks about the theory of the formation of mineral veins by infiltration. This theory he considered untenable, since there is an evident connection between this phenomenon and some action of the walls of the fissures in which veins are found:

"Thus at the vein we examined this morning at Prince's Location, we found each wall of the fissures covered with quartz crystals whose axes were perpendicular to the walls: those inside were crystals of calc-spar disposed in the same way. An electro-magnetic action, (which has been proposed by some geologists,) would fully account for this arrangement. If we suppose an electro-magnetic current passing through the fissure, this may have brought together similar particles scattered through the rock, and disposed them in the manner we see. In order to settle this point, however, it would be necessary to ascertain whether there is any constant relation in the arrangement of substances found in veins of different localities:—whether the minerals always follow each other in the same succession. If this be the case, it will give great probability to the supposition of an electro-magnetic current, over that of any merely mechanical agency like infiltration. Such an examination might probably also distinguish the cases where veins are formed by sublimation or deposition from vapors or gases from below.

Where the vein is composed of minerals not found in the surrounding rock, the probability would be in favor of sublimation: where the minerals occur, though in small quantities, in the rock, there the effect may have been produced by electro-magnetism. There has been as yet no sufficient investigation of this point.

"It may be remarked here that even where the vein is composed of hydrates, in whose composition *water* occurs, it is not necessary to suppose them deposited by infiltration, since it has been proved that hydrates may be formed by sublimation."

We remained here until half past three o'clock P.M., when, the weather continuing unfavorable, and even threatening a storm, we decided to give up our visit to Isle Royale, and to turn our faces homewards.

The distance of this, our westernmost point, from the Sault, was about four hundred and forty miles by the way we came; as we returned, rather more.

The wind was fresh from the southward, and when we got outside of the islands there was so much sea that the other canoe, although within a short distance of us, often disappeared, sail and all. It was rather a long swell for the lake, however, and we did not experience any difficulty from it, as we were nearly before the wind. We encamped on an island to the southward of the Pâté, in a deep bay with steep sides, overshadowed by trees of unusual size.

July 28th.— Started before sunrise. Weather calm and pleasant. We passed under the south-east side of Pie Island, a vertical cliff several hundred feet in height, presenting much the same appearance as Thunder Cape, viz: basaltic columns, across which may be traced the marks of an horizontal stratification. These columns in some places have fallen out, leaving hollows, like flues, in the side of the cliff. In other places single columns stand out alone, like chimneys; in others, again, huge flat tables of rock have scaled off from the face of the wall, and stand parallel and a little separated from it. The metamorphosed strata in one place were unconformable, exhibiting a sudden fault.

In the course of the forenoon several trout were caught, and the diversity of color led to some discussion. The men said there were three varieties, all of the same species: 1. the trout of the open lake, (*truite du large*,) of a gray silvery color, with inconspicuous spots and a white belly; 2. Those of the rocky ground, (*truite des battures*,) more yellowish, with large distinct spots; 3. Those of the sandy bottom, which are simply mottled. All the trout family spawn late; the lake trout in October, on the sandy beaches, when they are taken in abundance in nets, and with ground-lines having forty or fifty hooks.

The white-fish are everywhere scarce in August, (we could not learn why,) so that the Professor found some difficulty in getting specimens on our return. In October they spawn, on pebbly ground, and are then taken in great numbers. They are always seined; we did not hear of their ever taking the hook, though I have seen one take a fly from the surface. The lake herring spawns on similar ground, but in November; the siskawet in the latter part of August. Suckers, cat-fish and sturgeon in the spring; the sturgeon in swift streams; the sucker at the mouths of the rivers; the cat-fish on muddy flats; the dory (*Lucioperca*,) in bays.

We stopped at a little rock around which a great number of gulls (*Larus argentatus*,) were circling, and found there a few young ones and an addled egg. The young birds were about half grown, covered with grayish down, with irregular darker spots. None of them could fly, but they swam very well; indeed, as it seemed to me, better than the old birds. They were crouched in crevices of the rock, and we saw no appearance of nests. The egg was coffee-colored, with brown spots.

A fresh and fair breeze to-day, almost for the first time. We passed this morning several canoes of Indians, running before the wind with sails of birch bark. About noon, in threading a narrow passage among the islands we saw a smoke on shore, and directly afterwards the bateau, moored at the wharf of a deserted mining establishment, the buildings of which were still standing.

We kept on with the same fair wind until sunset, when we encamped on one of an extensive group of islands. As we glided rapidly into the little cove where we were to encamp, the water shoaled so suddenly, that looking down over the side of the canoe we seemed to be rushing against the side of a mountain. These coves shoal rapidly and have the bottom covered with huge rounded bowlders, like a gigantic pavement, whilst there are rarely large detached rocks on the beaches, doubtless owing to the violence of the waves, clearing out the smaller stones from the bottom, and heaping them up on the beach, and at the same time rounding the rocks below. We made about fifty miles to-day.

July 29th. — We started at sunrise, the weather clear and autumnal; the wind northerly. Breakfasted on a barren island terraced with ancient beaches, strewn with drift-wood, all of it showing strong action of the waves. Some logs of a foot or more in diameter had been thrown to the distance of fully a hundred and fifty yards from the water's edge, and thirty or forty feet above its level. Soon afterwards we entered a straight, narrow, river-like channel, some twelve or fifteen miles long, leading inside of Fluor Island and St. Ignace, whose dark wooded sides made a purple background to the vista. The banks were covered with birch, presenting an unbroken fringe of green; not a glimpse of the rock, and hardly, at intervals, the white line of sand at the edge of the water.

After passing through this channel we came out into Neepigon Bay, and had to keep round to the left to a deserted mining station at Cape Gourgan, before we could get a good camping ground. There we found a clearing and a convenient landing place. One of our companions two years before, in the month of October, had seen a large party of miners set ashore here from the propeller, to open the works. The marks of their labors, with the approaching winter before them, were everywhere visible. Wood had been cut and piled up; several log-cabins built and the cracks stuffed with moss and mud; and the paths through the woods showed where they went for fuel or to hunt. The ground was strewed with fur and bones of hares, and several lynx skulls were picked up by the men. Hunting must have formed the principal occupation of their days, since their mining operations had not been carried further then a few shallow pits, which doubtless soon convinced them of the fruitlessness of their errand.

It rained hard in the night, and we were somewhat incommoded by the leaking of our tent.

July 30th. — The rain continuing this morning, we did not think it worth while to start. The Professor took advantage of the opportunity to make the following remarks on the causes that influence the outlines of continents:

"The outlines of continents are not to be considered as fixed, immovable limits, but are variable, and dependent upon the degree of elevation above the level of the sea. For instance, were we to depress certain parts of South America or of the United States, even for a few feet, their outlines would be entirely changed, and immense tracts submerged; and vice versa, a slight elevation would produce corresponding changes.

"The west of Asia, comprising Palestine and the country about Ararat and the Caspian Sea, &c., is below the level of the ocean, and a rent in the mountain chains by which it is surrounded, would transform it into a vast gulf.

"Continents are in fact only a patch-work formed by the emergence and subsidence of land. These processes are still going on in various parts of the globe. Where the shores of the continent are abrupt and high, the effect produced may be slight; as in Norway and Sweden, where a gradual elevation is now going on without much alteration of their outlines. But if the continent of North America were to be depressed a thousand feet, nothing would remain of it except a few islands; and any elevation would add vast tracts to its shores.

"Elie de Beaumont, who has occupied himself much with tracing the changes wrought in continents by geological phenomena, has shown that chains of mountains elevated at the same time agree in direction. Thus the mountains of Scandinavia, the Ural chain and the Alps, &c. Before the elevation of the Alps, Europe was not divided into two great climatic regions. In this country the north and south direction of the mountains has a great influence. Animals migrate more extensively, and the cold winds, penetrating further south, influence the temperature.

"It would be very interesting to ascertain in detail the dependence of the forms of continents on geological phenomena. I have been struck with the possibility of this in running along the shore of this lake. The general shape of Lake Superior is that of a crescent. But it would be a great mistake to suppose it bounded by curved lines. Its shores are combinations of successive sets of straight parallel lines, determined in each instance by a peculiar system of trap-dykes. These dykes have five general directions, and the outlines of the shores are determined by their combinations. One of these directions is east, 30° north. This we find in the islands off Prince's Location, in Isle Royale, &c., and then again in Point Keewenaw and White-Fish Point. This is cut across by one east, 20° north: these two we have seen in several places together. Another is north, a little east. Another nearly E. to W. The last has a direction north and south, which we see in Neepigon Bay, where are the only inlets on the lake running north and south. Of these various sets of dykes each has its peculiar mineralogical character."

In looking round after the lecture for some more comfortable shelter than the tent, we espied a smoke rising from the chimney of a cabin at some distance in the clearing on the hill. Going thither we found one of the men very comfortably established on a sort of bench before a fire-place of stones and mud which occupied one of the corners. This was the only one of the houses that had a fire-place, and it was in all respects in much better condition than the rest, whether originally so, or from its remoteness having suffered less since its erection. Perhaps part of their company left the place when all hopes of copper vanished, and the rest then collected together in this building, leaving the other cabins to fall to pieces.

However this may be, the signs of habitation were still fresh here, and likewise unmistakable traces of the severity of the climate. Not only were the interstices between the logs carefully stuffed with moss and mud, but even the chinks between the two rooms into which the little hut (not over twenty feet by ten in the whole,) was divided, were filled throughout with hares' fur, large quantities of which were also piled up in a loft above and on a rude bedstead in the further room; a little circumstance which told not only of cold, but also of the listlessness and ennui of the poor devils shut up here, who could find time to pull to pieces skins enough to make such a quantity of loose fur. This was shown also by the caricatures scrawled all over the walls wherever the wood would show a mark, and an attempt apparently to make out an alphabet, some characters of which were entirely anomalous, and if inscribed on one of the rocks, might make work for some future antiquary. Each of the rooms had a fire-place occupying the corner, one still in good order, the other fallen to pieces from the softening of the mud cement. It was sad to think of the long days and nights they must have spent here, blocked up by the snow and crowding round the fire-places from the keen air rushing in at the chinks of door and window. Yet they were not destitute of provisions, as the remains of hares, and of sundry bean-barrels marked "Montreal Mining Company," testified; — they no doubt had cards, and perhaps, if they were Canadians, led pretty much the sort of life they liked best. The question of copper or no copper might be indifferent to them, if they were mere day-laborers, and for the rest, perhaps our commiseration was groundless.

One of the men having broken the stem of his clay pipe to-day, repaired it as follows; having cut a chip from a spruce log, he whittled it round, and cut a notch about the middle, leaving the ends connected by a thin spindle of wood. Then after burying it for some time in the hot ashes under one of the fires, he withdrew it, and twisting it in his hands one side came loose, and he drew it off, leaving a tube several inches in length, into which he inserted the stump of his pipe-stem. I afterwards saw this repeated, and both times, I may remark, the division of the wood had nothing to do with the annual rings, for the piece was taken near the outside of the log.

Towards sunset it seemed to clear off, and some of the party paid a visit to a deserted shaft, a mile or two distant, where they found small quantities of copper associated with chlorite, which from its greenish color had probably been mistaken for ore. In returning they got a ducking from a sudden shower.

July 31st.—We got off at five o'clock, the weather unsettled, and the wind high from N.N.W. We were in hopes to get round the point of St. Ignace, and then keep away before the wind. The prospect to windward was grand and striking. We were enclosed in an inner sea, a lake within the Lake: St. Ignace behind us, and on each side ridges of granite a thousand feet high. A sea of hills, rising from the rocky islands a few miles off, one over the other to the mountain chain far behind in the bottom of the bay. It was in fact an epitome of all the most remarkable scenery of the lake. The wind however increased so much that we judged it prudent to return. Accordingly we hoisted sail, and the canoe, right before the wind, swaying gently from side to side, like a sea-bird changing wings, made a comparative calm by its rapid flight; occasionally we struck a wave as it drew back, and then some care was required to keep from running bows under.

We encamped this time somewhat beyond the place we had left, more under the lee of the point. It continued windy and rainy all day, the wind going down at sunset.

Aug. 1st. — Started at four o'clock. Hazy, but soon cleared off, with westerly wind. We stopped to breakfast at a little sheltered cove on St. Ignace. The water here was filled for many rods with the larva-cases of a Phryganea, in such numbers that it was impossible to dip a cup of water without bringing up several of them. The insects themselves were flying about in swarms. This was the only time that we met any considerable number of these insects, which abound about the muddy flats of the lower lakes; the clear cold water of Lake Superior, and the pebbly bottom, are probably unfavorable to them. We continued coasting along St. Ignace, here a continuous cliff of red sandstone occasionally showing through its covering of forest. The wind was exceedingly variable to-day, shifting suddenly from one point of the compass to the opposite. I think we might sometimes have counted ten distinct directions in as many minutes.

Neepigon is said to signify "dirty water," and to-day it certainly deserved its name, being exceedingly turbid, and strongly in contrast with our experience of the other parts of the lake. But whether this is a constant phenomenon, or was an effect of the gale, I am unable to say. The bottom, in several places where I could observe it, was muddy, and the water unusually shallow.

We now approached the northern shore of the bay, a majestic line of rounded hills, the highest bare at the top, but in general covered with vegetation. A rocky cove where we stopped had been taken possession of by the Montreal Mining Co., who had made their mark on one of the trees, but apparently had not been encouraged to proceed farther. At our camping-ground this evening we found strawberries, still unripe.

Aug. 2d. — Hazy, wind east and strong, the Fates having seemingly determined that we should have head winds in whichever direction we steer.

At Turtle Island we looked for limestone, but were unable to find any. At this place an immense trap-dyke, running east and west across the point of the island, had tilted the sandstone 10° — 12°, and for some thirty feet on each side of it the rock was shivered into innumerable vertical fissures, of a line or two in width, and on an average not more than an inch apart. These fissures were filled with calc-spar.

We had now got back to the line of our westward course, and came this forenoon to the terraces spoken of July 15th. This remarkable formation (see frontispiece,) consists of three main terraces with several subordinate ones, rising one above the other by steep slopes. They occupy the whole bottom of the bay, (which has here an apparent width of a mile or more,) having the slope and curve of ordinary sand beaches, which indeed they evidently are. The slopes and widths of each respectively are, according to the Professor's measurements, as follows:—First the sand beach, rising from the water 11° for about twenty yards, then for a short distance 7°. Above this a ridge of pebbles 15°, beyond which was a belt of trees, and then a scanty growth of grass and a few low shrubs, extending about two hundred and fifty paces, with an ascent of 6°. From this an abrupt ascent of 20°, with a flat of fifty paces; then an ascent of 10° for a short distance, then sixty paces of 7°, and one hundred and fifty paces of 5°. Then comes another steep ascent of 30° to 33° to a space fifty paces deep of 10°—12°. Then another ascent of 26°—30°, succeeded by a succession of low, indistinct terraces, and finally an ascent of 20° to the top, which is nearly level for several hundred paces. The total height above the lake, according to Mr. Logan,* is three hundred and thirty-one feet. It will be seen that the whole presents a succession of acclivities in some cases as steep as the laws of equilibrium allow, alternating with slopes like the ordinary lake or sea beaches.

The general direction of these terraces is perfectly parallel to the present beach, and at right angles with the sides of the bay, which are high and rocky, and run in the same direction for some distance inland. From the further side of the highest terrace there is a uniform slope to a valley, apparently not much elevated above the level of the lake, and filled by a marsh and a small pond. The appearance is that of a deep inlet dammed across by the lake. The material is a coarse sand, with gravel, supporting a scanty covering of

* Geol. Survey of Canada. [A report to the Gov. General, Montreal, 1847.] p. 31.

G. Elliot Cabot from nat. Lith. of A. Sonrel Löffler on stone.

FALLS OF BLACK RIVER.

grass, and a few stunted spruces. The almost perfect regularity of these terraces, rising one above the other like one side of a gigantic amphitheatre, is very striking even at a distance, and the effect is increased by the absence of trees, giving the appearance of a clearing.

As the day had grown very hot we refreshed ourselves, after our scramble up these steep sandy slopes, by a bath in the icy water of the lake, and had to wade out several hundred yards from the shore before getting out of our depth. On the smooth sand of the beach were tracks of a lynx that had evidently been prowling there since the wind fell this morning.

As we pulled out of the bay a boat was entering it at the other side. It proved to belong to some government surveyors who were marking out mining locations, for which it seems there is still an active demand. They were established at the mouth of Black River, where we also encamped this evening.

This place strikingly resembles the mouths of the Crapauds and Chienne Rivers. A broad beach of white sand, about a mile long, is cut through at the west by the stream. The entrance is narrow, with a bar across it on which is five feet of water. Inside there is a wide expansion, across which projects from east to west (the course of the river being south,) a sand-spit in the shape of a half-crescent, with a broad base and tapering to a point. The rapids within sight from the beach.

Aug. 3d.—Rain. Held up early in the forenoon, and we started off up the river to see some falls about two miles above. One of the surveyors was kind enough to accompany us as guide, but the woods were so thick, and the ground so rough along the bank, that we kept off to some distance, where it was more open, hoping to strike the river higher up. But after half an hour's hard work, hearing the noise of rapids and coming down to the stream again, we found ourselves precisely where we started from. We resolved next time to keep near the river. Here we had to scramble over rocks covered with black lichens, (*Gyrophora,*) and make our way through dense spruce thickets, but whenever we strayed away from it we came to open desert tracts. At length we struck the river again, and came out at about the middle of a sand bank sloping uninterruptedly to the water. The distance to the top of the bank seemed trifling, but once embarked we found it a very severe tug, for the average slope being 30° to 31° and the sand very loose, we slipped back at each step nearly as much as we advanced. The height of the plateau above the river here is not less than a hundred feet, and the bank seemed to be composed of mere sand and gravel, horizontally stratified. Sitting down at the top to recover our breath, we had before us an extensive view over the forest, through which the river opened a long lane northward and seemed to expand beyond into a lake. At this spot we struck a trail leading to some works opened a year or two since near the Falls. The supposed copper, however, proving to be iron pyrites, they were speedily abandoned.

We had little difficulty now in reaching our place of destination, and came out of the forest upon a chasm of nearly vertical slate rocks, on a level again with the river, which comes in from the northward in a mass of rapids and little preliminary cascades, and falls in one sheet fifty or sixty feet into the chasm, a sort of gigantic well-hole, its sides black and savage with the splintered edges of the slate-rock, and so steep and even overhanging that we could not from any position get a view of the bottom. Below, the stream turns sharply to the left and rushes out through a deep gorge not more than five or six yards wide at the bottom. From below the gorge there is a very wild and picturesque view of the river boiling out from between overhanging rocks.

On our way back we followed the miners' trail all the way to the lake, coming out about a mile to the eastward of our camp. In our course we had diverged considerably from the river, and found the ground much more open, the trees scattered so much that we sometimes had difficulty in tracing the line which was "spotted" or scored upon them; the ground dry and lichenous. We descended to the lake by a succession of well-marked terraces of large rough pebbles, and then through thickets and over irregular broken rocks in piles smoothed by a treacherous covering of moss.

In the evening the Professor made the following remarks upon the terraces and the drift formation about the lake:

"We have seen at various points along our route, large accumulations of loose materials, often in the form of terraces. These loose materials are usually called 'drift,' but it is necessary to distinguish among the various formations known by this name, the beaches thrown up by the lake upon its present shores, and the ancient terraces above the present level of the water. Nevertheless, the connection between these two kinds of drift is such as to show that the latter also were formed by the lake, but under different circumstances from the present beaches. The first question is, whether the lake was anciently higher; the elevation of the ancient terraces having been the same as now; or whether the land has been elevated. Either is possible, for we have examples both of elevation and of depression going on in our own day, as upon the eastern coast of Sweden and the western coast of Norway. This question cannot be settled by a simple inspection of the terraces, but only by a comparison of their elevation with the level of the surrounding region. Now the terraces we saw yesterday show a difference of level of over three hundred feet above the present lake beaches. If we add this to the present level of the lake, and suppose it formerly to have stood at the height which they now exhibit, it must have overflowed the whole United States and joined the ocean. But if this were so, we ought to find the remains of marine animals here, which is not the case. It is more probable, therefore, that the land has been elevated.

"The foundation on which these terraces rest is uniformly rounded and scratched rock. During our whole journey we have nowhere seen serrated peaks; everywhere the surface is smooth, grooved and scratched in a north and south direction, occasionally diverging east and west. And it is evident that the force that produced these appearances acted from north towards the south, for we generally find the south side of the rocks rough and precipitous, showing no abrading action, whereas they are smoothed off towards the north. Now it may be asked whether the loose materials before spoken of were the agents that produced these effects? I think we may say positively that they were not. We have found the rounding and grooving at the highest point we have visited, that is, over twelve hundred feet above the level of the lake. This is much higher than any of these loose materials are to be found. Moreover we see they are disposed according to the present form of the lake, and evidently in many instances have been heaped up by a force acting in a direction from south to north, directly contrary to that of the grooving force. It is clear that the formation of the terraces was subsequent. They overlie the grooved and rounded rocks.

"To ascertain the cause of this latter phenomenon we must find what are its limits. Now we find it occurring universally over the northern portion of the globe, and always having the same general direction. Its limits in elevation, as ascertained on the sides of mountains, is about five thousand feet above the sea. At about this height on Ben Nevis in Scotland, and on Mt. Washington in New Hampshire, the grooving and polishing ends. Below this level the whole northern surface of the earth as a general thing shows the marks of this agency. Some geologists attribute these effects to the action of *currents*. But currents extending over such a vast extent of the earth's surface must necessarily have been ocean currents, and these must have brought with them marine animals, of the existence of which no traces have been found. Moreover such extensive currents in one direction could not have existed: there would necessarily have been refluxes and counter-currents.

These and other difficulties have led me to attribute these effects to another cause. It has been ascertained that the glaciers of Switzerland formerly extended much farther than at present, reaching, without interruption, to the vicinity of Paris, and, near their origin, to the height of nine thousand feet above the sea. Similar indications are to be found in all the mountain chains of Great Britain, and in various parts of Europe. Now at the time when such glaciers existed in Europe, the temperature must have been much lower than at present. The mean annual temperature of Switzerland must have been 15° Fah. below the present. That such a depression of temperature actually took place is also indicated by other facts. Thus the fossils found in the glacial moraines are of an arctic character, and shells of the German Ocean are found in the moraine gravels of Sicily. This, however, is inconceivable without a corresponding depression all over the globe. Now if we suppose the mean annual temperature of this country to be reduced to 26° Fah., it would naturally be covered to a considerable depth with ice, which would move from north to south. Such a mass of ice moving over the country would produce these effects of rounding and scratching the rocks, and would remove the soil, except from the depressions. It is sometimes objected to this theory that we have here no slope which should cause such a mass of ice to move onward. But it is not necessary that there should be any slope in order that a glacier should move. In the Swiss glaciers the motion is often slowest on the steepest part of the slope, and some glaciers of 7° inclination move faster than others with a slope of 40°. The great motive force is not the gravitation of the mass, but the pressure of the water infiltrated into it. Then supposing the country to have been subsequently depressed, (as we see has been the case in Sweden and Norway, where marine shells have been found at the height of three or four hundred feet above the level of the sea,) and afterwards raised again, these various terraces would mark the successive paroxysms or periods of reëlevation. Such a depression would not cause an irruption of the sea, since the level of the lake is over six hundred feet higher than the sea-level. But these phenomena are exceedingly complicated, and cannot be sufficiently illustrated without further details.

"The east and west direction of the scratches at Spar Island, contrary to the general rule, I suppose to have been caused by the depth of the channel there, giving the glacier on its retreat a direction parallel to the shore of the lake. We had there two very distinct systems of striæ, one much more southerly in direction than the other. Probably the glacier when advancing from the north, having an enormous thickness, disregarded the shape of the ground over which it passed, but on its retreat, that is, when it began to contract, having meanwhile melted away considerably and thus become lighter, its direction would be more easily modified. Similar phenomena are observed in the present glaciers in Switzerland. In a little loch near Ben Nevis there is also a secondary system of scratches, *at right angles* with the general direction, which may be traced even on the bottom of the loch."

We learned from the surveyors that a brown bear, differing from the black and grizzly bears, is found in this region. It was said to be about the size of the black bear, and is probably the barren-ground bear, (*Ursus arctos americanus*,) of Richardson, though he says this species is not found so far south.

On coming out of the tent we observed that standing by one of the fires, so as to bring it between us and the rapids, the roar of the water was suddenly shut off, as if by a door, the sound being interrupted no doubt by the ascending column of heated air.

The weather looked threatening this evening, and in the night we had a violent shower accompanied by thunder and lightning. In the midst of one of the gusts we were awakened by several small rivulets playing down upon us from folds in the tent, which, on account of the sandy soil, was not properly stretched. Indeed, without some better contrivance than mere loops for the tent-pins, a tent like that we had cannot be stretched so as to be water-proof in a violent shower. One of the tents, brought by Mr. Marcou, of the kind used by the French officers in Algiers, was entirely water-proof, and in every way more convenient than ours. It was square, with nearly perpendicular sides, and stretched near the top by cross-pieces at right angles with each other, while the pole ran up in a point in the middle. The only help was to cover ourselves as far as possible with our water-proof cloaks, &c. But these in the pitchy darkness were not so easily found. We then attempted to light a candle, but the matches were damp, and with all our precautions could not be coaxed quite to the igniting point. Finally by the intervention of a flint-and-steel, (let not the traveller be seduced into placing his reliance in any new-fangled substitute for this trusty companion,) we managed to get a light and find our things, and therewith made ourselves tolerably comfortable.

Aug. 4th.—Weather still unsettled, and we did not start until after breakfast. It was calm at first, but the wind soon rose strong from the N.N.W., obliging us to creep round very near the shore.

We encamped at night on a point where the very wide and steep beach ascended by terraces to a long regular ridge. This ridge was covered, in one place in an unbroken patch of an acre or more, with a checkerwork of large tufts of yellowish gray and dark pinkish lichens, mingled with deep green juniper (*J. virginiana,*) and Vaccinia.

The beach was covered with drift-wood, large trunks of trees with the roots often attached, most numerous on the top of the beach close to the trees, although the distance from the water must be a couple of hundred yards, and the elevation not less than thirty or forty feet. We never met with any floating wood. Doubtless the trees are washed away and thrown up in the winter, and cast higher by each successive storm until they are out of the reach of the water.

The Professor found here, in place, the red porphyry of which we had found erratic blocks at many points to the southward on our way hither; it was perfectly stratified, and associated with chlorite.

Aug. 5th.—We reached the Pic early this morning. As we approached the wharf we saw our companions whom we had left behind here, waiting to receive us. The sick man had pretty nearly recovered, but still looked thin and pale.

In the low grounds here, as at Fort William, we found partridges, (*Bonasia umbellus;*) in the wettest part of the swamp, directly at the foot of the ridge, I came upon a female with a brood of young

G. Elliot Cabot from nat. Lith. of A. Sonrel Löffler on stone.

PIC ISLAND from CAMP PORPHYRY.

nearly fledged. It is remarkable that this bird which with us affects dry situations, about the lake seems, as far as our experience went, to prefer swamps; the spruce-partridge (*Tetrao canadensis*,) being found rather on the high ground. But this apparent anomaly is explained when we remember that in the White Mountains, for instance, where both species are also found, the spruce-partridge is met with only at considerable elevations, among the spruces or "black growth," from which its popular name is derived, and the other bird in the valleys or lower slopes. But here, where the spruces come down to the general level of the country, the difference of distribution is still expressed, though less distinctly, notwithstanding it necessitates a change in what would seem a more important point. In this instance a very decided habit of the bird is sacrificed to what many naturalists would call a mere abstraction.

In the night we were disturbed by the dogs, who swarmed as usual about the Indian lodges, and as usual were half-starved and dependent solely on their own exertions for support. A camp-kettle left outside of the tents attracted them into our neighborhood, and they made a great noise in rolling it over in their endeavors to get the cover off. Among this vagrant crew I was astonished to see Mr. Beggs' Esquimaux dog, who might be supposed to be too well fed to be tempted into such ways. This dog was said to be of the pure breed. He was of a yellowish-white color, of moderate size, with a small head, the nose pointed and the face rather wolf-like, though not at all savage in its expression. Round the neck was a ruff of hair, and the tail was bushy and curled upon itself, as we see in the representations of this species.

Aug. 6th.—Mr. Ballenden stopped here at sunrise this morning, on his way to the Red River settlement, of which we understood he had been appointed governor. He had come all the way from Otter Head this morning, a distance of forty or fifty miles, running before a strong S.E. breeze in his large two-sailed boat. But this wind which was so favorable to him was quite the reverse to us, and kept us *degradés* here until six P.M., when, there being a slight lull, we embarked.

Mr. Swanston had promised to send us up some provisions hither from Michipicotin, but they had not arrived, and the stock in the store-house was so small that Mr. Beggs at first thought he could not spare us any, but just before we left, taking compassion on our destitute condition, he gave us a supply that would last us to Michipicotin.

When we got outside of the bar the wind rose again. We soon lost sight of the bateau, and the two canoes kept on alone as well as they could against the wind and sea. We in the larger canoe could not help watching with some anxiety the other one under our lee, occasionally throwing half her length out of the water, and then pounding down so as to make it fly up on all sides. This thumping does not agree very well with the birch bark. The gum gets cracked and lets in the water, and there is not substance enough about the fabric to float when filled. It was fast growing dark, and the shore to leeward showed a horrid line of grim weather-beaten rocks and white breakers. At length the men in the other canoe called to us that they could stand it no longer, and kept away for a cove we had just passed. We followed them, but although only a few hundred yards behind, yet it was so dark that when we entered the narrow mouth of the bay, we could see nothing of them. The outline of the shore to leeward, however, was still distinguishable against the western sky, and we assured ourselves that they had not gone further to leeward. We kept, therefore, an anxious lookout as we ran rapidly up the narrow bay, so narrow that we could not pass them undiscovered if they were afloat, and fired off several guns, but without answer. Before long we came to what seemed the bottom of the bay, but here we found no signs of our companions, and seeing a further passage to the left, we supposed they had kept on. Accordingly we pushed on up a river-like inlet, with high mountainous ridges on each side half a mile or more before we came to the bottom.

Here we landed on a little sand-beach, heaped up with a great quantity of drift-wood. While the men were pitching the tent in an open space inside the fringe of bushes, we lighted a fire, and looked about with a torch made of a roll of birch-bark for a tree suitable for a signal-fire. We soon found a tall spruce well covered with lichen, and applying the torch below, the flames climbed and spread upward and horizontally from one branch to another until the whole burst upwards in a vast tongue of flame, crackling and whirling up sparkles of burning twigs and leaves to such a height that it seemed impossible that our friends should be in the bay and not see it. But the flames went out, the last sparks one after the other dropped away, and the dark walls of the bay came into sight against the sky, yet we listened and looked in vain for an answering signal. Next morning, however, namely:

Aug. 7th—We were early awakened by their voices on the beach They had landed in the outer cove, and thus did not see our fire, being cut off by a high intervening ridge. They had heard the gun, but were engaged in hauling up the canoe, and so could not answer it. Looking round upon the prospects of the day we found the wind still so strong from the S.E. that there was no chance of getting off at present. Of this we could feel no more where we were, than if we had been at the bottom of a well, but the men pointed to the breakers at the mouth of the bay, where, at the distance of a mile or more, the large and rapidly shifting masses of white against the black rocks showed that the surf was beating outside at least as violently as the night before. On listening, the roar of the waves could be distinctly heard. But immediately about us it was dead calm, with occasional eddies in the tree-tops from all points of the compass. A contrast such as the lake seems to love, as if it sought to break up the uniformity of its general features as much as possible by brisk and abrupt changes in the minor ones. Thus although the weather throughout our journey might be called settled, yet we very rarely had a steady wind, either as to direction or strength, and in the hottest day the shade of a rock, or a cloud passing over the sun was enough to make it cool. The range of clothing thus necessitated within the twenty-four hours was extraordinary.

Our little point was as silent as a piece of the primeval earth; not a living thing stirring except a few musquitoes, and an impudent moose-bird that perched down, with a jerk of the tail and a knowing turn of the head, among our very camp-kettles. A heavy stillness seemed to hang over it and weigh down every sound, so that a few paces from the tents one forgot that he was not alone. It was as if no noise had been heard here since the woods grew, and all Nature seemed sunk in a dead, dreamless sleep.

Yet it was clear we were not the first visitants, for the fire-weed had sprung up here, and close at hand we found lodge-poles, and the remains of fires. Here also was an Indian *sweating-house;* a skeleton dome of sticks, about four feet high and two in diameter. The patient squats inside, and by his side are placed some hot stones, on which are thrown various herbs, by way of "medicine." Then the whole is covered in with blankets and pieces of bark, and he is left to simmer for the requisite period.

Back of this a path led a short distance through the woods to the mouth of a sluggish stream some five or six yards wide that joined the bay north of our camp, which was thus cut off on three sides by it and the lake, and on the fourth by the mountain.

Our beach, as I said, was heaped with drift-wood, most of it arbor-vitæ, recognizable by its twisted stem. This tree loves the water, and grows in situations where it is most exposed to be washed off by the winter storms. Some of the logs were of large size, a foot or more in diameter, completely stripped of branches and bark, and in general of their roots; and exhibited marks of very rough handling, being deeply grooved and rubbed, perhaps by chafing together, partly perhaps from ice. Many of them were very regularly and smoothly tapered at the end. Driven into the bay by the westerly gales in the winter, they had doubtless drifted along its steep sides, and been successively piled up at the bottom.

Our men having such a store at hand did not spare fuel, and were mightily amused when we told them they had on five dollars' worth at once. But although cold morning and evening, it was very warm in the middle of the day, the temperature rising from about 40° to near 80° Fah.

The water was deepest close to the rocks at the end of the point, though even there it was hardly anywhere more than five feet deep. Beyond, it was so shoal that we very easily waded across to the other shore, about a quarter of a mile. The bottom was an even surface of mud, on which we met one or two large rounded pebbles half imbedded, but no sand or small stones. Various water-plants, namely, two species of Potamogeton, and an Echinodorus, with pretty white flowers, were growing abundantly here.

The wind and waves still high outside. Several times the men went to explore, but returned, reporting it still too rough to venture out.

Aug. 8th.—This morning we heard distant reports of guns, and the men thought it might be our friends of the bateau over in the next bay. As our provisions were getting very low (the bulk of the stores being as usual in the bateau,) they resolved to cross the ridge and fetch a supply. They reached the cove after a laborious climb, but found no traces of them, and so kept on to the Pic, where they found them reëstablished in their old quarters.

We now reconnoitred again, but with the same results as before. Towards evening, however, the men seemed to have made up their minds that we should get off to-morrow. Certainly "*la vielle*," the old woman, as they called her, (a personage corresponding to our "clerk of the weather,") had given us a long enough bout of it, and it was time to expect a lull. Accordingly, they made all their preparations, and being desirous no doubt to appropriate to themselves the largest possible share of the good things of the wilderness, piled such a huge quantity of wood upon the fire that we were driven back yard by yard to the distance of some rods.

Aug. 9th.—Calm, with a slight fog, and soon cleared up very warm. This afternoon, for the first time on the lake, the wind was strong from the south. We encamped in a cove under a hook projecting from the southward. The beach of large stones covered with lichens, whence the name of Campement du Pays de Mousse, which the cove bears. It is terraced up to a dividing ridge, and thence down in like manner to the lake on the other side.

We had been struck for some distance back, and particularly to-day, with a falling off in the luxuriance of the vegetation, as compared with the country further north. This may be owing to the greater exposure to the northerly winds; the more northern shore being protected on that side by a lofty and continuous barrier. In a very sheltered cove where we landed to lunch, the trees were of considerable size. One larch measured seven feet two inches in girth, three feet from the ground, and we judged its height to be at least sixty feet.

Aug. 10*th.*—Calm this morning, with a swell on the lake; an unusual occurrence, owing to the southerly wind of yesterday. We passed at a short distance the river Rideau, which falls in a succession of cascades (said to have ninety feet descent in all) directly into the lake. The final fall, of about thirty or forty feet, is divided in the middle by a large rock, part of the wall of the cove into which it falls. This river, the only one we saw where the never-failing falls descend directly into the lake, was also the only one that had no sand-beach at its mouth. All the others were indicated from a distance by an expanse of white sand.

Shortly after, the wind sprung up fresh from the south-west, of which we took advantage with our tarpaulin sails. It is a mistaken notion that a canoe will not sail on a wind. Ours sailed very well, with the wind somewhat forward of the beam. Only the sails are not braced up much, but just enough to keep full; since otherwise, having no keel, the canoe would make too much leeway.

Opposite Otter Island we counted ten parallel trap-dykes, running north, twenty-five degrees west. Here are several terraces, passing by regular gradations into the present beach. At the Riv. á l'Oiseau Vert are veins of epidotic trap. The bateau hove in sight outside of us this morning, with both sails set.

In the afternoon we came upon the bateaux from Michipicotin, moored under the lee of some rocks. They had been several days on the way already, being kept back by the wind, and thus it was that our stores had not arrived at the Pic. These were now handed over to us, consisting of pork and excellent ship-biscuit. The men in the boats were mostly half-breeds, with their families. Several of the women were very pretty; their complexion, indeed, a faded or bleached olive, as if they had never seen daylight, but with a spot of color in the cheek. We passed Michipicotin Island, having neither time nor favorable weather for visiting it, and encamped on a beach of coarse dark sand, where we observed the white flowering raspberry for the first time on our return.

Aug. 11*th.*—At half past five this morning when we got under weigh, it was dead calm and somewhat foggy. The fog soon lifted, and the sun shone out warm. The surface of the lake continued unruffled, reflecting, unbroken and scarcely dimmed in color, the full form of every rock and tree. Running along at a moderate distance from the shore in this calm weather, we were often struck by an apparent convexity of the surface, as if the water were higher between us and the rocks. It even seemed to hide the line where land and water met.

Suddenly the water was spattered by the rising of a shoal of lake-herring, and our men were immediately full of excitement, and must needs get the fish-spear from the bateau to have a stroke at them. By that time, however, the shoal had sunk again, and the men watched in silence and without dipping an oar, for them to rise. Looking down over the side of the canoe, we could trace the vast, simple lines of the rock, until lost in the green mist. Everything below the surface seemed to shine with a diffused phosphorescent light, like a green unclouded sky. All at once the shoal came in sight, under the boat, pressing steadily on with a broad front, a solitary white-fish rather in advance of the rest. Each kept his relative position to the rest, like a flock of waterfowl, and they glided easily onward without any apparent exertion except a tremulous motion of the tail. Yet they soon vanished ahead, and not long after a great trout came sullenly following in their wake, like a pirate hovering about a convoy of merchantmen.

Some Indians came off to sell us fish, and our men in their gossip discovered they had in their lodge a couple of young foxes, which the Professor thereupon demanded to see, and bought. The poor little fellows were about half grown, and seemed to suffer from the heat. The first thing they did when we took them aboard, was to seek out the shadiest corner. They appeared to be perfectly tame, or at least inoffensive.

We caught several trout ourselves in the course of the forenoon. I was struck with the life-like appearance of the bait, (a trout's stomach drawn over the hook, and tied to the line above,) visible at a great depth. Out of water it has rather a shapeless appearance, but jerked along at a sufficient depth it has precisely the look of a small fish that has been wounded, so as swim with difficulty and somewhat sideways.

In the afternoon a favorable breeze sprung up. Our men were

profuse in their thanks and compliments to the "old lady," and in addition to the tarpaulin, must needs rig a spritsail, which they made of a blanket extended between an oar and the fish-spear.

We reached Michipicotin at about five P. M. One of our first questions was as to the flies. Mr. Swanston said they were "all gone," which we found, comparatively speaking, true, but at the old camping-ground there were a few left to remind us of our former sufferings.

We held a council this evening as to the advisability of making an excursion to Michipicotin Falls, six miles up the river. The majority were decidedly in favor of pushing on, and the Professor did not like to leave them. So it was settled that two of us who wished to go, should remain behind with the small canoe, and endeavor to overtake the rest by forced marches.

On opening this evening a tin case in which bird-skins were packed, I found the inside covered with drops of water, and some of the skins so wet that I had much difficulty in drying them. As the case was surrounded by an India-rubber covering, and the whole put into a wooden box, which was perfectly dry, the moisture could have come only from the condensation occasioned by the great and sudden changes of temperature. Metal, therefore, is to be avoided here, if dryness is requisite.

The dogs disturbed us somewhat in the night by their antics with a frying-pan and a tea-kettle, which Henry had unfortunately omitted to place out of reach. A troop of mongrel curs seems to be a general characteristic of an Indian village, though they neither make use of them nor seem to take any care of them,* and one does not see why they should keep them, unless it be for an occasional dog-feast, an observance which, to judge by the lean condition of the dogs, is rather gone out of fashion.

Aug. 12th.—Warm and cloudy. While our friends were making ready for departure, we set off for the falls, with an Indian lad for guide, paddled a few hundred yards up the river, and having pulled the canoe up on the scanty beach on the opposite side, climbed up

* One Indian, however, who readily sold his dog for a trifle, revoked the bargain when he understood that the skeleton only was wanted. Whether this was from any feeling for the dog, or only from some superstition, we could not learn.

the steep sandy bank, twenty or thirty feet high, and found ourselves upon a wide plain, bounded by the river on the right, and some steep rocks in the distance, on the left.

The surface was level and barren, not a tree in sight, but only a uniform expanse of withered herbage, bearberry, lichens and great quantities of blueberries and huckleberries, now ripe, much to our satisfaction, for we had not tasted fruit of any sort for so long that even these humble kinds had a flavor unknown before. There were two sorts ; the most abundant was of a light lead color ; the other larger and of a dull blackish. We did not stop to gather them however, but pulled them by handfuls as we ran along the trail, to the annoyance of our little Indian, who had evidently calculated upon a deliberate feast.

The path was worn through the crust of superficial vegetation and the thin seam of mould that supported it, a foot deep into the sand below, and so narrow that we had to walk Indian fashion with toes turned in, and I had some trouble to avoid grazing my ankles with my shoe-soles. My companion wore moccasins, a much more comfortable gear for this ground.

The weather was very warm, and the flies exceedingly troublesome, rising in swarms from the blueberry bushes when we touched them. Whether from a presentiment of their coming end, or from some other cause, they were not flying abroad to-day, but collected on the ground. Once roused, however, they showed no backwardness in making an attack. Having for the first time open ground enough to observe their manœuvrings, we tried to outrun them, and easily left them behind, but in a short time the swarm, like a pack of wolves, and guided to all appearance in like manner by scent, came ranging up in a body and fell on afresh.

Continuing on for about a mile we came to a sudden depression in the plain. We stood on the edge of a steep bluff some forty feet high. Below, the broad level valley stretched off apparently to the river on the right, and on the left to some rocky hills several miles distant. It seemed perfectly level and sandy, and in all respects like the plateau on which we stood, except that it was still more barren and showed patches of bare sand. On the opposite side the bluff rose again as abruptly to about the level at which we stood.

It had all the appearance of a sudden and even depression across a previously unbroken plain. My companion thought it a former bed of the river, and that he could see an opening in the hills to the left (which direction we knew the river took above) through which it might have flowed. I could see nothing of this, nor did the valley seem to me to present the appearance of a river-bed, for it was perfectly level, free from stones, and nowhere less than half a mile wide, varying from this to perhaps three fourths of a mile, at least six times the present width of the river. In our haste nothing very satisfactory could be made out, but my general impression was that it was the bed of a former arm of the lake.

Crossing the valley and ascending the bluff, by an equally steep path on the other side, we came before long to scattered spruce trees, and at the distance of about three miles from the factory, to the river again. Here we were made aware that what had seemed to us a horizontal plain, was in truth a gradually ascending level, for we now stood sixty or seventy feet above the stream. A little brook scarcely deep enough to swim a trout came into the river here at the same level, having sawed through the sand to its very base, leaving on each side a steep slope of pure sand, excessively fatiguing to ascend. We were now surrounded by a tolerable growth of spruce and birch, occasionally forming thickets. The aspect of the country was not unlike that of the White Mountains at the elevation at which the forest begins to disappear, only more abounding in lichens and small shrubs.

There was no opportunity in the course of our hasty walk to observe the stratification of the sand. We saw no freshly broken surfaces, and in the paths the materials were of course displaced. In general terms, however, I may say that it was a coarse, reddish sand, mixed with gravel and with a few stones, which were somewhat rounded but not scratched as far as I observed. The general appearance was much the same as that of the bluff at the factory, which is very distinctly stratified.

Afterwards we came out into an open space whence we had a very extensive view over woods and barren ground, with occasional glimpses of the river far below, and on the edge of the horizon a peep of the lake.

About three quarters of a mile from the falls we struck the portage path, running through deep moist woods. Across it were laid logs, at short distances apart, so that it was like walking on a railroad where the sleepers have not been filled in. An explanation soon presented itself, in a smooth, narrow trench in the middle of the path, such as would be made by the keel of a vessel, and on each side the traces of a heavy body dragged over the ground; we conjectured that it was an arrangement for facilitating the transport of the heavy bateaux that come down from Hudson's Bay. When we reached the head of the portage we found we had guessed rightly, for here lay several large boats ready to be hauled across. These bateaux measure generally twenty-eight feet in the keel and near forty above, and are very heavily built, yet as Mr. Swanston afterwards told us, the voyageurs make nothing of the portage, and amuse themselves with racing the boats against each other over the path.

At the head of the portage we found ourselves a good way above the falls, but there was no appearance of a path, so we made our way down stream through the tangled arbor-vitæs, and soon came out in front of the upper fall.

Michipicotin Falls consist of three cascades of about equal heights, separated by short intervals of rapids; the total descent is upwards of eighty feet. At each fall the river is compressed to the width of a few yards between projecting points of rock, and below each expands again somewhat.

The rock is a gray sienite, broken into huge parallelograms, some lying about in loose fragments, in others, the cleavage lines indi cated on the face of the rock having a dip of about 20° southwest, that is, at right angles with the fall. These projecting points and detached fragments of hard rock in the bed of the cascade, give it a peculiar character. Thus at the foot of the second fall the whole mass of water is thrown upwards again in a vast fountain of spray, from the resistance of some obstacle below the surface.

The third or lower fall is very striking. Whether from the sudden expanse of the channel, which becomes somewhat wider here, or from the shape of its bed, it forms a regular half-dome of broken water, a most magnificent spectacle, not at all like any other large fall I ever saw, but resembling on a gigantic scale the bell of water so often formed by a projecting stone in small mountain streams.

This indeed might serve for a description of the whole scene. It is a mountain torrent on a large scale, and without the majesty of Niagara, or even of Kakabeka, it has a charm of its own in its exuberant life and freedom. Below, the river turned to the right, leaving at its outer angle a whirlpool, in which were revolving a great quantity of logs, as cleanly stripped of bark, roots, and branches, as if prepared for the saw-mill.

From what I could observe, the river-bed above the falls is not much below the general level of the country; as if it flowed there over a rocky plateau, covered with a scanty depth of soil, and abruptly falling away at the falls, forming a barrier against which the sand and gravel from the lake have been heaped. Below, the banks are high, of loose drift deposit. This may be the edge of a step in the descent from the height of land.

Reaching the factory again, we found all in readiness for departure, the men anxious to be off, and the lake so smooth that we could take the direct line for Cape Choyye, which we reached a little after sunset, while the air was still full of rosy light, the moon just peeping through the fringe of forest on the edge of the cliff above us.

Here the men proposed to stop for rest and refreshment, and then to keep on by moonlight.

At the place where our tents had been pitched, I found the evergreen boughs still undisturbed on the stones; the balsam twigs still retained most of their leaves, but the spruce were entirely bare. We hastily drank our coffee, and the men their tea, and then reëmbarked. About ten o'clock we were awakened by the cessation of motion, and found ourselves in a narrow cove near Cape Gargantua.

Aug. 13*th*.—It was warm and rainy this morning, with fog. We started early, and approaching the Rivière aux Crapauds, the men saw a *boucane*, namely, a smoke (whence, bye the bye, the term *buccaneer*), and said we should find our friends there, though we could not well distinguish it from fog. They were right, however, for there they were, just done breakfast.

I was struck with the unhesitating accuracy with which our men steered in the fog to-day; they evidently knew the way now, though by no other landmarks than rocks and islets, which to an ordinary observer seemed all alike.

In the afternoon it rained hard. We protected ourselves with the tarpaulin, elevated in the middle with a tin map-case by way of tent-pole. The rain stopped towards evening, and close before us lay Mica Bay, with its wharf and crane, and Capt. Matthews' cottage on top of the bank.

The Captain had gone to commence mining operations at Michipicotin Island. Mrs. Matthews, however, and Mr. Palmer, a young gentleman attached to the establishment, received us most hospitably. Mr. Palmer gave the Professor several valuable specimens, and showed us the commencement of a very elaborate survey of the location, in which even the trap-dykes (which here intersect at some points in the most intricate manner,) were laid down.

Aug. 14*th*.—Before starting this morning, Mr. Palmer carried us up to the mine to see some "pot-holes," that had been discovered there since we were here before. The spot where they are found is two hundred feet above the present level of the lake, in a narrow vein filled with rolled pebbles and gravel, lying directly over the lode which is now worked. This vein runs vertically through a considerable thickness of unstratified drift, with angular bowlders, and *scratched*, but no rounded pebbles. The rock slopes steeply towards the lake, and some of the holes are joined together like stairs, the stones that formed them having evidently worked by degrees down the slope, as we see them doing now at Cape Choyye.

We left with a favorable breeze, passed Mamainse, and were already expecting to reach the Sault to-day, but by the time we were abreast of the Sandy Islands, it blew so hard that it was thought prudent to put in and wait for a lull, the bay beyond being, according to the men, a dangerous place in foul weather. The other boats had disappeared; the bateau to windward, the canoe working in shore towards Goulais Point.

On the broad sandy beach, as we landed, we found the tracks of a fox, just made, for the wind had not filled them up. I set out to explore the island, without my gun, however, contrary to my wont, having unluckily left my powder in the other canoe. As I approached a fallen spruce tree that lay about thirty yards off, with its top in the water, I saw coming towards me from on the other side—a fox! The fellow was of the variety called "Cross Fox," lean and hungry-looking. He trotted leisurely on, as one sees a dog trotting along a pathway,—occasionally pausing to sniff at a dead craw-fish. I did not attempt to hide myself, but stood perfectly still. He came carelessly on, and cleared the tree with the lightest and gracefullest of leaps, but his black paws hardly touched the sand before he had whisked like lightning from his course, and disappeared in the wood.

As the island is not a mile long and only a few hundred yards across, it was a matter of wonderment how he got here, or what he could find here to live upon. The men said he had most likely come across on the ice from the main land (a distance of about four miles) in the winter, and had not dared to swim back again. We found marks of digging in various parts of the island, and conjectured he had been reduced to a partly vegetable diet. If he could have trotted undisturbed a few rods further, he would have found what I picked up in his stead, the dead body of a little warbler that had evidently been beaten down and drowned in the storm the day before, and lay on its back on the sand at the water's edge, the wings a little open, quite fresh, and the plumage hardly ruffled.

At dusk, two figures appeared on the beach of an island about half a mile off. Our men said they were "Français," that is, not Indians,* but more could not be made out. They proved afterwards to have been some of our friends of the bateau, but they had encamped on the opposite side of the island, and did not see us. It rained at intervals, and blew very hard in the night, the wind shifting from north-west to north-east. We had fears for our tent, but fortified ourselves by felling a few trees to windward.

Aug. 15*th.*—At five o'clock this morning it still blew hard, and although the wind was more off shore, and the waves accordingly not so high, yet the rollers were plunging along the beach with a violence that rendered embarkation somewhat hazardous. But we were all anxious to be off. To-day was the day fixed for reaching the

* These half-breed voyageurs are true creatures of tradition, and still divide the human race into but two classes, "*Français*" and "*Sauvages.*" Before I understood this, one morning we found on a beach where we landed tracks of men who, they said, were "Français." When I asked them how they knew this, they pointed to the marks of *boot-heels* in the sand.

Sault, and we could reach it easily from here. Our men were as eager to be gone as we, for they had worked long enough at one job to be glad of a change. Then at this season it was as like as not to blow for a week, and harder, and our provisions would not hold out many days.

So the canoe was set afloat, and held head to sea by a man on each side, standing up to his middle in the water. In this position it was carefully loaded, and we got on board over the stern. Finally the men contrived to get in and push off without serious accident, though not without shipping a good deal of water. As the wind was directly off shore, matters improved as we proceeded, and before long we were under the lee of Gros-Cap.

The thickets of white flowering-raspberry were now full of fruit; the berries averaging about three quarters of an inch in length, by two thirds in diameter, and rather firmer and more symmetrical than the common cultivated species. The taste is slightly acid, but agreeable. Probably they were not entirely ripe. There was also an abundance of the common wild raspberries.

From Gros-Cap to the mouth of the river, the water was not more than three or four feet deep; the bottom gravel. Farther out it is deeper, but the amount of water that leaves the lake is small, as is shown by the moderate rate of the current at the entrance of the river, notwithstanding the narrowness of the outlet. At the Pointe-aux-Pins, where the shores from being over two miles apart suddenly approach to within half a mile of each other, we did not perceive any acceleration of the current. The fact is the channel has only this width all the way down to the Sault; the rest being very shallow. The banks are low, so that a very slight elevation of the surface of the lake would give an outlet of five or ten miles in width down to the Sault, and expanding below.

Arriving at the head of the portage, we found some of our friends awaiting us. Both the boats had got in just before us, and they had hastened to get on their civilized costume and run back to meet us. Singularly enough, the "Dancing Feather" had arrived that morning, about two hours before us! So here we were all on the day appointed for meeting, although we had paddled four hundred miles, and they twice as far since we parted.

We had arranged to shoot the Rapids, instead of landing above. The men did not seem to think it much of an exploit, and made no change in the stowage of the canoe. The oars were taken in; the steersman and bowman furnished with paddles instead. We glided quietly down, the paddles just touching occasionally, with a few rapid and vigorous strokes at certain points.

The water is so little broken that we seemed not to be moving very fast, and it was startling on looking down over the side to see the bowlders on the bottom twitched by so quickly that it was impossible to see their forms. It was like looking down from a railway car upon the sleepers. Whether from bravado on the part of our men, or from the necessity of the case, we several times passed within a foot or less of rocks apparently just under the surface. We were not more than three or four minutes going down, though the distance is nearly three quarters of a mile.

CHAPTER IV.

FROM THE SAULT HOMEWARD.

LAKE SUPERIOR is to be figured to the mind as a vast basin with a high rocky rim, scooped out of the plateau extending from the Alleghanies to the Mississippi valley, a little to the south of the height of land. Its dimensions, according to Capt. Bayfield, are three hundred and sixty miles in length, one hundred and forty in breadth, and fifteen hundred in circumference. The mountainous rim is almost unbroken; its height varies from the average of about three or four hundred feet, to twelve or thirteen hundred; the slopes are gradual towards the north, and abrupt on the opposite side, so that on the north shore the cliffs rise steeply from the water, whilst on the south it is said the ascent is more gentle; the abrupt faces being inland.

This difference of formation, joined to the prevalence of northerly winds, has given very different aspects to the two shores; the southern showing broad sand-beaches and remarkable hills of sand, whereas on the north shore the beaches are of large angular stones, and sand is hardly to be seen except at the mouths of the rivers. The rivers of the southern shore are often silted up, and almost invariably, it is said, barred across by sand-spits, so that they run sometimes for miles parallel to the lake, and separated from it only by narrow strips of sand projecting from the west.

The continuity of this rim occasions a great similarity among the little rivers on the north and east shores, and no doubt elsewhere. They all come in with rapids and little falls near the lake, and more considerable ones farther back. These streams are said often to have in their short course a descent of five or six hundred feet.

This huge basin is filled with clear, icy water, of a greenish cast, the average temperature about 40° Fahrenheit.* Its surface is six hundred and twenty-seven feet above the level of the sea; its depth, so far as actual soundings go, is a hundred and thirty-two fathoms, that is, one hundred and sixty-five feet below the sea level; but Bayfield conjectures it may be over two hundred fathoms in some places.†

In geographical position the lake would naturally seem to lie within the zone of civilization. But on the north shore we find we have already got into the Northern Regions. The trees and shrubs are the same as are found on Hudson's Bay; spruces, birches and poplars; the Vaccinia and Labrador tea. Still more characteristic are the deep beds of moss and lichen, and the alternation of the dense growth along the water, with the dry, barren, lichenous plains of the interior. Here we are already in the Fur Countries; the land of voyageurs and trappers; not from any accident, but from the character of the soil and climate. Unless the mines should attract and support a population, one sees not how this region should ever be inhabited.

This stern and northern character is shown in nothing more clearly than in the scarcity of animals. The woods are silent, and as if deserted; one may walk for hours without hearing an animal sound, and when he does, it is of a wild and lonely character; the cry of a loon, or the Canada jay, the startling rattle of the arctic woodpecker, or the sweet, solemn note of the white-throated sparrow. Occasionally you come upon a silent, solitary pigeon sitting upon a dead bough; or a little troop of gold-crests and chickadees, with their cousins of Hudson's Bay, comes drifting through the tree-tops. It is like being transported to the early ages of the earth, when the mosses and pines had just begun to cover the primeval rock, and the animals as yet ventured timidly forth into the new world.

The lake shows in all its features a continental uniqueness and uniformity, appropriate to the largest body of fresh water on the

* Logan, and Dr. Charles T. Jackson. A recent letter from the lake, dated July 1, 1849, mentions the temperature of the surface, at eight o'clock, P. M., as 37°.

† According to Bayfield's paper in the Transactions of the Literary and Scientific Society of Quebec, (cited in Bouchette's "British Dominions in North America." I., 128, et seq.)

globe. The woods and rocks are everywhere the same, or similar. The rivers and the islands are counterparts of each other. The very fishes, although kept there by no material barrier, are yet different from those of the other lakes. Where differences exist between the various parts, they are broad and gradual.

Aug. 16*th*, 17*th and* 18*th.* — Principally employed in arranging and packing specimens. Prof. Agassiz' collection alone occupied four barrels and twelve boxes, mostly of large size.

In the meantime our party gradually dispersed. Some took the steamer for Mackinaw; others were to remain for a few days at the Sault, whilst another party determined to take the English steamer "Gore," to Sturgeon Bay, and return home through Upper Canada.

Aug. 19*th.*—We started at eight o'clock A.M. in the "Gore," a very well-arranged and comfortable boat. Our first move was to cross the river, where we took in the (English) Bishop of Toronto, with his chaplain and another clergyman. We understood they had been consecrating a church on the English side.

The scenery below the Sault is pleasing, and in many respects like that we had just left, as if the influence of the Great Lake extended beyond its shores. The trees seemed to be of the same species, and there was the same abundance of wooded islands and islets. The Professor observed that the scratches on the rocks were not parallel to the valley, but have a constant north and south direction. The high land forming the sides of the valley retreats gradually on each side, leaving a wide expanse of low shores which would be inundated by a slight elevation of the water. For some distance below the Sault the river is shallow, and the bottom distinctly visible, showing ripple-marks in many places which are constantly covered by several feet of water.

About three o'clock P.M., we reached the Bruce copper-mine, to the northward of St. Joseph's Island. The long wooden pier to which we moored was heaped with the most brilliant ore of the kinds the miners call "horseflesh" and "peacock ore," having every hue of blue, purple and golden. The first question the agent asked us when we landed, was, whether we had a medical man with us, for two of his men had just been injured by a premature explosion. Fortunately, there were two of the profession in our party, not to count the Professor, and the poor fellows were immediately attended to. They were dreadfully burnt and torn about the face, and were moaning with pain, and still more at the thoughts of losing their eyes, and thus their means of support. The doctors shook their heads at first, but afterwards, after proper washing, &c., their case looked better. They were taken on board to be carried to the hospital at Penetanguishene, and we had the satisfaction on landing them there of believing that they would come out with an eye apiece, at the worst.

This mine belongs to the Montreal Company, and the little settlement has a thriving look. The works that we saw were mostly open trenches, displaying a few feet of top-soil, consisting of unstratified drift, clay with scratched pebbles and bowlders. The metalliferous rock, which is sienite and metamorphic talc-schist, with veins of quartz, is also polished and scratched. The ore consists of various sulphurets of copper, particularly the yellow. At St. Joseph's, where we stopped to wood, the Captain, (a very intelligent man, abounding in information concerning the country,) took us to see a rock which he considered a great curiosity. It proved to be a large bowlder of the most beautiful conglomerate, presenting a great variety of brilliant colors; agates, jasper, porphyry, trap, &c., all polished down to an even surface. Other bowlders of the same kind were lying about near the beach. The rock in place is Trenton limestone, and full of the organic remains peculiar to that deposit. We observed great numbers of bowlders on all the islands we passed in Lake Huron.

There is a little settlement on this end of the island, which the captain called his, as the land belongs to him. He bought seven hundred acres, (no doubt of our friend the Major and his co-tenant,) at the rate of twenty cents an acre, for land said to be fertile, and certainly supporting a fine growth of hard-wood trees.

In the evening the Professor made the following remarks on occasion of the bowlder:

"This bowlder may be considered as an epitome of all the rocks we have seen. A complete examination of it would occupy a geologist many months. This conglomerate is associated with the oldest stratified formations, and must have been formed in the same epoch with them. Its component parts give us some insight into its age. It contains no fragment of fossiliferous rock; thus the pebbles of which it is composed must have been broken off, rolled by the waves and thereby rounded and smoothed, and afterwards cemented together, before the appearance of animal life on the earth. On the other hand it contains trap; thus trap-dykes must have been thrown up at that early period. Its other elements are jasper, porphyry, agate, quartz, and even mica; all belonging to the ancient rocks which we have seen on Lake Superior. In one of the bowlders the materials are slightly stratified, so that they had been arranged in layers before they were cemented together. In all of them the cement is more or less vitrified, showing a strong action of heat. This must have been derived from plutonic agencies, so that the plutonic action on the lake commenced before the introduction of animal life. The sandstone formations about Gros-Cap and Batcheewauung Bay indicate in all probability the beaches of the ancient continents from which these fragments were detached, and the outlines of the seas by which they were rolled and worn. Afterwards they were conglomerated, and then removed hither by other agencies. This bowlder does not show the marks of having been transported by the action of water. Its surface is smoothed and grooved in a uniform manner, without the slightest reference to the different hardness of its various materials. Had it been worn into its present shape by the action of water, the harder stones would be left prominent. I have no doubt, from similarity of its appearance in this respect to the rocks of the present glaciers of Switzerland, that it has been firmly fixed in a heavy mass of ice and moved steadily forward in one direction, and thereby ground down."

These remarks being made in the main cabin, in the presence of the Captain and the other passengers, one of the clergymen afterwards took the Professor to task for denying that the world and its inhabitants were all made at once, as if this was a well-understood thing, and got quite indignant, when he would not admit that the Bible had so settled it. His tone on this occasion, (for otherwise he appeared to be a well-bred and educated man,) seemed to indicate a different position of the old theologico-geological question here, a question one would have thought finally disposed of among men of liberal training.

Aug. 20th.—We stopped this morning at a little settlement on the Grand Manitoulin, whither the Indians come yearly to receive their "presents." A few soldiers are stationed here to keep order on these occasions. It is a significant fact that both here and at Mackinaw, the ground-rent paid by the British and United States governments to the original lords of the soil, goes under the name of a *present*, as if dependent on the mere good-will and pleasure of the tenants.

The Indians had been collected here a week or two before, it was said, to the number of three or four thousand; we saw the traces of their encampment on the beach. In general it is only those living in the neighborhood that come, since to journey hither from the more distant villages would cost more than the "present" would come to.

On one occasion, the Captain saw a general collection of the tribe from all quarters, as far as the Red River settlement on the one hand, and Hudson's Bay on the other. There were in all about five thousand six hundred persons, men, women, and children. As usual they carried little or no food with them, and such a multitude soon exhausted the fish and game of the neighborhood. Terrible want ensued, and as the English authorities for some time refused any assistance, many were near starvation. Some families, to his knowledge, went three days without food; others lived on small bits of maple sugar, which were divided with scrupulous accuracy. At last the officer in charge ordered some Indian corn and "grease" to be served out to them. The Captain was standing with the officers when this order was executed, and understood (though *they* did not,) the speech the chief made to his men on the occasion. "When strangers come to visit *us*," said he, "we look round for the best we have, to offer to them. But we must take this, or starve."

If it be said that the strict law of nations is not applicable to dealings with savages, any more than the municipal law to the management of children,—at least they should have the benefit of the principle. If we claim to stand *in loco parentis* with regard to them, we should show some parental solicitude for their welfare. But the poor savages fall between the two stools, and get neither law, equity, nor loving kindness at our hands. It is difficult to see, for instance, why the annual stipend should not be paid to the Indians at places in a measure convenient for them to receive it, say at La Pointe, on the American side, and Fort William, the Red River settlement, and the like on the Canadian, instead of practically cheating them out of it in this way.

The settlement consists of a store-house on the beach, and a few neat whitewashed cottages along the top of the high bank, with their fronts overrun with vines. A little way back from the bluff was a neat Gothic church, of wood, not quite finished; service was held in a small building beyond. The rock, which is Trenton limestone, and full of fossils, crops out everywhere in nearly horizontal strata.

Soon after leaving this place we entered the Georgian Bay, so called, the Captain says, ever since he has known it, though one sees it named Lake Manitoulin, or Manitoulin Bay, on some maps. He commanded the first steamboat that plied between Penetanguishene and the Sault. The trip occupied four or five days; they crept along the northern shore, stopping to cut wood where they wanted it, and lying by at night.

High land was now in sight to the northward; mountains of about twelve hundred feet elevation. The water is very deep, but from the number of islands and rocks, the navigation is dangerous, and it is necessary to anchor in case of fog. Sometimes no bottom can be had close to shore, and then they have to make fast to trees. Northern Lights this evening.

Aug. 21*st.*—We arrived at Penetanguishene early in the forenoon, and remained there a short time to wood, &c. The wounded men were carried on mattresses to the Military Hospital. Near the entrance was a war steamer, moored at one of the wharves. This vessel, in accordance with treaty, carries but one gun. The village is situated at the bottom of a deep narrow bay; the shores on the right going in are low and covered with wood; on the left, the ground rising and cleared for cultivation. The sight of fences and farm-houses here was more home-like than anything we had seen for some time. The place seems to be a thriving one, and it is thought the road from the lake to Toronto will ultimately commence here. The upper part of the bay, however, near the town, seems to be too shallow to favor navigation. Judging from a slip of paper offering a reward for certain Indian curiosities, which was stuck up in one of the shops, there would seem to be some one here who has the good sense to look after the remains of the aboriginal inhabitants.

The distance to Sturgeon Bay, where we were to leave the boat, is not great, but from the stop at Penetanguishene, and the crookedness of the course, it was two o'clock before we got there. On our arrival, we found some confusion. So large a number of passengers had not been expected, for the travel on this route is very inconsiderable; the boats being maintained principally by their contracts with the Post-office.

The place consists of a small gap cut in the forest, large enough for a single rather neat frame-house and out-buildings. From it a dark lane, cut straight into the woods, was the road we were to take, a highway in its most primitive stages, as we found when, after some delay, we got off in three large open wagons, into which we were stowed with our luggage, as close as cattle on the way to market.

We found on our first landing a marked change both in the Fauna and the Flora. The woods are like those of Western New York in the size and species of the trees. We saw again red and sugar maples, red and white oaks, hop-hornbeam, beech, ash, basswood, sumach, &c., and among the birds we recognized the red-headed woodpecker and blue jay.

The road for the first thirteen miles was as bad as could be found, at this season of the year, on the continent, and we had to keep all the way at a walk. In the spring I should think it could be hardly passable by heavy wagons. For this distance, we saw no signs of habitation except a few scattered ruinous log-cabins, built by Indians, who had been encouraged to settle here, but who had long since deserted them.

After that we began to meet clearings, growing more and more numerous as we approached Coldwater. At one of these we succeeded in getting some excellent bread and milk, after convincing the mistress, a canny North-country woman, of our solvency.

Coldwater is a decayed looking village, run to pigs, snake-fences and wide straggling streets. According to the Bishop, who as curator of things spiritual in this district ought to know, the inhabitants have a very general antipathy to the article after which the place is called, whence perhaps their unprosperous condition. Beyond Coldwater we got on to higher land, where the road is better, and we mended our pace, but it was dark before we reached Orilla Landing on Lake Simcoe.

Finding the steamer here, we went on board to engage our passage, and were so much pleased with the appearance of things, that we resolved to pass the night there rather than at the tavern.

Aug. 22*d.*—The Lake Simcoe District as it is called, is, it seems, already noted for its fertility, particularly as a wheat country, although a large part of it is still uncleared. Judging from the growth of timber, the portion on Lake Huron must be at least equal to any of it. Patriotic and enthusiastic Sir Francis Head pronounces it the best land in North America; but without going so far as this, it may probably approach that of the north-western part of New York. The immediate border of the lake is, as I understand, less fertile; for this reason, probably, the forest is but sparingly interrupted by clearings. The lake is too large, and its shores too low and flat, to be beautiful; but it is saved from monotony by numbers of wooded islets. Its height above Lake Huron is 152 feet.

About noon we came to a river-like strait, with wide sedgy shores, which are said to afford capital duck and snipe shooting. Even at this time there were a few ducks. Arriving at Holland Landing, we found the same difficulty about conveyances to St. Albans, and most of us walked thither, three miles, sending our luggage by a wagon.

The name St. Albans has an old-world sound, and the place itself had an old-world look, for, though a raw kind of village enough, yet there were, I think, five very nice saddlers' shops, a tailor "from London," with a very neat establishment, and other signs of a somewhat aristocratic element in the population, probably due to the number of retired British officers who have farms in the neighborhood, and still keep up the equestrian habits, and something of the attention to dress, that distinguish their nation and class. Even the public houses were not "hotels," but "inns."

After dinner we packed into two stages, which, however, would not contain our effects, so they had to follow after, whereby we were much delayed, and I lost my best Mackinaw blanket, faithful companion in the wilderness, purloined from the top of my trunk.

The road beyond St. Albans is everywhere excellently well built, but the first part of it had been but recently macadamized, which reduced our pace to a walk. The country all the way is very pretty, neat villages and farm-houses increasing in number as we approached

Toronto, and all filled with troops of the rosiest children, and surrounded by fine orchards and corn-fields. The hay seemed in many instances at least to be *stacked*, in the English fashion, instead of being stored in barns. Vines and ornamental trees were beginning to be cultivated about the houses, though the prevalence of balsam-firs showed that they had not got far in this direction. The houses are sometimes of a very agreeable cream-colored brick, made in the neighborhood ; most frequently, however, rough-cast, upon lath, with a mixture of plaster, lime and coarse sand, which is said to stand perfectly well. The forest trees are principally white pines, some very fine specimens of which we saw along the road. These afford employment to a number of steam saw-mills, and large quantities of lumber are exported from Toronto.

The government lands here, I was told, are divided off into strips two lots deep, by parallel roads, and these being joined at certain intervals by cross lanes, the division of farms is rendered very symmetrical. Probably, however, this necessitates the buying of an entire lot, or none at all ; at all events, we understood that for some reason or other the transfer of real estate is much hampered by the regulations of the Land Office.

We arrived at Toronto by gas-light, and found nobody awake but a train of geese who were solemnly waddling across the street. We went to the Wellington Hotel, a very dirty and uncomfortable place.

Aug. 23*d.*—Our baggage did not arrive until this morning, fifteen minutes before the boat for Queenston started. My companions contrived to get on board, but I was left to pass the day in Toronto. My first move was to transport my effects to the North American House, somewhat better than the other, but very far from good.

Toronto is very regularly built, of the cream-colored brick above-noticed, in some cases stuccoed. The streets are wide, and both carriageway and sidewalk made of plank, laid transversely. Many of the houses in the suburbs have extensive gardens and ornamental grounds, but in the city itself there are no buildings of much pretension to beauty, and very few attractive shops.

Aug. 24th.—Early this morning I took to the boat for Queenston, and thence by a very wretched railway reached Niagara to dinner.

Aug. 25th.—We went by the railroad to Lockport, to pay a visit to Colonel Jewett, the most warm-hearted of collectors of fossils. He showed us his collection as far as it was accessible, gave the Professor several specimens, and showed us where to pick up more for ourselves. At the quarry of hydraulic limestone we saw an interesting document for the geology of the drift-period. The soft rock was abundantly furrowed, from a direction a little west of north. One of these furrows gradually deepened, until it was interrupted by a succession of horseshoe shaped hollows, sloping from the north, and deep and abrupt towards the south, showing that the furrowing mass was moving from north to south, and from some interruption had chipped out these bits.

From Lockport we drove to the line of the railroad, and returned home by the same way as we came.

END OF THE NARRATIVE.

LAKE SUPERIOR.

PHYSICAL CHARACTER, VEGETATION, AND ANIMALS,

COMPARED WITH THOSE OF OTHER AND SIMILAR REGIONS.

LAKE SUPERIOR.

I.

THE NORTHERN VEGETATION COMPARED WITH THAT OF THE JURA AND THE ALPS.

It is now universally known that living beings, animals and plants, are not scattered at random over the surface of the whole globe. Their distribution, on the contrary, is regulated by particular laws which give each country a peculiar aspect. We call climate the physical conditions which seem to regulate this distribution, however diversified the causes thus acting may be. The distribution of heat all the year round; the mode of succession of temperature, either by sudden or gradual changes; the degree of moisture of the atmosphere; the pressure of the air; the amount of light; the electric condition of the atmosphere; all these and perhaps some other agents continually influence the growth of plants and the development of animals. The nature of the soil is no less powerful in its influence upon organized beings, though here also very different agents are considered under one head; as the chemical properties of the ground are evidently as efficient as the physical.

Let us for a moment examine these circumstances. Temperature seems to be the all-ruling power. With the returning smile of spring, vegetation bursts out with new vigor, and dies again as the cold of winter brings back its annihilating rigors. Under the hot sun of the tropics the beauty and variety of vegetation exceed all that is known in more temperate regions, whilst as we approach the polar plains we see it grow gradually less diversified and more dwarfish, thus exhibiting all over the globe a close connection between the modifications of temperature from the equator to the poles, and the geographical distribution of vegetable and animal life. The more powerful influence of temperature upon vegetation does not, however, preclude the influence of other agents; even the manner in which the same amount of heat is distributed over the earth in a given time, will produce differences. It is well known, that countries in which the summers are short but very warm, and the winters very long and cold, have a vegetation totally different from those where the seasons are more equable and succeed each other by gradual changes, although the mean annual temperature of both be the same. Next in importance we may perhaps consider the degree of moisture of the atmosphere, which differs widely in different regions; the damp valleys of the Mississippi, for instance, present the most striking contrast with the rolling country farther west. Again, the swamps and the sandy plains, the rocky hills and the loamy soils, the snow-clad barrens and the frozen gravel of the North, even under circumstances otherwise most similar, afford the greatest diversity of vegetation. There is still another way in which moisture may act in a particular manner; as vegetation is not influenced simply by the annual amount of moisture, but also by the quantity of water that falls at one time, and the periods at which it falls. A low temperature in a moist climate will indeed produce some remarkable peculiarities; for instance where early winters cause an extensive sheet of snow to be accumulated over the ground, and to protect vegetation from the destroying influence of frost; as is the case in the Alps, where the most delicate flowers prosper admirably under their white blankets, and show themselves in full development as soon as the snow melts away, late in the spring, when the warm season is already fairly set in. Light, again, independently of heat, will also show its influence; shaded places are favorable to plants which would be killed under the more direct influence of the rays of light.

Atmospheric pressure would at first seem to have only a very subordinate influence upon vegetation. But comparing Alpine vegetation with that of higher latitudes, which from their situation must have climates otherwise very similar, we shall be led to the conclusion that atmospheric pressure has its share in bringing about the diversity of plants; for though analogous, the flora of the high North is by no means identical with that of the most elevated Alpine ridges, over which vegetation continues to extend. The influence of atmospheric pressure seems to me particularly evinced in the great, I may say the prevailing number of Alpine species endowed with a volatile fragrance which adds so much to the sweet and soothing influence of mountain rambles; whilst the northern species, however similar to those of the Alps, partake more or less of the dullness of the heavy sky under which they flourish.*

Whatever may be the intensity of other causes, and even when they are most uniform, the chemical nature of the soil acts perhaps as powerfully as the physical conditions under which the plant may grow. To be fully impressed with the important influence of the soil we need only be familiar with the differences noticed in the growth of wheat or other grains in different soils, or with the different aspect of pastures on rich or poor grounds, and to trace the same modifications through any small tract of land with the view to understand similar changes over wider countries.†

* It would be a mistake to ascribe to reduced atmospheric pressure the peculiar aspect of most plants in the higher Alps, as they are undoubtedly more influenced by the temperature, and especially by the pressure of the snow of those high regions. These plants are commonly covered with a thick and close down, which reminds us of the soft fur of the northern animals; they creep for the most part attached to the compact and tenacious soil among the clefts of rocks, where their roots can penetrate and where they find shelter. Several of them have fleshy and succulent leaves, filled with liquid, derived rather from the atmosphere, than from the stony and dried soil upon which we generally find them. These phenomena of Alpine vegetation occur successively at a less considerable elevation the more we advance northwards, and show themselves on the plains towards the polar regions, where the temperature agrees with that of the high Alpine summits. The fact that many plants of the highest summits live very well at the foot of the glaciers which descend into the lower valleys, would seem to show that atmospheric pressure has only a limited influence upon Alpine plants; but the moment we have satisfied ourselves that the most fragrant of these species never prosper below, we must admit that the relation between fragrance and atmospheric pressure to which I have alluded above, is well sustained. The Alpine plants are, it is well known, very difficult to cultivate; Mr. Vaucher, at Fleurier, assisted by Mr. Lesquereux has however succeeded in bringing together a magnificent and numerous collection of species of the high Alps. In order to preserve them, they took care to harden and press the soil, or to introduce small blocks of limestone into it, and to cover them with snow in the spring, but especially to press the roots very often into the ground in the spring, as they are otherwise pushed out after every frost, and perish in a single day if care be not taken to put them again without delay into the ground.

† The chemical elements of the soil seem, however, to have less influence upon the geographical distribution of the large vegetables or phænogames, than upon the cryp-

To satisfy ourselves of the powerful influence of electricity upon vegetation, we need only remember the increased rapidity with which plants come forth, during spring, after thunder storms.

Many other causes still more intimately connected with the aspect of our globe have also a great influence upon the distribution of the animals and plants which live on its surface. The form of continents, the bearing of their shores, the direction and height of mountains, the mean level of great plains, the amount of water circumscribed by land and forming inland lakes or seas, each shows a marked influence upon the general features of vegetation. Small low islands, scattered in clusters, are covered with a vegetation entirely different from that of extensive plains, under the same latitudes. The bearing of the shores again, modifying the currents of the sea, will also react upon vegetation. Mountain chains will be influential not only from the height of their slopes and summits, but also from their action

togames. The attempts made to group the former according to the nature of the soil upon which they grow, have afforded no satisfactory results. It is otherwise when we consider the hydrodynamic capacity of the soil, that is to say, the property which it has to retain the water for a longer or shorter time. Tracing our investigations in this direction we arrive, on the contrary, at very important conclusions. A sandy desert and a peat-bog for instance, as the two extremes, have quite peculiar floræ, which stand completely isolated from the vegetation of soils whose essential component material is humus. This fact is in perfect accordance with recent discoveries in vegetable physiology, which seem to prove that plants extract nothing from the soil except water, or nourishment in a liquid state, and that their other components, the carbon in particular, are furnished them from the atmosphere.

As we descend the scale, and arrive at the cryptogames, the chemical influence of the soil is gradually more and more felt in the distribution of the genera, and even of the species. The mosses even may be readily grouped according to the localities where they live. The Orthotrichœ occur almost exclusively upon the bark of trees, and upon granite and limestone; the Phascaceæ inhabit clayey soils, with the Gymnostomeæ, Pottieæ, Funarieæ and some Weissiæ. The Sphagneæ occur only in peat-bogs, or in waters charged with ulmic acid; the Splachneæ generally upon animal substances in decomposition; the Grimmieæ upon granitic rocks; whilst the greatest number of the Hypnums and Dicranums cover large surfaces of rotten vegetables. And if we take into consideration the modifications which temperature introduces in the habitation of some mosses, we are enabled to account even for the cosmopolitism of some species which, like the Bryums, would seem to be less influenced than others by the nature of the soil upon which they grow.

The examination of the lichens which attach themselves commonly to the surface of woods and rocks leads to conclusions still more striking. Some species live exclusively upon limestone; others upon mica schist; others upon various kinds of granite; and others finally upon certain species of trees or other vegetables. The analysis of the substances upon which lichens live, has, if not completely explained, at least led to the understanding of the causes of the remarkable distribution of these plants.

upon the prevailing winds. It is obvious, for instance, that a mountain chain like the Alps, running from east to west, and thus forming a barrier between the colder region northwards, and the warmer southwards, will have a tendency to lower the temperature of the northern plains, and to increase that of the southern, below or above the mean which such localities would otherwise present; while the influence of a chain running north and south, like the Rocky Mountains and the Andes, will be quite the reverse, and tend to increase the natural differences between the eastern and western shores of the continent, and, laying open the north to southern influences and the south to those of the north, render its climate excessive, i. e., its summer warmer and its winter colder.

Again, the equalizing influence of a large sheet of water, the temperature of which is less liable to sudden changes than the atmospheric air, is very apparent in the uniformity of coast vegetation over extensive tracts, provided the soil be of the same nature, and also in the slower transition from one season into the other along the shores; the coasts having less extreme temperatures than the main land. The absolute degree of temperature of the water acts with equal power; as the aquatic plants of the tropical regions, for instance those of Guyana, differ as widely from those of Lake Superior, as the palms differ from the pine forests. *

* One of the most prominent causes of the dispersion, not to say of the distribution of plants, is certainly the direction and the swiftness of water-courses. On one hand the rivers bring down from the summits or the elevated parts of the country a large number of plants and seeds, which are stopped and take root farther below, on their banks; on the other, they spread in their neighborhood a greater or less amount of moisture. This is, I think, the best cause to assign to the uniformity of vegetation over large plains, traversed by rivers, or to that of the sea-shores, or especially to that of the low islands and peninsulas of little extent. We must also admit, however, that there are along the course of rivers a great variety of stations, which we may find nowhere else, valleys, abrupt rocks, shaded places, constantly or alternately lighted by the sun according to their bearing; and that in this manner secondary agents may have their influence in varying greatly the aspect of vegetation.

It is also a curious but positive fact, that high mountain chains have a direct influence upon the dissemination of the species over the neighboring secondary chains, even at a considerable distance. This fact is plainly shown in the Jura for instance, where from the summits of the Dole to those of the Chasseral we observe a true Alpine vegetation, less and less abundant the more we recede from the Alps in one or another direction. At an equal elevation the summits of the northern Jura lose every trace of Alpine plants which we find so abundantly upon its southern summits, especially upon the ridges near the Alps, as the Dole, the Mount Tendre, for instance. The same takes

But however active these physical agents may be, it would be very unphilosophical to consider them as the source or origin of the beings upon which they show so extensive an influence. Mistaking the circumstantial relation under which they appear, for a causal connection, has done great mischief in natural science, and led many to believe they understood the process of creation, because they could account for some of the phenomena under observation. But however powerful may be the degree of the heat; be the air ever so dry, or ever so moist; the light ever so moderate, or ever so bright; alternating ever so suddenly with darkness, or passing gradually from one condition to the other; these agents have never been observed to produce anything new, or to call into existence anything that did not exist before. Whether acting isolated or jointly, they have never been known even to modify to any great extent the living beings already existing, unless under the guidance and influence of man, as we observe among domesticated animals and cultivated plants. This latter fact shows indeed that the influence of the mind over material phenomena is far greater than that of physical forces, and thus refers our thoughts again and again to a Supreme Intelligence for a cause of all these phenomena, rather than to so-called natural agents.

Coming back from these general views to our special subject, it will be observed that North America must, *a priori*, be expected to have, in some parts, a very diversified vegetation, owing to the peculiarities of its natural geographical districts, and in others, viz., over its extensive tracts of uniform plains, a vegetation as uniform as anywhere in the world.

The physical agents whose influence upon organized beings we have just examined, show a regular progression in their action, which agrees most remarkably with the degrees of latitude on one side, and the elevation above the level of the sea on the other. Hence the difference in the vegetation as we proceed from the tropical regions towards the poles, or as we ascend from the level of the

place westwards. The list of Alpine species found upon the Dole amounts to one hundred, whilst upon the Weissenstein, where even the Anemones have disappeared, we find no other representative of that beautiful flora of the snow regions, than the sole *Erinus Alpinus*.

sea to any height along the slopes of a mountain. In both these directions there is a striking agreement in the order of succession of the phenomena, so much so, that the natural products of any given latitude may be properly compared with those occurring at a given height above the level of the sea; for instance, the vegetation of regions near the polar circles, and that of high mountains near the limits of perpetual snow under any latitude. The height of this limit, however, varies of course with the latitude. In Lapland, at 67° north latitude, it is three thousand five hundred feet above the level of the sea; in Norway at lat. 60° it is five thousand feet; in the Alps at lat. 46° about eight thousand five hundred; in the Himalaya at lat. 30° over twelve thousand; in Mexico at lat. 19° it is fifteen thousand; and at Quito under the equator, not less than sixteen thousand. At these elevations, in their different respective latitudes, without taking the undulations of the isothermal lines into consideration, vegetation shows a most uniform character, so that it may be said that there is a corresponding similarity of climate and vegetation between the successive degrees of latitude and the successive heights above the sea. As a striking example I may mention the fact of the occurrence of identical plants in Lapland in lat. 67° at a height of about three thousand feet and less above the level of the sea, and upon the summit of Mount Washington in latitude 44° at a height of not less than six thousand feet, while below this limit, in the wooded valleys of the White Mountains, there is not one species which occurs also about North Cape.

There is nevertheless one circumstance which shows that climatic influences alone, however extensive, taking for instance into account all the above-mentioned agents together, will not fully account for the geographical distribution of organized beings, as their various limits do not agree precisely with the outlines indicating the intensity of physical agents upon the surface of the earth. A few examples may serve to illustrate this remark. The limit of forest vegetation round the Arctic Circle, does not coincide with the astronomical limits of the Arctic zone; nor does it agree fully with the isothermal line of 32° of Fahrenheit; nor is the limit of vegetation in height always strictly in accordance with the temperature, as the Cerastium latifolium and Ranunculus glacialis, for instance, occur in the

Alps as high as ten, and even eleven thousand feet above the level of the sea. Again, eastern and western countries within the same continent, or compared from one continent to the other, show such differences under similar climatic circumstances, that we at once feel that something is wanting in our illustrations, when we refer the distribution of animals and plants solely to the agency of climate. But the most striking evidence that climate neither accounts for the resemblance nor the difference of animals and plants in different countries, may be derived from the fact that the development of the animal and vegetable kingdoms differs widely under the *same* latitudes in the northern and in the southern hemispheres, and that there are entire families of plants and animals exclusively circumscribed within certain parts of the world; such are, for instance, the magnolia and cactus in America, the kangaroos in New Holland, the elephants and rhinoceros in Asia and Africa, &c. &c.

From these facts we may indeed conclude that there are other influences acting in the distribution of animals and plants besides climate; or perhaps we may better put the proposition in this form: that however intimately connected with climate, however apparently dependent upon it, vegetation is, in truth, independent of those influences, at least so far as the causal connection is concerned, and merely adapted to them. This position would at once imply the existence of a power regulating these general phenomena in such a manner as to make them agree in their mutual connection; that is to say, we are thus led to consider nature as the work of an intelligent Creator, providing for its preservation under the combined influences of various agents equally his work, which contribute to their more diversified combinations.

The geographical distribution of organized beings displays more fully the direct intervention of a Supreme Intelligence in the plan of the Creation, than any other adaptation in the physical world. Generally the evidence of such an intervention is derived from the benefits, material, intellectual, and moral, which man derives from nature around him, and from the mental conviction which consciousness imparts to him, that there could be no such wonderful order in the Creation, without an omnipotent Ordainer of the whole. This evidence, however plain to the Christian, will never be satisfactory to the man of science, in that form. In these studies evidence must rest upon direct observation and induction, just as fully as mathematics claims the right to settle all questions about measurable things. There will be no *scientific* evidence of God's working in nature until naturalists have shown that the whole Creation is the *expression of a thought*, and not the *product of physical agents*. Now what stronger evidence of thoughtful adaptation can there be, than the various combinations of similar, though specifically different assemblages of animals and plants repeated all over the world, under the most uniform and the most diversified circumstances? When we meet with pine trees, so remarkable for their peculiarities, both morphological and anatomical, combined with beeches, birches, oaks, maples, &c., as well in North America as in Europe and Northern Asia, under most similar circumstances; when we find again representatives of the same family with totally different features, mingling so to say under low latitudes with palm trees and all the luxuriant vegetation of the tropics; when we truly behold such scenes and have penetrated their full meaning as naturalists, then we are placed in a position similar to that of the antiquarian who visits ancient monuments. He recognizes at once the workings of intelligence in the remains of an ancient civilization; he may fail to ascertain their age correctly, he may remain doubtful as to the order in which they were successively constructed, but the character of the whole tells him that they are works of art, and that men, like himself, originated these relics of by-gone ages. So shall the intelligent naturalist read at once in the pictures which nature presents to him, the works of a higher Intelligence; he shall recognize in the minute perforated cells of the Coniferæ, which differ so wonderfully from those of other plants, the hieroglyphics of a peculiar age; in their needle-like leaves, the escutcheon of a peculiar dynasty; in their repeated appearance under most diversified circumstances, a thoughtful and thought-eliciting adaptation. He beholds indeed the works of a being *thinking* like himself, but he feels at the same time that he stands as much below the Supreme Intelligence in wisdom, power and goodness, as the works of art are inferior to the wonders of nature. Let naturalists look at the world under such impressions and evidence will pour in upon us that all creatures are expressions of the thoughts of Him whom we know, love and adore unseen.

After these general remarks let us consider more closely the vegetation of the temperate and of the colder parts of North America, and compare it with that of the elevated regions forming in Central Europe the ridge which separates the nations of German tongue from the Roman. In these notes I shall, however, limit myself mostly to trees and forest vegetation, as this is the characteristic vegetation of those tracts of land, and only introduce now and then occasional remarks upon the other plants. It is indeed a peculiarity of the northern temperate regions all over the world, to be wooded, and to afford room for an extensive development of other plants only in those places where permanent accumulations of water exclude forests, where a rocky soil does not afford them a genial ground, or where artificial culture has destroyed them, introducing in their place agricultural products.

A few families, however, constitute the whole arborescent vegetation of temperate regions, and the uniformity of the forests all over that zone in the Old and New World is quite remarkable. In the first rank we find the Amentaceæ and Coniferæ, with their various subfamilies and tribes; next to them maples, walnut, ashes, linden, wild cherries, &c., &c. In the special distribution of each of these families, we observe, however, some peculiarities which will equally claim our attention.

There is, for instance, a striking contrast within these limits, between the vegetation of Coniferæ, which are evergreen, and that of Amentaceæ, Juglandeæ, Fraxineæ, Acerinæ, Tiliaceæ, &c., which lose their foliage in the fall. Again taken as a natural assemblage, the plants which constitute the northernmost forests are farther remarkable for covering extensive tracts of land with one and the same species, to the exclusion of others. Or else a few species are combined together in various ways, the Coniferæ generally excluding the trees with deciduous leaves, or occurring together but rarely, and vice versâ.

In the warmer parts of the temperate regions, the diversity of forest trees with deciduous leaves is greater than farther north, where Coniferæ appear almost exclusively. Another difference is observed in the more continuous distribution of northern forests, while in the warmer climates of the temperate zone they alternate more frequently with shrubs or grazing grounds, with smaller plants growing among them. Whatever may be the peculiarities which we observe in the details of this arrangement, there is, nevertheless, a remarkable coincidence between the vegetation of the plains from the middle latitudes northwards, and the vegetation of mountainous districts, especially in the Alps, as we ascend from the plains towards their snowy summits; the same variety of Amentaceæ, Fraxineæ, Juglandeæ, Acerinæ, Pomaceæ, interspersed with corresponding shrubs, occur in the lower regions, while in the higher the Coniferæ come in more extensively, to the almost entire exclusion of the others.

The correspondence between this ascending forest vegetation, and the distribution of trees over the whole extent of the temperate zone, is so great, that it may be considered as a most positive and universal law. The Juglandeæ and various forms of Amentaceæ, especially those which produce eatable fruit, as the chestnuts, occur in the lower latitudes under the influence of a more genial climate, and disappear entirely below the parallels where agriculture ceases. So also we find them in the lower regions of mountainous countries. Farther north we have a variety of poplars, oaks, willows, maples, ashes, etc., interspread with pines, which begin to form more continuous forests, till they make room northwards for the almost uniform pine and birch forest, which covers in unbroken continuity the northern countries as far as tree vegetation extends; and again in a similar succession we observe Amentaceæ, Acerinæ, &c., &c., in ascending higher and higher on the slopes of mountains, the coniferous trees gaining gradually the ascendency over those with deciduous leaves, until these disappear below the limit of perpetual snow. A more detailed comparison of this resemblance between northern and Alpine vegetation, will show that they agree in almost every respect, and that there are corresponding species under similar circumstances in different parts of the Old and New Worlds, following each other in the same succession from south to north, or from the plains to the mountain summits, modified only by those influences which constitute the contrasting peculiarities of the eastern and western shores of America, Europe and Asia; but in the main agreeing most extensively over the whole range of forest vegetation throughout both continents. The tabular view of these plants which is given below, will at once show the correspondence and divergence.

From these facts it might be inferred that the aspect of wooded lands, whether mountainous or level, would be very similar; that in the northern regions, it compares in every respect with that of high mountain chains. Such an impression is almost universally prevalent among those who are conversant with these laws of the geographical distribution of plants, without having had an opportunity actually to compare such countries. It having been my good fortune, after having been for years familiar with the vegetation of the Alps, to visit the northern regions of this continent within the limits of the temperate zone, I was at once struck with the great difference in the general aspect of their vegetation. Indeed, the picturesque impression is an entirely different one, and nevertheless the above-mentioned laws are correct; but the fact is that the changes of mean annual temperature in this country take place at the rate of about 1° of Fahrenheit for every degree of latitude, or for every sixty miles; or in other words, as we travel north or south, we reach successively every sixty miles, localities the mean annual temperature of which is 1° Fahrenheit lower or higher; while in the Alps we meet, in ascending or descending, the same change of 1° Fahrenheit in mean annual temperature, for every three hundred feet of vertical height; so that we pass within the narrow limits of between six to seven thousand feet, from the vine-clad shores of the lakes of Northern Italy and Switzerland, to the icy fields of snow-mountains, whose summits are never adorned by vegetation; a journey which can easily be performed in a single day. Whilst on the other hand from the 40th degree of northern latitude, where the mean annual temperature is nearly the same as that of the foot of the Alps, we find towards the northern pole a diminution of one degree of temperature for every degree of latitude, or for every sixty odd miles; so that we should travel over twenty degrees of latitude, or more than twelve hundred miles from south to north, for instance, from Boston to Hudson's Bay, before passing over the same range of climatic changes as we do in one day in the Alps; thus causing a narrow vertical stripe of Alpine flora to correspond to a broad zone of northern vegetation stretching over a widely-expanded horizon. So that notwithstanding the correspondence of species, we have in the first case, in the Alps, a rapid succession of highly-diversified vegetation, whilst in the other case, in northern latitudes, we have a monotonous uniformity over extensive tracts of land, although the elements of the picture are the same. But it is a picture seen in a different perspective: in one case we have a simple vertical profile, which in the other case is drawn out into disproportionate horizontal dimensions; like the far-reaching shade of a steeple cast under the light of the setting sun, which may change all proportions, and destroy all resemblance between the shade and the object itself, simply because it is so much elongated. Fantastic images presented at various distances before a light falling at various angles, may prepare us to understand these different aspects of the landscape, be it a wooded plain along a gentle slope, or a forest along a more abrupt mountain chain.

There is another feature in the geographical distribution of organized beings which deserves to be particularly noticed, and which contributes to increase the diversity of aspect of vegetation in any given part of the world. There are in all continents remarkable differences between the vegetation of the shores of a continent, east and west, within the same latitude or the same isothermal line. The forests of the Atlantic and Pacific coasts of temperate America are not altogether composed of the same plants; we remark that in the East there will be a tendency in the different families to develop in different proportions, and perhaps with the addition or disappearance of one or two peculiar types; for instance, the walnut family contains several more representatives on the eastern side of the continent than on the western, and they prosper here in latitudes where in Europe there is only one introduced species of that family growing wild. Again, we find Liquidambar on the American side of the Atlantic, which has no representative either on the Pacific coast, or in Europe. This comparison might be traced farther, and we should see the same correlation even among the shrubs.

But these indications will be sufficient for my object, which is to show that, although there is an intimate correlation between climate and vegetation, the temperature and other influences which constitute climate do not reveal the whole amount of causes which produce these differences, as they are repeated under the same isothermal lines, between the eastern and western shores of the Old World in the same order as along the eastern and western shores of North America; so much so that the northern Chinese and Japanese vegetation coincides very closely with that of the Atlantic States, whilst that of the Pacific coasts of America and that of Europe agree more extensively.

This picture would be incomplete did I not institute a farther comparison between the present vegetation of those regions and the fossil plants of modern geological epochs. If we compare, namely, the tertiary fossil plants of Europe with those living on the spot now, we shall be struck with differences of about the same value as those already mentioned between the eastern and western coasts of the continents under the same latitudes. Compare, for instance, a list of the fossil trees and shrubs from Oeningen, with a catalogue of trees and shrubs of the eastern and western coasts, both of Europe, Asia, and North America, and it will be seen that the differences they exhibit scarcely go beyond those shown by these different floræ under the same latitudes. But what is quite extraordinary and unexpected, is the fact that the European fossil plants of that locality resemble more closely the trees and shrubs which grow at present in the eastern parts of North America, than those of any other part of the world; thus allowing us to express correctly the differences already mentioned between the vegetation of the eastern and western coasts of the continents, by saying that the present eastern American flora, and I may add, the fauna also,* and probably also that of Eastern Asia, have a *more ancient character* than those of Europe and of Western North America. The plants, especially the trees and shrubs growing in our days in this country and in Japan, are, as it were, old-fashioned; they bear the mark of former ages; a peculiarity which agrees with the general aspect of North America, the geological structure of which indicates that this region was a large continent long before extensive tracts of land had been lifted above the level of the sea in any other part of the world.

The extraordinary analogy which exists between the present flora

* The characteristic genera Lagomys, Chelydra and the large Salamanders with permanent gills remind us of the fossils of Oeningen, for the present fauna of Japan, as well as the Liquidambar, Carya, Taxodium, Gleditschia, etc. etc.

and fauna of North America, and the fossils of the miocene period in Europe, would also give a valuable hint with respect to the mean annual temperature of that geological period.

Oeningen, for instance, whose fossils of all classes have perhaps been more fully studied than those of any other locality, could not have enjoyed during that period a tropical or even a sub-tropical climate, such as has often been assigned to it, if we can at all rely upon the indications of its flora, for this is so similar to that of Charleston, South Carolina, that the highest mean annual temperature we can ascribe to the miocene epoch in Central Europe must be reduced to about 60° Fah.; that is to say, we infer from its fossil vegetation that Oeningen had, during the tertiary times, the climate of the warm temperate zone, the climate of Rome, for instance, and not even that of the northern shores of Africa. We are led to this conclusion by the following argument:—The same isothermal line which passes at present through Oeningen at the 47th degree of northern latitude, passes also through Boston, lat. 42°. Supposing now, (as the geological structure of the two continents and the form of their respective outlines at that period seem to indicate,) that the undulations of the isothermal lines which we notice in our days existed already during the tertiary period, or in other words, that the differences of temperature which exist between the western shores of Europe and the eastern shores of North America, were the same at that time as now, we shall obtain the mean annual temperature of that age by adding simply the difference of mean annual temperature which exists between Charleston and Boston, (12° Fah.,) to that of Oeningen, which is 48° Fah., as modern Oeningen agrees almost precisely with Boston, making it 60° Fah.; far from looking to the northern shores of Africa for an analogy, which the different character of the respective vegetations would render still less striking. The mean annual temperature of Oeningen during the tertiary period would not therefore differ more from its present mean, than that of Charleston differs from that of Boston.

This old-fashioned look of the North American forests goes also to show the intimate connection there is all over the globe between the physical condition of any country, and the animals and plants peculiar to it. But far from supporting the views of those who believe that there is a causal connectión between these features of the creation, we must, on the contrary, conclude from the very fact that there are so many special thoughtful adaptations for so long successive periods in their distribution, that those manifold relations could only be introduced, maintained and regulated by the continuous intervention of the Supreme Intelligence, which from the beginning laid out the plan for the whole, and carried it out gradually in successive times.

What is true of plants is also true of animals; we need only remember that it is in North America that Lepidosteus and Percopsis are found ; that species of Limulus occur along the Atlantic shores; and that Trigonia and Cestracion live in New Holland along palæozoic rocks.

II.

OBSERVATIONS ON THE VEGETATION OF THE NORTHERN SHORES OF LAKE SUPERIOR.

THE vegetation of the Northern shores of Lake Superior agrees so closely with that of the higher tracts of the Jura, which encloses the lower and middle zone of the subalpine region, that on glancing at the enumeration below, one is astonished to find so great a number of species entirely identical. Making full allowance for the influence of the lake, and leaving out of consideration a small number of species peculiar to North America, there remains about Lake Superior a subalpine flora which is almost identical with that of Europe, with which it is here compared. Although this fact is very striking, it is nevertheless in accordance with the general laws of botanical geography, and is another proof that the vegetation of the two continents becomes more and more homogeneous the more we advance northwards.

I have divided the catalogue of the phænogamous plants collected about Lake Superior into four lists: The first containing such plants as are really subalpine in their character, or correspond to those of the forests of the lower Alps; * the second containing the plants of the lake proper, or the aquatic plants; † the third comprising the plants purely American,‡ and the fourth the cosmopolitan plants, or those which extend beyond the subalpine region. In the different

* Only such plants are introduced in the first list as have true representatives in Central Europe.

† Lacustrine Floræ and Faunæ present so many peculiarities that it has been thought best to separate the plants of the lake, which are aquatic, from those of the main land enumerated in the first list.

‡ Besides the plants which have true analogues in Europe, there are some about Lake Superior which are truly American types; these constitute the third list.

lists I have indicated as nearly as possible the analogous species whose location is the same in Europe.*

SUBALPINE PLANTS OF LAKE SUPERIOR.	EUROPEAN PLANTS OCCURRING IN THE SUBALPINE REGION.
RANUNCULACEÆ.	
Anemone parviflora *Michx.*	Anemone sylvestris *L.*
" multifida *DC.*	In Europe the Anemones are for the most part alpine plants, but those only whose carpels are plumose, and which ought to be generally considered as a peculiar genus. Anemone sylvestris, the only European species which agrees with the American ones, occurs in the plains.
" pennsylvanica *L.*	
Ranunculus repens *L.*	Ranunculus repens *L.*
" micranthus *Nutt.*	Jura and Alps. In the Alps it rises to the height of 4,000 feet.
Thalictrum Cornuti *L.*†	Thalictrum minus *L.* Creux du Vent.
Actæa rubra *Willd.*	Actæa spicata *L.* Woods of the higher Jura.
" alba *Bigel.*	
CISTACEÆ.	
Helianthemum canadense *M.*	Helianthemum vulgare *J.* Pastures of the lower Alps and Jura.

* All the plants enumerated below were collected by me and some of the gentlemen of our party, who took particular interest in the study of botany, as C. G. Loring, Jr., T. M. Lea, J. E. Cabot and Dr. Keller. They were for the most part determined on the spot with the excellent work of my friend Prof. Asa Gray on the Botany of the Northern United States. Afterwards my collection was revised by Dr. Gray himself, and by Messrs. Leo Lesquereux and Ed. Tuckerman; the latter of whom examined the lichens with particular care, while Mr. Lesquereux revised more particularly the mosses, and furnished me with very minute information about the distribution of plants in Switzerland, to which I had myself paid a good deal of attention in former years. I owe it nevertheless to his contributions upon this particular point, that I have been able to carry my comparisons of the plants of Lake Superior and Central Europe so much into detail as I have done. Prof. Gray has also furnished me with very important documents respecting the distribution of many species, beyond the regions I have examined myself. The general views, however, derived from this study, as I have expressed them in the preceding and following pages, so far as they are new, are my own.

† This and several other plants of this list have a rather extensive range southwards; but this seems to be in accordance with the general direction of the mountain chains and the form of the American continent itself, in w ich both animals and plants peculiar to the arctic and temperate zones extend farther south, than their analogues in the Old World.

Lake Superior.	Europe.
CRUCIFERÆ.	
Arabis petræa *L.*	Arabis petræa *L.* Mts. of Auvergne.
" lyrata *L.*	
Sysimbrium canescens *Nutt.*	Sysimbrium pinnatifidum *DC.* Central Alps.
Draba arabisans *Mx.**	Drabra incana *L.*
Turritis glabra.	Turritis glabra *L.*
DROSERACEÆ.	
Drosera rotundifolia *L.*	Drosera rotundifolia *L.* } Peat bogs of the higher Jura.
" longifolia *L.*	" longifolia *L.* }
OXALIDEÆ.	
Oxalis acetosella *L.*	Oxalis acetosella *L.* Woods of the mountains.
PARNASSIEÆ.	
Parnassia palustris *L.*	Parnassia palustris *L.* Meadows of the mountains.
HYPERICINÆ.	
Hypericum ellipticum *Hook.*	Hypericum Elodes *L.* In peat bogs in Central Europe.
CARYOPHYLLACEÆ.	
Stellaria longipes *Gold.*	Stellaria graminea *L.* Subalpine pastures.
" borealis *Bigel.*	" uliginosa *Murr.* Peat bogs.
Cerastium arvense *L.*	Cerastium arvense *L.*
Sagina nodosa *L.*	Sagina nodosa *L.* } Lower Alps, and the higher Jura.
Alsine Michauxii *Fenzl.*	Alsine stricta *Wahl.* Peat bogs; Jura and Alps.

It is a remarkable fact, that the family of Caryophyllaceæ, so extensive in the alpine regions of Europe, has so few representatives about Lake Superior. The reason is, that the Caryophyllaceæ, like the Cruciferæ, belong for the most part, to the alpine flora properly, and to the flora of the plains, and are missing in the subalpine, or intermediate regions.

* A small species of Draba with yellow flowers, found at Michipicotin, was lost.

LAKE SUPERIOR.	EUROPE.
ANACARDIACEÆ.	
Rhus Toxicodendron, and several other species which were not collected.	Rhus Cotinus *L.* does not correspond to any of the North American species.
ACERINACEÆ.	
Acer saccharinum *Wang.* " spicatum *Lam.*	Acer Pseudoplatanus *L.* Pastures of the higher Jura. This truly subalpine species ascends as high as the Pines (Abies excelsa and pectinata.)
GERANIACEÆ.	
Geranium carolinianum *L.*	Geranium dissectum *L.* Meadows of La Chaux de Fonds.
" robertianum *L.*	" robertianum *L.* Everywhere.
LEGUMINOSÆ.	
Vicia americana *Muhl.*	Vicia sylvatica *L.* Higher Vosges.
Hedysarum boreale *Nutt.*	Hedysarum obscurum *DC.* Alpine pastures.
Lathyrus ochroleucus *Hook.*	Lathyrus pratensis *L.* Common.
ROSACEÆ.	
Cerasus pumila *Mx.* " pennsylvanica *Lois. and var.* borealis *Mx.* " serotina *DC.*	Cerasus avium *L.* Marks in the Jura the limit between the region of the beech, (Fagus sylvatica,) and that of the pines.
Prunus americana *Marsh.*	Prunus insititia *L.* Cultivated.
Spiræa opulifolia *L.*	Spiræa aruncus *L.* Mts. of the Jura.
" salicifolia *L.*	" salicifolia *L.* Mounts of Auvergne.
Agrimonia Eupatoria *L.*	Agrimonia Eupatoria *L.* Mid. Jura.
Geum rivale *L.*	Geum rivale *L.*
" macrophyllum *Willd.* " strictum *Ait.*	" montanum *L.* Alpine.
Potentilla norvegica *L.*	Potentilla aurea *L.* Subalpine.
" tridentata *Atl.* " fruticosa *L.*	" caulescens *L.* } Creux du Vent.
" simplex *Michx.*	" salisburgensis *DC.* "
" arguta *Pursh.*	" rupestris *L.* Jura and Alps.

Lake Superior.	Europe.
	ROSACEÆ.
Comarum palustre *L.* Very abund't.	Comarum palustre *L.* Abounds in the peat bogs of the higher Jura.
Fragaria vesca *L.*	Fragaria vesca *L.* Middle Jura.
Rubus triflorus *Rich.*	Rubus saxatilis *L.* Higher Jura.
" strigosus *Mx.* Everywhere.	" Idæus *L.* Everywhere in the Jura.
" canadensis *L.*	
Rosa stricta *Lindl.*	Rosa alpina *L.* } Pastures of the higher Jura.
" blanda *Ait.*	" rubrifolia *DC.* }
	" tomentosa *L.* }
Sorbus americana *DC.*	Sorbus Aucuparia *L.* The higher limit of the trees in the Jura.
Amelanchier canadensis *Torr. & Gr.*	Amelanchier vulgaris *DC.* Middle Jura.

The Malvaceæ are generally plants of warm countries. This family is not represented about Lake Superior by a single species, nor are the intermediate families between this and the Leguminosæ. The Leguminosæ themselves are very rare, since they are, like the Caryophyllaceæ, plants of the higher Alps, or of the plain. The Rosaceæ, on the contrary, generally extensive in the sub-alpine regions of Europe, are also abundant around Lake Superior.

	ONAGRARIÆ.
Circæa alpina *L.*	Circæa alpina *L.* Woods of the higher Jura.
Epilobium angustifolium *L.*	Epilobium angustifolium *L.* Forest.
" coloratum *Muhl.*	" tetragonum *L.* Moist places.
" palustre *L.*	" palustre *L.* Peat bogs.

	RIBESIEÆ.
Ribes prostratum *L. & Ait.*	Ribes petræum *Jacq.* Higher Jura.
" hirtellum *Mx.*	" alpinum *L.* " "
" lacustre *Pers.*	" Uva-crispa. " "
" oxyacanthoides *H.*	" Grossularia *L.* In rocky places.

	SAXIFRAGEÆ.
Saxifraga Aizoon *Jacq.*	Saxifraga Aizoon *Jacq.* Higher Jura
" tricuspidata *Retz.*	" aizoides *L.* Alps, and lower Alps.
" virginiensis *Mx.*	

LAKE SUPERIOR.	EUROPE.
	SAXIFRAGEÆ.
Mitella nuda *L.* / " diphylla *L.*	These two species have no other analogues in Europe than the Saxifraga rotundifolia, and the species similar to it. In general, the Saxifrageæ, which have few representatives about Lake Superior, belong to the alpine region, so that in order to meet them in the plain, we have to go as far as Greenland, where they are numerous. The species of the plains are represented in America by the genera Sullivantia, Heuchera, Mitella, and Tiarella.
	UMBELLIFERÆ.
Sanicula marilandica *L.*	Sanicula europæa *L.* Creux du Vent.
Archangelica atro-purpurea *Hoff.*	Archangelica officinalis *Hoff.* Jura, also in the Valtellina.
Osmorrhiza brevistylis *DC.*	Chærophyllum hirsutum *L.* Jura.
Sium lineare *Michx.*	Sium latifolium *L.*
	ARALIACEÆ.
Aralia hispida *Michx.*	This family has but one representative in Central Europe, Hedera Helix *L.*
	CORNACEÆ.
Cornus stolonifera *Mx.*	Cornus sanguinea *L.* Middle Jura.
	CAPRIFOLIÆ.
Linnæa borealis *Gron.*	Linnæa borealis *Gron.* Lower Alps: Valais.
Symphoricarpus occidentalis *R. Br.*	
Lonicera parviflora *Lans.* / " hirsuta *Eaton.* Var. Douglasii.	Lonicera Caprifolium *L.* / " Periclimenum *L.* } In the region of the vineyards.
" involucrata *Spr.* Saskatshewan, Oregon, Rocky Mountains, California.	" involucrata, *Spr.* Siberia *L.* alpigena which resembles it somewhat, occurs in the Jura and the Alps.
Sambucus pubens *Mx.*	Sambucus racemosa *L.* Cr. du Vent.
Viburnum Opulus *L.*	Viburnum Opulus *L.* Belongs in Europe to the region of the beech. (Fagus sylvatica.)
" pauciflorum *Pyl.*	

LAKE SUPERIOR.	EUROPE.
RUBIACEÆ.	
Galium trifidum *L.*	Galium rotundifolium *L.* } Characteristic of the subalpine flora.
" triflorum *Mx.*	Asperula odorata and " taurina *L.* } Characteristic of the subalpine flora.
COMPOSITÆ.	
Eupatorium purpureum *L.*	Eupatorium cannabinum *L.* Common in wheat places.
Aster corymbosus L.	
" macrophyllus *L.*	
" puniceus *L.*	
" laxifolius Nees.	
" ptarmicoides *Torr.* et *Gray.*	
" graminifolius *Pursh.*	Aster alpinus *L.* Creux du Vent.

Of these six American species, the last is exclusively northern, and occurs as far as Labrador, to the pine region. It has its analogue in the fine Aster alpinus of the Creux du Vent, and of the lower Alps. The other species, more widely distributed, are represented in Europe by the Aster Amellus and A. salignus, *L.*, which are plants of the plains.

Lake Superior	Europe
Erigeron philadelphicum *L.*	Erigeron alpinum *L.* Creux du Vent.
" strigosum *Mühl.*	
Diplopappus umbellatus *Torr. & Gr.*	
Solidago stricta *At.*	Solidago virgaurea *L.* Var. alpestris, which grows at Chasseron, and in the lower Alps.
" bicolor *L.*	
" thyrsoidea *E. Meyer.*	
" arguta *Ait.* Var. juncea.	
" canadensis *L.*	
" lanceolata *L.*	

The genera Aster and Solidago are exceedingly numerous in America, where, on the contrary, the Inula and the Hieracium, which abound in Europe, are very rare. The same is the case with the Senecionidæ, the Centaureæ, and the Carduaceæ, which are as few in America as they are numerous in Europe.

Lake Superior	Europe
Achillæa Millefolium *L.* Var. setacea.	Achillæa Millefolium *L.* Var. setacea. Declivities of the lower Alps, in the Valais.
Tanacetum huronense *Nutt.*	Tanacetum vulgare *L.* Chaux de Fonds.

Lake Superior.	Europe.
	COMPOSITÆ.
Artemisia canadensis *Mx.*	We might take as analogous of that plant in the subalpine flora of Europe, the Artemisia pontica, which grows in the Valais. But this approaches more the Artemisia maritima *L.*, and belongs thus to the flora of the shores.
Antennaria margaritacea *R. Br.* " plantaginifolia *Hook.*	Antennaria margaritacea *R. Br.* Mt. Cenis.
Senecio aureus *L.* " " var. Balsamitæ	Senecio viscosus *L.* " sylvaticus *L.* " sarracenicus *L.* } Three species of the subalpine flora of the Jura.
Cirsium horridulum *Mx.* " muticum *Mx.*	Cirsium spinosissimum *Scop.* Subalpine Alps. Cirsium rivulare *DC.* " acaule *L.* " eriophorum *L.* } Subalpine Alps, with several other species.
Hieracium canadense *Mx.* " scabrum *Mx.*	Hieracium umbellatum *L.* " amplexicaule " Jaquini *DC.* } Sub Alps and higher Jura, with many other species.
	CAMPANULACEÆ.
Campanula rotundifolia *L.* " " var. linifolia. " aparinoides *Pursh.*	Campanula rotundifolia *L.* " rhomboidalis *L.* This plant is one of the most extensive and the most characteristic of the subalpine region of the whole of Europe, and agrees in its habitat with the Campanula aparinoides, but not in its forms.
	ERICACEÆ, VACCINICEÆ, ERICINEÆ, AND PYROLEÆ.
Vaccinium Oxycoccus *L.* " macrocarpon *At.* " Vitis Idæa *L.* " uliginosum *L.* " pennsylvanicum *Lam.* " cæspitosum *Mx.* " canadense *Kalm.*	Vaccinium Oxycoccus *L.* Subalpine peat bogs. Vaccinium Vitis Idæa *L.* " uliginosum *L.* " Myrtillus *L.* } Forests of the higher Jura.

LAKE SUPERIOR.	EUROPE.
	VACCINIEÆ.
Chiogenes hispidula. *Torr. & Gr.*	
Arctostaphylos Uva-Ursi *Spreng.*	Arctostaphylos Uva-Ursi *Spreng.* La Tourne, higher Jura, and lower Alps.
Loiseleuria procumbens *Des.*	Loiseleuria procumbens *Des.* Pastures of the Alps.
Andromeda polifolia *L.*	Andromeda polifolia *L.* Peat bogs of the higher Jura.
Ledum latifolium *At.*	Ledum palustre *L.* Peat bogs of the North.
Pyrola rotundifolia *L.*	Pyrola rotundifolia *L.* Pastures and forests of the Jura.
" asarifolia *Mx.*	" rosea *L.* Forests.
" chlorantha *Sw.*	" chlorantha *Sw.* Forests.
" secunda *L.*	" secunda. *L.* Woods of the higher Jura.
Monotropa uniflora *L.*	Monotropa hypopythys *L.* In the forests of the Jura.
Moneses uniflora *Salisb.*	Moneses uniflora *Salisb.* Woods of the Vosges.
Chimaphila umbellata *Nutt.*	Chimaphila umbellata *Nutt.* Forests of the Vosges.

No family is more homogeneous in its distribution, or more equally spread in the North of America and Europe, than that of the Ericaceæ, which characterizes rather the region of the pines than the subalpine flora; for these species follow the pine forests in their more or less elevated stations.

LAKE SUPERIOR.	EUROPE.
	PLANTAGINEÆ.
Plantago major* *L.*	Plantago major *L.* Rich, moist soil.
	PRIMULACEÆ.
Primula mistassinica *Michx.*	
" farinosa *L.*	Primula farinosa *L.* Marshes of the North. Higher Jura.
Trientalis americana *Pursh.*	Trientalis europæa *L.* Damp forests.
	OROBANCHEÆ.
Aphyllon uniflorum *Torr. & Gr.*	Orobanche epithymum *L.* And several other species abundant on the declivities of the Jura.

* Can scarcely have been introduced where it was found.

Lake Superior.	Europe.
	UTRICULARIEÆ.
Pinguicula vulgaris *L.*	Pinguicula vulgaris *L.* Sub-Alps and Jura.
	SCROPHULARINEÆ.
Veronica scutellata *L.*	Veronica scutellata *L.* Peat bogs, Jura, and Sub-Alps.
Euphrasia officinalis *L.*	Euphrasia officinalis *L.* Pastures of the Jura.
Rhinanthus Crista-galli. Var. minor. *L.*	Rhinanthus Crista-galli. Var. minor. Pastures of the Sub-Alps and high Jura.
Melampyrum pratense *L.*	Melampyrum pratense *L.* Pine forests.
	LABIATÆ.
Clinopodium vulgare* *L.*	Clinopodium vulgare *L.* Dry declivities of the Jura.
Prunella vulgaris *L.*	Prunella vulgaris *L.* do.
Scutellaria galericulata *L.*	Scutellaria galericulata *L.* Shores of the Lake Etaillères, high Jura.
" lateriflora *L.*	
Stachys aspera *Mx.*	Stachys alpina *L.* Subalpine.
Mentha canadensis *L.*	Mentha arvensis *L.* Moist grounds.
Dracocephalum parviflorum *Nutt.*	Dracocephalum Ruyschiana *L.* In Wallis.
	ASPERIFOLIÆ.
Cynoglossum virginicum *L.*	Cynoglossum montanum *L.* Creux du Vent.
Mertensia pilosa *DC.*	Pulmonaria angustifolia *L.* High Jura.
	GENTIANEÆ.
Gentiana alba *Mühl.*	Gentiana punctata *L.*
	" rubra *L.*
" saponaria *L.* Var Frölichii.	" acaulis *L.*
	" Pneumonanthe *L.* And several other species of Gentiana, which characterize the subalpine declivities.
Menyanthes trifoliata *L.*	Menyanthes trifoliata *L.* Marshes of the mountains.
Halenia deflexa *Griseb.*	Swertia perennis *L.* Peat bogs of the high Jura.

* Probably native where it was found.

LAKE SUPERIOR.	EUROPE.
OLEACEÆ.	
Fraxinus sambucifolia *Lam.*	Fraxinus excelsior *L.*

The Ash (Fraxinus excelsior) and the Sycamore (Acer pseudoplatanus) are, with the Pines, the trees which ascend highest in the mountains of Central Europe.

LAKE SUPERIOR.	EUROPE.
CHENOPODEÆ.	
Corispermum hyssopifolium* *L.*	Corispermum hyssopifolium *L.* Caucasus.
POLYGONEÆ.	
Polygonum viviparum *L.*	Polygonum viviparum *L.*
" cilinode *Mx.*	" Convolvulus *L.*
" sagittatum *L.*	

Polygonum viviparum is the most extensively spread in the subalpine pastures, and the most characteristic of that region. It is also very common about Lake Superior. The same is also true of Empetrum nigrum *L.*, which marks the higher limit of the pine region.

LAKE SUPERIOR.	EUROPE.
EMPETREÆ.	
Empetrum nigrum *L.*	Empetrum nigrum *L.* Region of the pine trees. — Higher Jura and Sub-Alps.
CUPULIFERÆ.	
Quercus rubra *L.* A few dwarfish specimens occur south of Michipicotin.	
Fagus ferruginea *Mx.* Begins to lose its majestic appearance, and forms only meagre forests as far north as Mackinaw.	Fagus sylvatica *L.* Grows dwarfishly and disappears in the subalpine regions of Europe.
Corylus rostrata *Ait.*	Corylus Avellana. *L.* Forests of the Jura. Everywhere.

* I found this plant on the northernmost shore of Lake Superior, near the entrance of Nepigon Bay. Sir W. Hooker mentions it from the Saschatchewan, Athabasca, and Red River.

Lake Superior.	Europe.
	BETULACEÆ.
Betula papyracea *Ait.*	Betula pubescens *Pall.* High Jura.
" excelsa *Ait.*	" nana *L.* Peat bogs of the high Jura.
" pumila *L.*	
Alnus incana *Willd.*	Alnus glutinosa *L.* Valleys of the Jura.
" viridis *DC.*	" viridis *DC.* The Handeck, in the Bernese Alps.
	SALICINEÆ.
Salix pedicellaris *Pursh and others.*	For the willows and poplars, which are rather extensively distributed aquatic plants, see the second list.

About Lake Superior the Amentaceæ are represented only by species of cold countries, or subalpine regions, and are, with a few exceptions, the same as those of Europe. The Quercus rubra is scarcely an exception, since the Quercus pedunculata ascends the valleys of the high Jura; we find very large trunks of it in the marshes of the Verrieres, on the frontier of France and Switzerland.

Lake Superior.	Europe.
	ULMACEÆ.
Ulmus fulva *L.*	Ulmus effusa *Willd.* Banks of the Doubs.
" americana *L.*	
	URTICACEÆ.
Humulus Lupulus *L.*	Humulus Lupulus *L.* Hedges of Val de Travers.
Urtica canadensis *L.*	Urtica dioica *L.* Everywhere. These two species spread diversely in various regions, and have nothing characteristic.
	CONIFERÆ.
Pinus Strobus. *L.*	Pinus sylvestris *L.* Declivities of the Jura.
" resinosa *L.*	" Pumilio *Clus.* Peat bogs of the higher Jura.
" Banksiana *Lamb.*	" Cembra *L.* Declivities of the Alps. Handeck. Glacier of the Aar.

LAKE SUPERIOR.	EUROPE.
CONIFERÆ.	
Abies alba *Mx.*	Abies excelsa *DC.* Forests of the Jura.
" canadensis *Mx.*	
" nigra *Poir.*	
" balsamea *Marsh.*	" pectinata *DC.* Forests of the Jura.
Larix americana *Mx.*	Larix europea *DC.* High Jura.
Thuja occidentalis *L.*	
Juniperus communis *L.*	Juniperus communis *L.* } Forests of the Jura.
" virginiana *L.*	" Sabina *L.* } Forests of the Jura.
Taxus canadensis *Willd.*	Taxus baccata *L.* } Forests of the Jura.

The resemblance of the Coniferæ of Lake Superior to those of the subalpine region is very striking, for though they are not of the same species, the analogy of the forms is so great, that it requires the eye of a botanist to be satisfied positively that these forests are not composed of identical trees in the two hemispheres.

ALISMACEÆ.

Triglochin elatum *Nutt.*	See also the second list.

ORCHIDEÆ.

Microstylis ophioglossoides *Nutt.*	Microstylis monophyllos *Lindl.* In the Sub-Alps.
Corallorhiza multiflora *Nutt.*	Corallorhiza innata *R. Br.* Pine forests in the Sub-Alps. Creux du Vent.
" Macræi* *Gray.*	
Gymnadenia tridentata *Lindl.*	Gymnadenia conopsea *L.*
Platanthera psycodes *Gr.*	Platanthera bifolia *Rich.*
" orbiculata *Lindl.*	
" Hookeri *Lindl.*	
" dilatata *L.*	
" obtusata *Lindl.*	
Goodyera repens *R. Br.*	Goodyera repens *R. Br.*
" pubescens *R. Br.*	
Listera cordata *R. Br.*	Listera cordata *R. Br.* Sub-Alps.
Cypripedium pubescens *Willd.*	Cypripedium Calceolus *L.*
" acaule *Ait.*	

* "CORALLORHIZA MACRÆI (sp. nov.): scapo multifloro; floribus (pro genere maximis) brevissime pedicellatis; petalis ovali-oblongis; labello ovali integerrimo basi utrinque auriculato-inflexo, palato prominulo subbilamellato in plicam antice productam desinente; calcare plane nullo; columna subalato-triquetra; capsula ovoidea. In umbrosis humidis ad 'Caledonia Springs,' Canada Occidentali detexit beatus *W. F. Macrae*, ann. 1843, exemp. fructif. Nuper in insula 'Mackinaw' floriferam legerunt celeb. Agassiz et C. G. Loring, Jr.—Radix ignota. Scapus pedalis. Flores purpurascentes: sepala et petala semiunciam longa!" *A. Gray.*

These Orchideæ, and several more which correspond by their forms to those of Europe, or are even identical with them, characterize all the subalpine regions. The Orchideæ are among the most characteristic plants, in a geographical point of view, for their forms vary in a striking manner, the more we descend towards the warmer latitudes, where they assume more and more brilliant colors, whilst their flowers become larger and more diversified.

Lake Superior.	Europe.
	SMILACINEÆ.
Smilacina racemosa *Desf.*	Convallaria multiflora *L.* } Middle Jura.
" stellata *Desf.*	" Polygonatum *L.* } Middle Jura.
" bifolia *Ker.*	Smilacina bifolia *Ker.* } Middle Jura.
	LILIACEÆ.
Allium schoenoprasum *L*	Allium schoenoprasum *L.* Common in the Alps to the height of 7000 feet.
Lilium philadelphicum *L.*	Lilium Martagon *L.* Pastures of the Sub-Alps.
Streptopus amplexifolius *DC.*	Streptopus amplexifolius *DC.* High Jura.
Tofieldia glutinosa *Willd.*	
" calyculata *Wahl.*	Tofieldia calyculata *Wahl.* Pastures of the Sub-Alps and high Jura, Creux du Vent, &c.
	CYPERACEÆ.
Scirpus cæspitosus *L.*	Scirpus cæspitosus *L.* Peat bogs of the higher Jura.
" Eriophorum *Mx.*	
Eriophorum alpinum *L.*	Eriophorum alpinum *L.* This plant and the preceding are very characteristic of the peat bogs of the high Jura.
" virginicum *L.*	
Carex trisperma *Dew.*	
" canescens *L.*	
" straminea *Schk.*	
" oligocarpa *Schk.*	
" aurea *Nutt.*, var.	
" bicolor *All.*	Carex bicolor *All.* In the highest Alps, in grazing places, occurs also in Labrador.
" Vahlii *L.* Var. elata.	" Vahlii *L.* Found in Lapland Occurs also in Greenland.

LAKE SUPERIOR.	EUROPE.
GRAMINEÆ.	
Alopecurus aristulatus *Mx.*	Alopecurus pratensis *L.* Meadows of the Jura.
Phleum alpinum *L.*	Phleum alpinum *L.* Pastures of the Sub-Alps.
	" Micheli. *L.* Summit of the Chasseron. Highest ridge of the Jura.
Cinna pendula *Trin.*	
Agrostis scabra *Willd.*	Agrostis vulgaris *Willd.* " alba, et } High Jura.
Muhlenbergia sylvatica *T. et Gr.*	
Calamagrostris arenaria *Trin.*	Calamagrostis arenaria *Trin.* Northern shores.
" canadensis *P. de Beaur.*	" baltica. *Skr.* Baltic.
Oryzopsis canadensis *Torr.*	
Rebóulea pennsylvanica *Gr.*	
Spartina cynosuroides *Willd.*	
Glyceria fluitans *R. Br.*	Glyceria fluitans *R. Br.* Brooks of the Jura.
" aquatica *Smith.*	" aquatica *Smith.* Brooks of Jura.
" nervata *Tr.*	
Poa alpina *L.*	Poa alpina *L.* One of the most characteristic plants of the subalpine regions.
" serotina *Erh.*	
Festuca ovina *L.*	Festuca ovina *L.* Peat bogs.
Bromus secalinus* *L.* (Introduced?)	Bromus secalinus *L.* Fields of the Jura.
Triticum repens *L.*	Triticum repens *L.* In sandy places.
" dasystachyum *Gray.*	
Elymus canadensis *L.* Var. glaucifolius.	Elymus europæus *L.* Forests of the high Jura.
" mollis *R. Br.*	Judging from its form, this species is rather a plant of the shores.
Hordeum jubatum *L.*	Hordeum murinum *L.*

* I could not discover indications of this plant having been introduced where it was found. However, even an accidental landing might account for the presence of a plant which can scarcely be a native of the northern shores of Lake Superior.

Lake Superior.	Europe.
GRAMINEÆ.	
Aira flexuosa *L.*	Aira flexuosa *L.* Sub-Alps.
Trisetum molle *Kunth.*	Avena flavescens *L.* Subalpine meadows.
Phalaris arundinacea *L.*	Phalaris arundinacea *L.* Banks of the brooks of the Jura.
Hierochloa borealis *Röm. & Sch.*	Hierochloa borealis *Röm. & Sch.* Northern Europe.
Milium effusum *L.*	Milium effusum *L.* Characterizes the subalpine forests.
EQUISETACEÆ.	
Equisetum sylvaticum *L.*	Equisetum sylvaticum *L.* Woods of the high Jura.
" arvense *L.*	" arvense *L.*
" limosum *L.*	" limosum *L.* Brooks of the Jura.
FILICES.	
Struthiopteris germanica *Willd.*	Struthiopteris germanica *Willd.* Mountains of the Vosges.
Polypodium Dryopteris *L.*	Polypodium Dryopteris *L.* Creux du Vent.
Pteris aquilina *L.*	Pteris aquilina *L.* Woods of the Jura.
Allosorus gracilis *Presl.*	Allosorus crispus *P.*
Cystopteris bulbifera *Bernh.*	Cystopteris fragilis *B.*
Woodsia ilvensis *R. Br.*	Woodsia ilvensis *R. Br.*
Dryopteris dilatata *Gray*	Dryopteris dilatata *Gray.* Higher Jura.
" intermedia *Gray.*	
Botrychium virginicum *Swartz.*	
" Lunaria *L.*	Botrychium Lunaria *L.* Summit of the Jura.
LYCOPODIACEÆ.	
Lycopodium lucidulum *Mx.*	Lycopodium Selago *L.* Higher Jura.
" inundatum *L.*	" inundatum *L.* Marshes of the higher Jura.
" annotinum *L.*	" annotinum *L.* Summit of the Jura, Creux du Vent, etc.
" dendroideum *Mx.*	
" clavatum *L.*	" clavatum *L.* Higher Jura.
" complanatum *L.*	" complanatum *L.* Higher Vosges.

Lake Superior.	Europe.
Selaginella selaginoides *Spring.*	Selaginella selaginoides *Spr.* Pastures of the lower Alps and the higher Jura.
" rupestris *Spring.*	

The Equisetaceæ, the Ferns, and the Lycopodiaceæ of Lake Superior are almost absolutely the same species as those of the subalpine region of Europe. As we descend the scale of the vegetable kingdom under higher latitudes, vegetation seems to follow the sides of an angle, as it were, which become convergent about the zone of pine forests. Thus the Lichens and the Mosses are already entirely the same species here as in Europe, and it will be sufficient to make a single list of them, without indicating the corresponding European species, since all are identical. Few *Hepaticæ* are also enumerated.

Mosses of Lake Superior.	Localities in the Jura.
Sphagnum capillifolium *Brid.*	Peat-bogs of the high Jura.
" cuspidatum *Brid.*	" " " " " "
" squarrosum *Hedw.*	Peat-bogs of the Vosges and Hartz. This species belongs to the granitic peat-bogs.
Funaria hygrometrica *L.*	Grows everywhere.
Grimmia apocarpa Var. rivularis *B. et S.*	Dripping rocks in the Alps and Jura.
Hedwigia ciliata *Hedw.*	Everywhere on granite.
Orthotrichum Hutchinsiæ *H. et T.*	" " "
" strangulatum *Beauv.*	Is missing in Europe, but replaced in the forests by a great number of analogous species.
" leiocarpum *B. et S.*	Forests.
" anomalum *Hedw.*	Stones.
Ceratodon purpureus *Brid.*	Everywhere.
Dicranum scoparium *Hedw.*	Forests.
" undulatum *Ehrh.*	Moist forests.
" congestum *Brid.*	Forests of the higher Jura; descends never in the middle region of the pine forests.
" Schraderi *W. et M.*	Peat-bogs of the higher Jura.
" fulvum *Hook.*	Forests of the Alps.
" longifolium *Ehrh.*	Granitic blocks.
" virens *Hedw.*	Forests of the Alps and higher Jura.
" polycarpum *Brid.*	Fissures of rocks, and the forests in the Alps.
" majus *Turn.*	Higher Jura; descends never in the middle region.
" glaucum *L.*	Peat-bogs of the Jura.

Mosses of Lake Superior.	Localities of the Jura.
Distichum inclinatum *B. et S.*	Summits of the Jura. Declivities of the Alps.
" capillaceum *B. et S.*	Fissures of the rocks. Subalpine regions.
Encalypta ciliata *Hedw.*	On the ground in the higher Jura.
Pogonatum alpinum *Brid.*	Sub-Alps.
Polytrichum formosum *Hedw.*	Woods of the mountains. Everywhere.
" piliferum *Hedw.*	
" juniperinum *Hedw.*	
Bartramia pomiformis *Hedw.*	Granite in the Vosges and Alps.
" Oederi *Brid.*	Rocks of the Jura.
" fontana *L.*	Everywhere near springs.
Aulacomnium palustre *Br.*	Peat-bogs of the higher Jura.
Bryum pseudo-triquetrum *L.*	Moist places in the forests. Everywhere.
" nutans *L.*	
Var. elongatum *B. et S.*	Elevated peat-bogs.
Mnium cuspidatum *Hedw.*	Skirts of the forests.
Hypnum Schreberi *Willd.*	Pine forests.
" tamariscinum *Hedw.*	" "
" splendens *Hedw.*	" "
" aduncum *L.*	" "
" uncinatum *Hedw.*	" "
" cupressiforme *L.*	" "
" Crista-castrensis *L.*	" "
" abietinum *L.*	" "
" nitidulum *L.*	" "
Neckera intermedia *Hedw.*	" "
Marchantia polymorpha *L.*	Moist places.
Jungermannia barbata *Hook.*	Pine forests.
Ptilidium ciliare. *Nees.*	" "

Enumeratio Lichenum a D. Prof. Agassiz ad Lacum Superiorem, anno 1848, lectorum, ab Edvo. Tuckerman, Cantabr.

Vidi olim in Museo Parisiensi aliquot plantas a D. Comite de Castelnau in itinere suo ad Lacum Superiorem decerptas, inter quas Lichenes decem insequentes reperi: —

Usneam barbatam, Var. pendulam.
Everniam jubatam *Fr.*
Ramalinam calicarem *β. Fr.*
Cetrariam islandicam *Ach.*
C. glaucam *Ach.*
C. lacunosam *β* Atlanticam *Tuck.*

Stictam pulmonariam *Ach.*
Parmeliam saxatilem *Ach.*
P. caperatam *Ach.*
Cladoniam rangiferinam *Hoff.*

Hisce primitiis incrementum attulit, quantum scio, nemo usque donec oras insulasque Lacus perlustrans Professor noster illustriss. Agassiz, dum plantarum nobiliorum distributionem geographicam persequitur, Lichenum etiam, hac in re multum adjuvantibus discipulis ejus commilitonibusque, viris amicissimis J. E. Cabot, J. M. Lea, C. G. Loring, and Dr. Keller,—messem satis largam fecit.

Has igitur opes Lichenosos, mihi benevolentia V. ill. mandatos, pro viribus explicare pergam.

LICHENES.

USNEA.

1. *barbata* Fr. var. *dasypoga*, Fr., infert.
2. *longissima* Ach., cum cephalodiis.
3. *cavernosa* Tuckerm. mss. Thallo pendulo laxo molli glaberrimo tereti-compresso plus minus cavernoso ochroleuco, ramis primoribus simpliciusculis subventricosis attenuatis ad apices dichotome ramosis, ramulis ultimis tenuissime capillaceis; apotheciis sessilibus radiatis disco albido-pruinoso demun subcarneo margine obscuriori evanescente.——— HAB. ad arbores in oris Lacus Superioris; *C. T. Jackson*, 1845; *Agassiz*, 1848. Ipse legi sterilem in Montibus Albis, anno 1843. Specimen habeo omnibus conveniens e Madras, Ind. Orient., ex *Hb. Hook.*

   Thalli rami majores e subtereti demum compressi, angulati annulatim rupti, lacunis regularibus subellipticis plus minus insignes, apicibus dichotomis elongatis teretiusculis tenuissime demum capillaceis. Apothecia omnino Usneæ, at discus strato gonimo viridi impositus! albido-pruinosusque! Hos characteres Usneis a Friesio plane denegatos, iis primum tribuit Montagne (Annales 1834, t. 2, p. 2, p. 368, and Cryptog. Canar. in Webb & Berth. Hist. Nat. d. Iles Canar., p. 93). Ex observationibus Montagnei U. ceratina discum habet pruinosum, et U U. Jamaicensis Ach., et Ceruchis Montag., discum pruinosum strato gonimo impositum. Species nunc descriptá pluribus notis cum U. Ceruchi (Americæ tropicæ adhuc privæ, a Montagneo (Ann. l. c.) luculentissime llustratæ) convenit; distat facie, statuque (normali ut videtur) pendulo. Disci characteribus jam laudatis facillime distinguenda est U. cavernosa ab omni (ni fallor) Usnea borealiamericana.

EVERNIA.

1. *jubata* Fr. *β. chalybeiformis* Ach., inf.
   γ. implexa Fr., infert.
2. *Prunastri* Ach., infert.

RAMALINA.

1. *calicaris* Fr. β. *fastigiata* Fr., fert.
δ. *farinacea* Sch., fert.

CETRARIA.

1. *islandica* Ach. γ. *crispa* Ach., fert.
2. *nivalis* Ach., infert.
3. *glauca* Ach. β. *sterilis* Fr., infert.
4. *ciliaris* Ach., fert.
5. *lacunosa*, β. *atlantica* Tuck., fert.
6. *Oakesiana* Tuckerm., infert.
7. *Pinastri* Sommerf., infert.

PELTIGERA.

1. *aphthosa* Hoffm., fert.
2. *canina*, Hoffm. fert.
3. *rufescens* Hoffm., fert.
4. *polydactyla* Hoffm., infert.
5. *horizontalis* Hoffm., infert.

SOLORINA.

saccata Ach., fert.

STICTA.

1. *pulmonaria* Ach,. infert.
2. *linita* Ach., infert.
4. *glomerulifera* Delis., fert.

PARMELIA; subsect. *Imbricaria*.

1. *perlata* Ach., infert.
2. *tiliacea* Ach., fert.
3. *Borreri* Turn. β. *rudecta* Tuckerm., infert.
4. *saxatilis* Ach., fert.
5. *aleurites* Ach., infert.
6. *physodes* Ach, infert.
7. *olivacea* Ach., fert.
8. *caperata* Ach., fert.
9 *conspersa* Ach., fert.
10. *centrifuga* Ach., fert.
11. *parietina*, γ. *rutilans* Ach., fert.

Subsect. *Physcia*.

12. *speciosa* Ach., fert.
13. *stellaris* Ach. *a*. fert.

Subsect. *Placodium*.

14. *saxicola* Ach., fert.
15. *chrysoleuca* Ach., fert.
16. *elegans* Ach., fert.

Subsect. *Patellaria.*

17. *subfusca* Fr. β. *distans* Fr.
18. *albella* Ach.
19. *ocrina* Ach.

Subsect. *Urceolaria.*

20. *oncodes*, Tuckerm. mss. Thallo crustaceo tartareo (farinoso-pulverulento) contiguo rimoso-areolato ambitu verrucoso-subplicato glauco-albicante; apotheciis innatis mox protrusis sessilibus disco pruinoso demum protuberante nigro margine proprio tenui erecto thallodem tumidum demum obtegente.——*Turner Island*, in rupe porphyritico; *Agassiz.* P. Glaucomæ, Ach. Fr. et P. repandæ, Fr. affinis. Distincta videtur crusta tenui, apotheciisque nigris infantia solum conspicue pruinosis, margine proprio erecto persistente.

STEREOCAULON.

1. *tomentosum* Fr., fert.
2. *paschale* Laur., fert. Adsunt quoque specimina *S. coralloidi* forsan referenda.

CLADONIA. Ser. *Glaucescentes.*

1. *turgida* Hoffm. *a.* fert.

   β. *grypea*, Tuckerm. mss. Podetiis majoribus fastigiato-ramosis glauco-viridibus, scyphis obscuris in ramos fastigiatos radiato-dentatos, v. subulatos abeuntibus.——Major, pulchre glauco-viridis. Formis majoribus americanis C. uncialis β. similis et analoga, reipsa vero C. turgidæ omnino referenda. Thallo foliaceo destituta sunt specimina; squamulæ tamen (iis C. turgidæ similes) hic illic apparent.

Ser. *Fuscescentes.*

2. *pyxidata* Fr. *a.* fert.
3. *gracilis* Fr. γ. *hybrida*, Fr., fert.
4. *degenerans* Fl. *a.* fert.
5. *cornuta* Fr. *a.* fert.
6. *squamosa* Hoffm. *a.* fert.
7. *furcata*, Fl. δ. *subulata* Fl. infert.
8. *rangiferina* Hoffm. *a.* fert.
   β. *sylvatica* Fl., fert.
   γ. *alpestris* Fl., infert.

Ser. *Ochroleucæ.*

9. *amaurocræa* Fl., fert.
10. *uncialis* Fr. β. *adunca* Ach., fert.
    γ. *turgescens* Sch., fert.

Ser. *Cocciferæ.*

11. *cornucopioides* Fr., fert.
12. *Floerkeana* Fr., fert.
13. *deformis* Hoffm., fert.

BIATORA.

1. *rufonigra* Tuckerm., fert.
2. *icmadophila* Fr.
3. *vernalis* Fr.

LECIDEA.

1. *parasema* Fr. Specimina in Betula aliquantum differe videntur.
2. *geographica*, *a*. Schær.

UMBILICARIA.

1. *pustulata*, *β*. *papulosa* Tuck., fert.
2. *hirsuta* Ach., fert.
3. *Dillenii* Tuckerm., infert.
4. *Muhlenbergii* Ach., fert.

OPEGRAPHA.

scripta Ach. Schær. *a*.

ENDOCARPON.

1. *miniatum*, *β*. *complicatvm*, Sch. Status pusillus, teneritate etiam a Lichene Novæ Anglicæ distans.
2. *Manitense* Tuckerm. mss. Thallo cartilagineo-membranaceo tenui fragili lævi lobato ex olivaceo-nigricante, lobis ambitus rotundatis incisis planis margine subplicatis crenatis, cæteris flexuosis irregularibus, subtus e fusco-nigrescentibus; ostiolis prominulis nigris pertusis.——Proxima E. fluviatili, at colore, superficie nitidiuscula, lobatione fere Imbricariæ, apotheciisque diversa.

PERTUSARIA.

pertusa Ach. *a*.

COLLEMACEÆ.

COLLEMA *saturninum*, Ach., infert.

Fungi were not collected, except a few of the more solid ones, which have not yet been determined. The softer species are very difficult to preserve during such a journey, when travelling constantly upon water in birch-bark canoes.

To this first enumeration of the species of plants occurring about Lake Superior, and which belong to the subalpine region as such, we subjoin a list of species, which cannot strictly be referred to this one, though they occur in it. They are few in number and still fewer of them belong to the Cryptogamous plants.

II. *Plants of the lake and shores, which have or have not their analogous representatives in Europe.**

LAKE SUPERIOR.	EUROPE.
Ranunculus aquatilis *L.*	Ranunculus aquatilis *L.* Everywhere.
" reptans *L.*	" reptans *L.* Sand of the lake shores.
Cardamine hirsuta *L.*	Cardamina hirsuta *L.* Moist places.
Barbarea vulgaris *R. Br.*	Barbarea vulgaris *R. Br.* Along ditches.
Nuphar lutea *Smith.* Var. Kalmiana.	Nuphar pumila *Sp.* Black forest. Meadows and margin of lakes.
Cakile americana *Nutt.*	Cakile maritima *L.* Baltic Sea.
Callitriche linearis *Pursh.*	Callitriche autumnalis *L.*
" verna *L.*	" verna *L.* In brooks.
Lathyrus maritimus *Bigel.*	Lathyrus maritimus *B.* Marine plant.
" palustris *L.*	" palustris *L.* Marshes of the lakes.
Oenothera biennis *L.*	Oenothera biennis *L.* Lake of Neuchatel. Introduced into Europe.
Myriophyllum spicatum *L.*	Myrioph. spicatum *L.* Quiet waters.
Sium lineare *Mx.*	Sium angustifolium *L.* In brooks.
Bidens cernua *L.*	Bidens cernua *L.* Ditches.
Lysimachia stricta *Ait.*	Lysimachia vulgaris *L.* Marshes.
" ciliata *L.*	" ciliata *L.* Marshes.
Naumburgia thyrsiflora *L.*	Naumburgia thyrsiflora *L.* Near St. Blaise, Lake of Neuchatel.
Veronica americana *Mx.*	Veronica Beccabunga *L.* Brooks and lakes.
Lycopus virginicus *L.*	Lycopus europæus *L.* Margins of waters.
" sinuatus *Ell.*	
Polygonum amphibium *L.*	Polygonum amphibium *L.* Margins of quiet waters in diverse regions.
Myrica Gale *L.*	Myrica Gale *L.* Shores of the Baltic.
Salix candida *Willd.*	In Europe the same species of willows are found at the margin of waters in diverse latitudes, but most of them differ from the American species. The extensive distribution of these trees along the shores of lakes and rivers at various
" lucida *Mühl.*	
" discolor *Mühl.*	
" angustata *Pursh.*	
" pedicillaris *Pursh.*	
" pumilis *Marsh.*	

* The number of aquatic plants found along the shores of Lake Superior, is so small, that I have put them all together in this list, whether they have, or not, their analogies in Europe.

LAKE SUPERIOR.	EUROPE.	
	latitudes, shows their closer connection with the nature of the ground than with the temperature of the country where they grow.	
Populus balsamifera *Mx.*	Populus nigra *L.*	Jura.
" tremuloides *Mx.*	" tremula *L.*	
Sparganium natans *L.*	Sparganium natans *L.*	Quiet waters, lakes and rivers of Europe.
Potamogeton natans *L.*	Potamogeton natans *L.*	
" lucens *L.*	" lucens. *L*	
" prælongus *Wulf.*	" perfoliatus *L.*	
" heterophyllus *Schreb.*		
" pectinatus *L.*		
" pauciflorus *Pursh.*		
Triglochin elatum *Nutt.*	Triglochin palustre *L.* This species occurs also in N. America.	
Alisma Plantago *L.*	Alisma Plantago *L.*	"
Sagittaria variabilis *Engl.*	Sagittaria sagittifolia *L.*	
Echinodorus subulatus *Engl.*	Echinodorus is an aquatic type peculiar to the American flora.	
Udora Canadensis *Nutt.*	Udora occidentallis *Pursh.* Northern Germany.	
Vallisneria spiralzis *L.*	Vallisneria spiralis *L.* Lombardy and Tessino.	
Iris versicolor *L.*	Iris pseudc-acorus *L.* Margins of waters. Everywhere.	
Juncus effusus *L.*	Juncus effusus *L.*	
" acuminatus *Mx.*	" acutiflorus *Ehrh.*	
" paradoxus *E. Meyer.*		
" nodosus *L.*		
" balticus *Willd.*	" balticus *Willd.* Northern Sea and Baltic.	
Eleocharis obtnsa *Schultz.*		
" palustris *R. Br.*	Eleocharis palustris *R. Br.* Marshes.	
" tenuis *Schult.*		
" acicularis *R. Br.*	" acicularis *R. Br.* Margin of lakes and marshes.	
Scirpus lacustris *L.*	Scirpus lacustris *L.* Common in all lakes of Switzerland.	
Carex stipata *Mühl.*	Many of these species are the same in the two continents; but there are at the margin of waters of the whole middle and northern Europe, many more Carices re-	
" scoparia *Schk.*		
" festucacea *Schk.*		
" vulgaris *Fries.*		
" stellulata *Good.*		

LAKE SUPERIOR.	EUROPE.
Carex crinita *Lam.*	sembling those of North America, which are however not identical.
" tentaculata *Muhl.*	
" hystricina *Willd.*	
" Ocderi *Ehrh.*	
" intumescens *Rudge.*	
" retorsa *Schwr.*	
Nitella flexilis *Agardh.*	Nitella flexilis *Agardh.* Lake of Geneva.
Fontinalis antipyretica.	Fontinalis antipyretica. In the brooks of the Jura.

It seems at the first glance to be a contradiction to unite in a separate table the aquatic plants of the lakes, leaving as characteristic of the subalpine region the aquatic plants of the peat-bogs. That is, however, not the case, for the peat-bogs and the plants which form them, (the peat-bogs with Sphagna at least,) never descend below the Pine region, which they follow in its whole extent, whilst lake and marine plants follow the shores in various latitudes. The former being of course under the direct influence of the temperature, the latter, on the contrary, being more dependent upon the moisture of the soil.

III. *American plants of Lake Superior, which have no analogous representatives in Central Europe.**

Sarracenia purpurea *L.* Hudsonia tomentosa *Nutt.*	} Truly American types.
Rubus Nutkanus *Moç.*	There are no Rubus of the type of odoratus and nutkanus in Europe.
Potentilla fruticosa *L.*	Cultivated in the gardens of Europe, where it succeeds very well in temperate plains and in the mountains.
Cornus canadensis *L.*	A charming little plant of which we find no other analogue in Central Europe than a few Umbelliferæ, for their general form, the Buplevrums for instance, which grow in the Sub Alps. But Cornus suecica *L.* is its strict analogue in Northern Europe.

* Besides the genera which have no representatives at all in Central Europe, there are several introduced in this list which have only remote analogues, or indeed, real representatives; but in such countries of the Old World which are far distant from the mountain chains, the vegetation of which has been compared here with that of Lake Superior.

Diervilla trifida *Moench.* Mitchella repens *L.*	Truly American types.
Coreopsis lanceolata *L.*	This genus, one of the finest of the Compositæ, is wanting in Europe.
Mulgedium leucophæum *D. C.*	Comes near the Mulgedium alpinum of Lapland.
Nabalus racemosus *Hook.*	Entirely wanting in Europe.
Lobelia Kalmii *L.*	The Lobeliœ are not numerous in Europe, being replaced there by the Campanulæ and Phyteumata, of which genera the first is scantily represented in America, and the second not all.
Dianthera americana *L.* Mimulus ringens *L.*	Truly American types.
Castilleja coccinea *Spr.*	Bartsia alpina *L.* Found upon the highest peaks of the Jura, is the nearest relative to Castilleja coccinea in Central Europe.
" septentrionalis *Lindl.*	Castilleja pallida *L.*, closely allied to C. septentrionalis, occurs on the N. E. confines of Russia.
Monarda fistulosa *L.*	
Calystegia spithamæa *Pursh.*	
Apocynum androsæmifolium *L.*	We cannot consider this plant as corresponding to the Apocynum Venetum, which belongs to the seashores of the Adriatic. These two species differ in form and habitat.
Polygonum articulatum *L.*	Of this type of Polygonum there is no analogous form in Europe.
Shephardia canadensis *Nutt.* Comandra livida *L.* " umbellata *Nutt.* Clintonia borealis *Raf.* Sisyrinchium bermudianum *L.*	Truly American types.

IV. *The few plants of Lake Superior, indicated in the following list, have either a very wide range or are perhaps introduced.*

Corydalis aurea *Willd.* " glauca *Pursh.*	Corresponds to Corydalis lutea *L.* Vauxmarcus. The Corydalis are cosmopolites of the middle region.
Capsella Bursa — Pastoris *D. C.* (Introduced ?)	Everywhere in Europe.

Astragalus canadensis *L.* (*Cosmopolite.*)	Corresponds to Astragalus glyciphyllos *L.* Equally cosmopolite.
Trifolium repens *L.* (Introduced?)	Everywhere in Europe.
Potentilla anserina *L.*	"
Mentha piperita *L.* (Introduced.)	Mentha piperita *L.* Everywhere in Europe, especially in the plains.
Galeopsis Tetrahit *L.* (Introduced.)	Everywhere in Europe.
Physalis viscosa *L.*	Corresponds to Physalis Alkekengi *L.*, cosmopolite like the Solaneæ in general, and all plants which attach themselves to man.
Blitum capitatum *L.*	Blitum capitum *L.* In Wallis.
Amaranthus albus *L.* (Introduced.)	The sands of Europe.
Polygonum dumetorum. *L.*	Grows in Europe in diverse latitudes.

From these various tables it is easy to see that the vegetation of the northern shores of Lake Superior is perfectly similar to the subalpine vegetation of Europe, at that zone which, in the Jura for instance, extends from 3,000 to 3,500 feet, and which in the Alps extends from 3,500 to 5,000 feet. Now removing some plants of the lakes, and some few peculiar American types, the subalpine flora remains in its integrity, and will be found to form chiefly the vegetation about the northern shores of Lake Superior.

SPECIAL COMPARISON.

Distribution of the Trees and Shrubs of Switzerland from the Plains to the Summit of the Mountains, compared with those of North America.

As it is easier to perceive the regular order of succession of the different growths which follow each other along the slope of a mountain, and to determine under such circumstances the precise limits of their distribution, than to ascertain the natural range of the corresponding vegetation northwards over extensive tracts of land, in level countries, I shall first introduce a general picture of the arborescent vegetation of the Swiss mountains, before I undertake to show that it agrees most minutely in its internal arrangement with that of the lake districts.

The vines which cover the margins of the Lake of Neuchatel, 1338 feet above the level of the sea, characterize, of course, the lower regions, which we call, for that reason, the region of vineyards. The trees which are cultivated there, the mulberry, peach, apricot, and even the fig in the warmest places, are all exotic. All fruits of the temperate zone, however, succeed there perfectly well, and among the wild trees and shrubs which characterize this zone, we find especially Rubus: Rubus corylifolius, Rubus fruticosus *L.*, Rubus tomentosus *W.;* some Roses: Rosa pimpinellifolia *L.*, Rosa eglanteria *L.*, Rosa alba *L.;* the Pyrus communis *L.*, the Cratægus torminalis *L.*, Mespilus germanica *L.*, and Mespilus eriocarpa *DC.* The most common ornamental shrubs which are cultivated there on level ground, are the Philadèlphus coronarius and the Lilac, which we find as far as the lower valleys of the Jura. This zone is almost entirely cultivated, and has few indigenous trees. We meet now and then with forests of oak trees (Quercus Robur *L.*,) and of chestnut trees (Castanea vesca).*

Immediately above this horizon, at an elevation of some hundred feet higher, from 1600 to 1700 feet begins the zone of oaks, which ascends somewhat into the valleys. The two species of this genus, the Quercus Robur *L.*, and the Quercus sessiliflora *Sm.*, grow in the same places; the latter ascends, however, a little higher, and occurs but very thinly, it is true, in the Val de Ruz, and in the Val de Travers. On the slopes of the Alps it ascends 1,500 feet higher, especially in sheltered valleys. The shrubs and trees which follow these are not numerous, (for the vegetation of the oak forests, like that of the pine trees, excludes other trees;) they are the hedge-plants, which are found as far as the region of the pines, (Viburnum Opulus *L.* et Viburnum Lantana *L.*); the yew, (Taxus baccata *L.*); the box-tree, (Buxus sempervirens *L.*); the hornbeam, (Carpinus betulus *L.*,) very rare; the alder, (Alnus glutinosa *Gærtn.*) At the margins of the brooks, some briars, the honeysuckle, (Lonicera Caprifolium,) cultivated; the buckthorn, (Rhamnus catharticus *L.*); the holly, (Ilex Aquifolium). The fruit trees cultivated with the greatest success in this zone, are the walnut, the apple, the pear, &c.

* Along the margin of the lakes grow the Populus nigra and several species of willows, which are characteristic, but have no direct affinity with the localities in which they occur. The Clematis Vitalba, on the contrary, attaches itself to the trees of the region of the vines and oak trees, but never ascends higher.

Between the region of the oak and that of the beech, we have at a height of 2,000 feet, as a transitory zone, a narrow tract characterized by the wild cherry tree and the Pinus sylvestris, which is, however, particularly adorned by a large variety of shrubs. To this zone belongs in the first place the linden tree, (Tilia microphylla *V.*, and Tilia platyphylla *Scop.*) ; three maples, (Acer opulifolium *L.*, Acer platanoides *L.*, and Acer campestre *L.*) ; the Evonymus europæus *L.*, Cerasus Padus *DC.*, Prunus spinosa *L.*, Cratægus Aria *L.*, Mespilus oxyacantha, Lonicera Periclymenum *L.*, Sambucus nigra *L.*, Cornus mas *L.*, Cornus sanguinea *L.*, Viscum album *L.*, Ligustrum vulgare L., Daphne Cneorum *L.*, Populus tremula *L.*, with the introduced Æsculus Hipocastanum, which succeeds in this zone better than anywhere else. This is the region of shrubs, properly speaking, with which is mingled the beech tree, whose zone, however, is more extended, and ascends in the Jura to 3,500 feet, and to 4,000 feet in the Alps.

To the region of the beech tree, which extends over a thousand feet of vertical height, from 2,500 to 3,500 feet, belong the following shrubs:—Rhamnus Frangula *L.*, Cytisus Laburnum *L.*, Rubus saxatilis *L.*, Rubus cæsius *L.*, Rubus idæus *L.*, Rosa eglanteria *L.*, Rosa villosa *L.*, Rosa canina *L.*, Rosa rubiginosa *L.*, Cratægus Amelanchier *L.*, Lonicera Xylosteum *L.*, Sambucus Ebulus *L.*, Daphne Mezereum *L.*, Daphne alpina *L.*, Daphne laureola L., Ulmus campestris *L.*, Corylus Avellana *L.*

The region of the pines or Coniferæ extends from 3,500 feet to 4,500 feet in the Jura, and to 6,000 feet in the Alps. It is well characterized in its lower and middle parts, where we find Fraxinus excelsior *L.*, Abies excelsa *DC.*, Abies pectinata *DC.*, Juniperus communis *L.*, and in the higher part the Pinus Cembra *L.*, Pinus Pumilio *Clus*, Larix europæa *DC.* In this zone live the Betula alba *L.*, Betula pubescens *Ehr.*, and Betula nana *L.*, and some bushes which never leave it, the Ericineæ especially; Vaccinium Myrtillus *L.*, Vaccinium uliginosum *L.*, Vaccinium Oxycoccos *L.*, Vaccinium Vitis-idæa *L.*, Andromeda polifolia *L.*, Arbutus Uva-ursi *L.*, Arbutus alpina *L.*, Pyrola rotundifolia *L.*, Pyrola minor, *L.*, Pyrola chlorantha *Sn.*, Pyrola secunda *L.*, Pyrola umbellata *L.*, Pyrola uniflora *L.*, Linnæa borealis *L.*, Lonicera alpigena *L.*, Lonicera cærulea *L.*, Rosa rubrifolia *Willd.*, Rosa alpina *L.*, Rhamnus alpinus, *L.*, and in the higher parts, Cratægus Chamæmespilus *L.*, Azalea procumbens *L.*, Empetrum nigrum *L.*, Acer pseudoplatanus *L.*

Above all these we meet already in the Jura the Rhododendrons and the Salix herbacea, which belong truly to the alpine flora characterized by all those handsome plants covered with a light cotton down, which we find along the margin of the glaciers in the Alps, and as high as the uppermost limit where all vegetation ceases somewhat suddenly, at a level of about 8,000 feet above the level of the sea.

Trees of the Lake Superior Region.

We may place at about 40° northern latitude the zone of vegetation, which in America corresponds to the upper limit of the cultivation of the vine, as we observe it on the banks of the Swiss lakes. At about this latitude the family of the Magnoliaceæ dies out, though we may still meet the Magnolia glauca in the swamps, as far as the 43° N. lat., and though the tulip tree still flourishes there. This is also the northern limit of the Anonaceæ, Melastomaceæ, Cactaceæ, Santalaceæ, and Liquidambar; and though in Europe we have no representatives of these families, it is easy to perceive, on reflecting upon the examples just mentioned, that the limits of vegetation under consideration are natural, and correspond to each other, though characterized in the two continents by different plants. Again, the numerous species of wild vines which America produces, although they do not extend farther northwards than the cultivation of the vine in Europe, yet prosper on this continent in a colder climate.

The State of Massachusetts, with its long arm stretched into the ocean eastwards, or rather the region extending westward under the same parallel through the State of New York, forms a natural limit between the vegetation of the warm temperate zone, and that of the cold temperate zone, whose forests G. B. Emerson, Esq., has so well described in his admirable Report upon the Trees and Shrubs of Massachusetts. With this book, we may become well acquainted with the arborescent vegetation of the zone which corresponds to the horizon of oaks and shrubs in the Jura; so that I need not enumerate these characteristic species. Not only is this also the northern limit of the culture of fruit trees, but this zone is equally remarkable for the great variety of elegant shrubs which occur particularly on its northern borders, where we find so great a variety of species belonging to the genera, Celastrus, Cratægus, Ribes, Cornus, Hamamelis, Vaccinium, Kalmia, Rhodora, Azalea, Rhododendrum, Andromeda, Clethra, Viburnum, Cephalanthus, Prinos, Dirca, Celtis, &c. I shall only add, that in the latitude under which the St. Lawrence winds its course from the great Canadian lakes, and takes a more independent course north-eastwards, we perceive already great changes in the growth of trees. About Niagara, or rather somewhat farther north along the northern shores of Lake Ontario, and the hills which rise above Toronto, the following species begin to disappear: Sassafras officinale, (I have not seen this species north of Table Rock,) Juglans nigra and cinerea, Carya alba and amara, Castanea americana, Quercus alba and Castanea, Platanus occidentalis, Tilia americana, (this species occurs, however, as far north as Sturgeon Bay, on Lake Huron,) Rubus odoratus. Though the Beech is extensively distributed among the forests of this zone, we cannot but be struck with their splendid growth further north, where the Elm, Red Oak, Hornbeam, Hop-hornbeam, several species of Birches, various Maples, Ashes, Wild Cherries, &c., &c., more or less mixed with Coniferæ, form the most beautiful forests of the temperate zone, particularly remarkable for their diversified shades of green and dark foliage, and which almost uniformly cover the ground along the shores of the Great Lakes as far as Lake Superior, the Coniferæ gradually coming in in a larger proportion to the successive exclusion of the trees with deciduous leaves. As soon as we reach Mackinaw we find the Beech has almost entirely disappeared, or become so dwarfish as no longer to be a handsome tree, while Ostrya, Carpinus, Betula populifolia, Quercus rubra, and indeed all Cupuliferæ are entirely gone, and the Canoe-Birch, the Black Ash, with Pinus balsamifera, alba, nigra, Larix americana, Pinus Strobus, Sorbus americana, and some Poplars on the lake shore, form the mass of forests, with a few low shrubs among them, such as Arctostaphylos Uva-ursi, Vaccinium, Chiogenes, &c. This zone, which corresponds to the horizon of Pines in the Jura, extends all along the northern shores of Lake Superior. North of Fort William are extensive forests of Pinus Banksiana, with Pinus resinosa and Strobus. We noticed no Cupuliferæ beyond Batcheewauaung Bay, and we learnt that but a few dwarfish Red Oaks are seen in the Island of Michipicotin; but the Elm is still handsome about Fort William, though it is very scarce in other parts of the northern shores.

The shores of Nipigon Bay, the northernmost point we visited, are covered with Pine forests, with a few Ashes and Maples, and here and there a Sorbus americana among them. At this latitude, the 49°, we had therefore not yet reached the zone of the true alpine vegetation, and remained for the whole extent of our journey within the limits of the sub-alpine flora.

The highest point which we visited, the summit of a mountain upon St. Ignace Island, which we called Mount-Cambridge, afforded the following harvest for our herbarium:—Abies balsamea, Abies alba, Betula papyracea, Alnus viridis, Sorbus americana, Amelanchier canadensis, Acer montanum, Diervillea trifida, Sambucus pubens, Rhus Toxicodendrum, Vaccinium uliginosum, Corylus rostrata, Linnæa borealis, Cornus canadensis, Spiræa opulifolia, Salix, Corydalis glauca, Epilobium angustifolium, Polygonum ciliare, Melampyrum, Clintonia borealis, Stereocaulon paschale, Gyrophora hirsuta, Cladonia pyxidata, and rangiferina, Parmelia tiliacea and Sphagnum acutifolium.

From this list it is obvious, that even a thousand feet of height will introduce very slight differences in the vegetation of these regions. For, though Mount Cambridge is about a thousand feet above the level of the lake, its whole slope is covered with the same vegetation which occurs at the very level of the lake.

This fact would seem in flat contradiction with the general laws of the geographical distribution of plants, to which we have alluded above, but for the presence of the lake itself and its peculiar character.

So large a sheet of so deep water as Lake Superior, preserving all the year round a very equable and low temperature even on its shores, which are generally very precipitous, must of course influence greatly the temperature of the main land in its immediate vicinity, at considerable heights above its surface.

There is, therefore, nothing very surprising in our finding so uniform a vegetation at rather considerable heights above the surface of the lake and on its immediate shores.

This fact is to be attributed to the equalizing local influence of the lake, and does not form an exception to the law of distribution, and change of the character of vegetation in the interior of continents, upon the slopes of high mountains; for we have, even a few degrees farther south, in the same continent, a striking example of the fixity of these laws, in the White Mountains, which are sufficiently distant from the sea-shore, and not surrounded by any large sheet of fresh water, so that the zones of vegetation are very well marked on their slopes, and can be traced in gradual succession beyond the range of the Mountains proper to the level, where the vegetation has the character which distinguishes it, in this latitude, near the level of the sea.

In the vicinity of the White Mountains, the changes of vegetation are rather conspicuous, owing to their gradual elevation above the surrounding flat country, and also to the more sudden rise of several of their peaks. We no sooner begin to ascend the head waters of the Connecticut valley towards Littleton, than the forest vegetation begins to assume a different character from what it has lower down in the main valley nearer the sea. Juglans cinerea and Carya porcina disappear in that village. The oaks also are fewer and smaller. The mountain maple, which is not found below, here makes its appearance. The following trees may be seen between Windsor and Littleton: — Abies Canadensis, Pinus strobus, Thuya occidentalis, Larix Americana, Platanus occidentalis, Fagus ferruginea, Comptonia asplenifolia, Betula populifolia, B. lenta, B. excelsa, B. papyracea, Quercus alba, Q. rubra, Q. bicolor, Ulmus Americana, Carpinus Americana, Ostrya Virginica, Fraxinus alba, Populus tremuloides, Tilia Americana, Acer saccharinum, A. montanum, A. Pennsylvanicum. The chestnut has already disappeared at Windsor, where the height above the level of the sea is three hundred feet.

From Littleton, eight hundred and thirty feet above the sea, to Fabyan's, which is fifteen hundred feet,* we notice Abies alba, A. balsamifera, A. Canadensis, Pinus strobus, Larix Americana, Tilia Americana, Fraxinus alba, Acer saccharinum, A. montanum, A. Pennsylvanicum, Ulmus Americana, Sorbus Americana, Betula excelsa, B. papyracea, B. populifolia, Alnus incana, Comp-

* This and the following measures were ascertained barometrically by Professor A. Guyot.

tonia asplenifolia, &c. The Cupuliferæ have disappeared; Pinus rigida, also, is no longer observed, and thus vegetation continues from Fabyan's to a level of two thousand and eighty feet, where the pine vegetation forms the larger proportion of the features of the forest.

This height of two thousand and eighty feet is a very natural level in the chain of the White Mountains, and especially on the slope of Mount Washington. It indicates the horizon where the slope begins to be much steeper, and where the variety of trees combined in the forests is greatly reduced; for above this level to the height of four thousand three hundred and fifty feet we may say that the vegetation consists entirely of Abies alba and balsamea and Betula excelsa and papyracea, which grow gradually more and more stunted, till at the height of four thousand three hundred and fifty feet, those species even, which form tall, splendid trees one or two thousand feet lower, appear here as mere shrubs, low bushes, with crooked branches so interwoven as almost entirely to hedge up the way, excepting in places where a bridle-path has been cut through.

Above this level the mountain is naked, and many fine plants make their appearance which remind us of the Flora of Greenland, and many of which grow on the northern shores of Lake Superior, such as Arenaria Grœnlandica, Vaccinium cæspitosum, uliginosum, &c.

The summit of the mountain, at the height of six thousand two hundred and eighty feet, produces several plants which have no representatives south of Labrador. Such are Andromeda hypnoides, Saxifraga rivularis, Rhododendron lapponicum, Diapensia lapponica.

Before leaving this subject I ought to make an additional remark about the identity of so many plants which are common to both continents. It is a general fact, that the farther north we proceed, the greater is the primitive uniformity of the plants, as well as the animals, in both hemispheres; so much so, that the arctic flora and the arctic fauna are identical, not only in their general character, but also in almost all the species which characterize that region as a natural botanical and zoölogical province. But there are a great many plants and animals occurring in the temperate zone, which are equally identical in Europe and America, and which, nevertheless, do not belong originally to both hemispheres,* but were introduced into America since the settlement of Europeans in this part of the world, many of which, though foreigners, have spread so extensively, as to be generally considered as natives of this country. But if we carefully examine their distribution, we soon perceive that they follow everywhere the tracks of civilization, and occur nowhere except in those districts and in those soils where the hands of white men have been at work. In such localities, however, they have almost completely replaced the native weeds, which have disappeared before them as completely as the Indian tribes have disappeared before the pressing invasion of the more civilized nations. These plants are chiefly such as occur in Europe by the road-sides, or near the habitation of man, and which to a certain degree may be considered as satellites of the white race. Their occurrence is particularly striking along the new lines of railroads, where they settle almost as soon as the tracks are marked out, and increase in a few years so rapidly within the enclosure of the roads, as to suppress the primitive vegetation almost completely, with the exception of a few hardy natives which resist the new invaders. Several of these plants occur naturally, in America, in more northern latitudes. Nevertheless, I have no doubt that in most cases they were introduced into the more temperate and cultivated latitudes from Europe, rather than from their northern residence in America.

The following list of these plants was chiefly made from an examination of the railroad tracks between Boston and Salem, in company with that liberal cultivator of botany, Hon. John A. Lowell, and also from materials collected during an excursion made with

* I do not wish by this remark to be understood as intending to deny the identity of any native plant in the temperate zone of Europe and America. I know that many species which occur very far north, and are there truly identical in both continents, are also found among the plants of the temperate zone on the two sides of the Atlantic; but there still remains a large number, the identity of which ought to be ascertained by direct comparison of authentic specimens from the two continents, before it can be finally admitted that there is no specific difference between them. As such, I may mention Hepatica triloba, Geranium Robertianum, Oxalis Acetosella, Spiræa Aruncus, Circæa lutetiana, Calystegia sepium, Agrimonia Eupatoria, Majanthemum bifolium, and many aquatic plants. The identity of these with European species seems to me the more questionable, as the freshwater animals, the fishes, mollusks and insects differ specifically throughout.

the same gentleman to Niagara Falls and the White Mountains. The European weeds which are limited to cultivated ground, as Lychnis Githago, Centaurea cyaneus, are entirely omitted in this list, as well as plants escaped from gardens, which are found only occasionally, in an apparently wild condition, in the United States, as Abutilon Avicennæ, Althæa officinalis, &c.

Ranunculaceæ.

Ranunculus acris.
" bulbosus.
" sceleratus.

Berberideæ.

Berberis vulgaris.

Papaveraceæ.

Chelidonium majus.

Fumariaceæ.

Fumaria officinalis.

Cruciferæ.

Nasturtium officinale.
Lepidium ruderale. Often side by side with Lepid. virginianum.
Barbarea vulgaris.
Sisymbrium officinale.
" thalianum.
Draba verna.
Sinapis nigra.
" arvensis.
Capsella Bursa-Pastoris.
Raphanus Raphanistrum.

Hypericineæ.

Hypericum perfoliatum.

Caryophyllaceæ.

Saponaria officinalis.
Silene inflata.
Arenaria serpyllifolia.
Stellaria media.
Cerastium vulgatum.
Spergula arvensis.
Scleranthus annuus.

Portulacaceæ.

Portulaca oleracea.

Malvaceæ.

Malva rotundifolia.

Geranieæ.

Erodium cicutarium.

Leguminosæ.

Trifolium pratense.
" arvense.
" repens.
" procumbens.
Medicago lupulina.
Vicia sativa.
" cracca.
Melilotus officinalis.

Crassulaceæ.

Sedum Telephium.

Umbelliferæ.

Daucus Carota.
Pastinaca sativa.
Conium maculatum.

Rubiaceæ.

Galium Aparine.
" verum.

Valerianeæ.

Fedia olitoria.

Compositæ.

Tussilago Farfara.
Inula Helenium.
Achillæa millefolium.
Xanthium strumarium.
Leucanthemum vulgare.
Tanacetum vulgare.
Lappa major.
Cichorium Intybus.
Leontodon autumnale.
Maruta cotula.
Anthemis arvensis.
Taraxacum Dens Leonis.
Senecio vulgaris.
Sonchus oleraceus.
" arvensis.

Plantagineæ.

Plantago major.
" lanceolata.

Primulaceæ.

Anagallis arvensis.

Scrophularineæ.

Linaria vulgaris.
Verbascum Thapsus.
Veronica officinalis.
" serpyllifolia.
" arvensis.
" agrestis.

Labiatæ.

Lycopus Europæus.
Nepeta Cataria.
Leonurus cardiaca.
Prunella vulgaris.
Origanum vulgare.
Clinopodium vulgare.
Lamium amplexicaule.
Galeopsis Tetrahit.
" Ladanum.
Marrubium vulgare.
Ballota nigra.

Borragineæ.

Echium vulgare.
Lycopsis arvensis.
Symphytum officinale.
Lithospermum officinale.
" arvense.
Echinospermum Lappula.
Cynoglossum officinale.

Convolvulaceæ.

Convolvulus arvensis.

Solaneæ.

Solanum Dulcamara.
" nigrum.
Datura Stramonium.
Hyoscyamus niger.

Oleaceæ.

Ligustrum vulgare.

Chenopodiaceæ.

Chenopodium album.
Agathophytum Bonus-Henricus.

Polygoneæ.

Polygonum Hydropiper.
" aviculare.
" Convolvulus.
" Persicaria.
Rumex Acetosella.
" obtusifolius.
" crispus.

Urticaceæ.

Urtica urens.
" dioica.

Euphorbiaceæ.

Euphorbia helioscopia.
" platyphylla.
" Peplus.

Euphorbia Esula.

Salicineæ.

Salix purpurea.
" viminalis.
" alba.
" fragilis.

Liliaceæ.

Allium vineale.

Gramineæ.

Alopecurus pratensis.
Phleum pratense.
Agrostis canina.
" vulgaris.
" alba.
Cynodon Dactylon.
Dactylis glomerata.
Poa pratensis.
" annua.
Festuca duriuscula.
" elatior.
" pratensis.
Bromus secalinus.
Triticum repens.
" caninum.
Lolium perenne.
Arrhenatherum elatius.
Holcus lanatus.
Anthoxanthum odoratum.
Panicum Crus-galli.
Setaria viridis.

It is still a question whether all these plants originate from Europe, as many of them occur there in the same circumstances as in this continent, under the immediate influence of agricultural improvements, and might have followed the Caucasian race of men from farther east, in his migrations over the temperate zone of Europe. Various other remarks respecting the vegetation of this continent may be found above, in the course of the Narrative, pp. 10, 13, 19, 89. Many interesting remarks upon the foreign vegetation of this continent may also be gathered in Kalm's Travels in North America. Quite a number of European insects have also been introduced into this country with those plants, among which I may mention some showy butterflies, as Vanessa atalanta, Cardui and Antiopa, which are very erroneously considered by some entomologists as native Americans.

III.

CLASSIFICATION OF ANIMALS FROM EMBRYONIC AND PALÆOZOIC DATA.

FOR several years I have been in the habit of illustrating, in my public lectures and elsewhere, principles which have not yet been introduced in our science, and to which I feel it my duty to call attention in a more formal manner on this occasion, as during our excursion we had several opportunities to discuss them at length. These remarks will form an appropriate introduction to the lists of the animals found about Lake Superior, which are given below.

The principle which has regulated our classifications for the last half century, is that which Cuvier worked out by his anatomical investigations; I mean the arrangement of the whole animal kingdom according to the natural affinities of animals as ascertained by the investigation of their internal structure. This fruitful principle, applied in various ways, has produced a series of classifications, agreeing or differing more or less in their outlines, but all resting upon the idea, that a certain amount of anatomical characters may be easily ascertained, expressing the main relations which exist naturally among animals, and affording a natural basis for classification. Structure, therefore, internal as well as external, is, according to the principles of Cuvier, the foundation of all natural classifications; and undoubtedly his researches and those of his followers have done more, in the way of improving our natural methods, than all the efforts of former naturalists put together; and this principle will doubtless regulate, in the main, our farther efforts.

Nevertheless, so much is left in this method to the arbitrary decision of the observer, that it would be in the highest degree desirable to have some principle by which to regulate the internal details of the edifice.

We may indeed form natural divisions simply from structural evidence, bring together all fishes as they agree in the most important details of their structure, and combine all reptiles into one class, notwithstanding the extreme differences in their external form. We may also recognize the true affinity of whales, and bring them together with other Mammalia, notwithstanding their aquatic habits and their fish-like form; we may even subdivide those classes into inferior groups upon structural evidence, and thus introduce orders, like the Quadrumana, Carnivora, Rodentia, Ruminantia, &c., &c., among Mammalia. But we are at once at a loss how to determine the relative value of those groups, and to find a scale for the natural arrangement of further subdivisions. After having, for example, circumscribed the Carnivorous Mammalia into one natural family, how are we to group the minor divisions like that of the swimming Carnivora, the Plantigrada and the Digitigrada; or, after circumscribing the reptiles into natural groups like those of Chelonians, Saurians, Ophidians and Batrachians, how shall we, for instance, arrange the various types of Batrachians? To those who have been familiar with our proceedings in all these attempts, it must be evident that the grouping of our subdivisions has been almost arbitrary and entirely left to our decision without a regular guide. We have, it is true, subdivided the Batrachians into the more fish-like forms which preserve their gills and tails, or at least their tails; and into another group, containing those which undergo a complete metamorphosis; but it has not yet occurred to naturalists to take this metamorphosis as the regulating principle of classification, to arrange genera according to their agreement with certain degrees of development, in the natural order of changes which the higher of these animals undergo. Now it is my firm belief, that such a new principle can be introduced into our science; that methodical arrangement may be carried into the most minute details, without leaving any room for arbitrary decision. Proteus, Menobranchus, Amphiuma, Triton, Salamandra will hereafter have a natural place in our classification, which will be commanded by embryology, and no longer be left to a vague feeling that aquatic animals are lower than amphibious and terrestrial ones, and that the retaining of the gills indicates a lower position than their disappearance.*

Of course, in the outset, we do not find sufficient data to trace this arrangement throughout the animal kingdom, and to make the principle which I have just mentioned the ruling law of nice classical arrangement. But until such sufficient knowledge is acquired, let me show that my principle does in fact apply to all classes of the animal kingdom, and will at once contribute to improve all their subdivisions. Among Mammalia, for example, we shall continue to give the aquatic carnivorous animals a lower position among Carnivora, but no onger simply because they are aquatic, but because they are webfooted, as the webfoot is the earlier form of the limbs in all Mammalia whose embryonic development has been traced. We shall be led, for similar reasons, to deny the bats the high position which has been assigned to them, and to combine them closer with the Insectivora. We shall separate the manatees from their present relations and combine them with tapirs, elephants, &c., as they are rather webfooted Pachyderms, than true Cetaceans.†

* These views were fully illustrated in a series of twelve lectures upon *Comparative Embryology*, delivered before the Lowell Institute during the last winter, and reported for the Daily Evening Traveller, and afterwards published as a separate pamphlet.

† These aphorisms will be justified by a more elaborate illustration of the peculiar changes which the limbs of Mammalia undergo during their embryonic growth, as far as I have been able to trace them, in various animals. It may suffice, for the present, for me to say here, that in all young embryos of Mammalia which I have recently had an opportunity to examine, I have found the extremities arising as oblong tubercles, flattened at their extremities, spreading more and more into the form of hemispherical paddles, in which the changes in the cellular growth gradually introduce differences upon the points where the fingers are to be developed. But for a longer time they remain combined in a common outline, and the microscopic structure of the tissues alone indicates the points of growth; and even after the fingers have been fully sketched out, they remain for a certain time united by a common web, which is successively reduced as the fingers grow longer and thicker.

It is very remarkable how uniform, and indeed how identical in form and structure the anterior and posterior extremities are in the beginning, whatever may be the difference at a later period of growth. Thus, for instance, there is not the slightest difference between the anterior and posterior extremities of the bat, in the early stages of development. The wing is then a very short limb, terminated by a flat, webbed paddle, of a semicircular form, identical in development, size and form with the hinder extremity, and differing in no respect from the appearance of the hand and foot in young human embryos, or in embryos of cats, dogs, squirrels, hares, rabbits and pigs, and bearing

Among birds we shall also avail ourselves of the discovery I made last year, that embryos of birds have web-feet and web-wings, and no longer consider Palmipedes as forming a natural group by themselves, but allow the possibility of having several natural groups of birds, beginning each with web-footed forms. Every one who is conversant with the natural history of birds must have been struck with the great diversity of features in birds united in our systems under the head of Palmipedes. Taking all birds together, we hardly notice among them greater differences than those which exist between the various families of Palmipedes, which are, confessedly, brought together upon no other character than the webbed form of their feet; though among them we have birds of prey, such as the gulls, and others, which seem to stand by themselves unconnected and without any analogy with any other family, such as the swans, geese, and ducks; and again, the pelicans and the genera allied to them, and also the divers. It can hardly be understood why birds so widely different should be brought together; and indeed, their reunion would long ago have been given up, had it not been for the difficulty of finding characters to separate them, and for the strong impression, that the similarity of the structure of their feet should overrule the other characters.

But now, since it is known that birds of the most heterogeneous character in the structure of their legs, in their adult form, have, when very young, identical legs, whether they belong to the type of hawks, or to that of crows, or to that of sparrows, or to that of swallows, or to that of pigeons, or to that of hens, or to that of waders, or to that of true Palmipedes,—when we know all these types to have an identical development of their legs, and, I may add also, of their wings,—for the young wing is equally a small, webbed fin,—there can be no longer any doubt left upon the impropriety of combining any two families of adult birds solely on the ground of their legs having webbed feet.

It is a fact, too well known in zoölogy, that different families will

the same relation to the extremities of birds, in which also legs and wings are developed according to the same pattern.

These facts have been partly described in my Lectures on Comparative Embryology, and more extensively illustrated in a paper laid before the American Association for the advancement of Science, in Cambridge, August, 1849. See also Narrative, p. 35.

repeat, in the same class, the characteristic changes which are peculiar to the whole family, to require any further argument to show that Palmipedes are not, necessarily, a natural division; and though we may fail for the present in reärranging the families of this class into natural orders, I trust after these remarks, more importance will yet be attached, and more attention paid in future, to the fact that Palmipedes, as they are now characterized, have very different types of wings and bills. I have, for my own part, been strongly impressed with the resemblance which exists between gulls and frigate birds, and the birds of prey, of the hawk and vulture families, in which the toes are by no means so completely distinct as they are among other birds. And, far from considering birds of prey as the highest family among birds, I would only consider them as highest in the series which includes simultaneously Procellaridæ and Laridæ. Whether the family of pelicans belongs to this group or not, I am not prepared to say; but, at all events, the fact of their preserving their four toes in one continuous web shows them to rank lowest among birds.

Again, among reptiles there will no longer be a foundation for any arrangement resting merely upon impressions; thus the terrestrial turtles will stand higher than the freshwater, and these again higher than the marine; and among Batrachians, which are best known in their embryology, we can already arrange all the genera in natural series, taking the metamorphosis of the higher as a scale, and placing all full-grown forms in successive order, according to their greater or less resemblance to these transient states. Even the relative position of toads and frogs may be settled with as much internal evidence as any other question of rank in wider limits, merely upon the difference of their feet.

In my researches upon fossil fishes I have on several occasions alluded to the resemblance which we notice between the early stages of growth in fishes, and the lower forms of their families in the full-grown state, and also to a similar resemblance between the embryonic forms and the earliest representatives of that class in the oldest geological epochs; an analogy which is so close, that it involves another most important principle, viz., that the order of succession in time, of the geological types, agrees with the gradual changes which the animals of our day undergo during their metamorphosis, thus giving us another guide to the manifold relations which exist among animals, allowing us to avail ourselves, for the purpose of classification, of the facts derived from the development of the whole animal kingdom in geological epochs, as well as the development of individual species in our epoch. But to this most fruitful principle I shall have hereafter an opportunity of again calling attention.

At present there is some doubt among zoologists, as to the respective position of the classes of worms, insects and Crustacea, some placing the Crustacea, and others the insects uppermost. Embryonic data may afford the means of settling this question; we need only remember the extensive external changes which insects undergo from their earliest age, and the many stages of structure through which they pass, whilst Crustacea are less polymorphous during the different periods of their life, and never obtain an aërial respiration, but breathe through life with gills, which many larvæ of insects cast before they have accomplished their metamorphoses, to be satisfied that the affinity between Crustacea and worms is greater than between worms and insects, especially if we consider the extraordinary forms of some parasitic types of the former. As soon as the higher rank of insects among Articulata is acknowledged, many important relations, which remain otherwise concealed, are at once brought out. The whole type of insects in its perfect condition, contains only aërial animals, while the Crustacea and worms are chiefly aquatic. And if we compare these three classes in a general way, we cannot deny the correctness of the comparison as made by Oken, that worms correspond to the larval state of insects, Crustacea to their pupa state, and that insects pass through metamorphoses corresponding to the other classes of Articulata. The little we know about the embryology of worms will already satisfy us that the earlier stages of the higher of these animals agree most remarkably in character with such of them as, from other reasons, we have been in the habit of considering as the lowest, thus affording another prospect of regulating finally the arrangement of those curious animals entirely upon embryonic data.

If there is any internal evidence that the whole animal kingdom is constructed upon a definite plan, we may find it in the remarkable agreement of our conclusions, whether derived from anatomical evidence, from embryology or from palæontology. Nothing, indeed, can be more gratifying than to trace the close agreement of the general results derived from the study of the structure of animals, with the results derived from the investigation of their embryonic changes, or from their succession in geological times. Let anatomy be the foundation of a classification, and in the main, the frame thus devised will agree with the arrangement introduced from embryological data. And, again, this series will express the chief features of the order of succession in which animals were gradually introduced upon our globe. Some examples will show more fully that this is really the case. Resting more upon the characters derived from the nervous system, which in the crabs is concentrated into a few masses, zoologists have generally considered these animals as higher than the lobsters, in which the nervous ganglia remain more isolated. Now as far as we know, the embryos of brachyuran Crustacea, that is, of crabs, are all macrural in their shape, that is to say, they resemble at an early age the lobsters more than their own parents; and again, lobster-like Crustacea prevailed in the middle ages of geological times during the triassic and oölitic periods, that is, ages before crabs were created, as we find no fossils of that family before the tertiary period.

Of the class of insects I have for the present little to say, the diversity of their metamorphoses having not yet allowed an insight into their bearing. I will only mention that the predaceous character of the larvæ of most of the sucking insects, which are provided with powerful jaws in their early stages of growth, seems to indicate that the chewing insects rank lower than the sucking tribes. Investigations which I am tracing at present, will, I hope, throw some light upon this most important question.*

* Since the above remarks were written, I have devoted most of my time to the investigation of these metamorphoses in insects; and to my great satisfaction (but, I may say, as I anticipated,) I find that the metamorphoses of the higher insects throw such light upon the real relations of the different orders of that class, as to settle finally the question of their gradation. It has now become with me a matter of fact, that Coleoptera, Orthoptera, Neuroptera and Hymenoptera, rank below Hemiptera, Diptera and Lepidoptera. A careful investigation of the changes of Lepidoptera has shown to me that, prior to assuming its pupa form, the young butterfly assumes, under the last

In the department of Mollusca, if the above principles are correct, embryology is likely to introduce modifications in our systematic methods, which will entirely overthrow the views entertained at present respecting their systematical arrangement; not that we should ever be led to consider Acephala as higher than the Gasteropoda, or these as higher than the Cephalopoda; but within these classes, taken by themselves, I look for considerable changes, which, when once introduced, might explain why there is apparently so little agreement between the geological succession of their types and their systematic arrangement, especially among Gasteropoda. Now it is precisely among these, that I anticipate the greatest changes. It is indeed a remarkable fact, that so many, if not all naked branchiferous Gasteropoda should be provided with a shell in their early age, and lose this protecting envelop as they grow older, which would lead to the conclusion, that among these animals the fact of having a shell indicates a rather lower condition. The comparison of Octopus, Loligo, Sepia and Nautilus would lead to similar conclusions. Indeed it is scarcely any longer doubted, that Nautilus has many points of resemblance in common with the Gasteropoda, and from its numerous tentacles (multiplication being always an indication of a lower degree,) must be considered the lowest type among Cephalopoda; next we should place the Dibranchiate Cephalopoda, among which the Argonauta, with its external shell, ranks the lowest; next the naked Octopodidæ, while the Sepiadæ with their ten tentacles and internal shell or bone would be the highest in that class. Now if this arrangement be the real order of succession of the Cephalopoda accord-

skin of the caterpillar, (in which state the caterpillar is so seldom examined, from fear of disturbing it in its transformation) that under this last skin of the caterpillar, I say, the young butterfly assumes the characters of a Coleopteron. It has then an upper pair of wings, having the character of elytra, and a lower pair of membranous wings. At that time its jaws have not yet assumed the form of a sucker, and are still free, as are also the legs. But these parts, which are easily observed in caterpillars immersed in diluted alcohol at the very moment when they are casting their last skin, are soon soldered together to form the hard coating of the pupa, and are cast off before the perfect butterfly comes out. It is, therefore, correct to say, that the structural condition of Coleoptera, in their perfect state, answers to that stage of moulting of Lepidoptera which precedes their perfect development. Coleoptera are, therefore, one stage behind Lepidoptera; they rank below them; they are an inferior degree of development of the type of insects.

ing to their structure and development, is it not remarkable, does it not indicate the maintenance of the same plan throughout the creation, when we find chambered shells, so abundant throughout the ancient geological formations, and belemnites, the analogues of the cuttle-fish, beginning late in the secondary epoch in the lias; whilst fossil argonauts do not occur before the tertiary times? So that we might almost conclude, that in this class the order of succession of their fossil types is a safer guide for our classification, than anatomical investigation.

In the class of Acephala the low position of brachiopods in the order of appearance in time, as well as in our estimation of their structural standing, is another striking instance of the correspondence between the order of geological succession and the gradation in structure. I may add as a link for farther inference, that I have seen embryonic cyclas attached by a byssus to the gills of the mother.

There is perhaps no department in which we may expect more important results for methodical arrangement from embryological researches than that of the Radiata. Let us only consider the metamorphosis of the Medusæ, their first polyp-like condition, their division and the final transformation of their stem into several distinct individuals, exemplifying in a higher sphere the growth of compound Polypi, where the successive buds remain united upon a common stock. Let us remember the free Comatula growing from the egg upon a Crinoid-like stem; let us then remember, that there are animals of that class, which preserve throughout life this articulated support, and remind us of corals even in the highest class of Radiata; let us farther know, that even the arrangement of plates in those Crinoids agree in some respects with the first formed calcareous granules in free moving starfishes; let us finally and above all here remember, that those Crinoids with stems are only Echinoderms of earlier ages, which die out gradually, to be replaced by new and free forms, and there will not be the slightest doubt left in our minds, that besides the structure, there is no safer guide to the understanding of the plan of the creation of the animal kingdom, as it has been in former ages and as it is in our days, than embryological and palæontological researches.

The internal arrangement of these classes as I now conceive it, would require that we introduce Bryozoa among Acephala and place them lowest in that class, next the compound and simple Ascidiæ, and then the Brachiopoda and true Acephala. Among Gasteropoda I would introduce Foraminifera as their lowest type, exemplifying, in a permanent condition, the embryonic division of their germ, next the Pteropoda would follow, also as an embryonic form of Gasteropoda, in which the lateral fin-like appendages and the symmetrical shell remind us of the deciduous shell of naked Gasteropoda with their vibrating wheels, and next the Heterobranchia, the common branchiferous Gasteropoda, and uppermost the Pulmonata, in some of which the embryo is not even aquatic, nor provided with fringed appendages. As for the Cephalopoda, I have recently had sufficient evidence from embryonic investigations that the Octocera stand below Decacera.

IV

GENERAL REMARKS UPON THE COLEOPTERA OF LAKE SUPERIOR.

BY JOHN L. LECONTE, M.D.

THE materials which form the basis of the present catalogue, were not altogether derived from explorations made during the expedition which produced this volume. They embrace the results of my collections during three journeys made to Lake Superior, and were procured at various points around the entire circumference of that sheet of water, and during various months from June to October.

The distribution of species does not appear to differ materially on the two sides of the lake; nevertheless many species occurred on the north shore, which were not found on Point Kewenaw, while many water beetles were taken at the last mentioned place, which were not seen during the present voyage. Still in each case the delay at particular localities was so short, that necessarily many even of the most common species would be overlooked. We may therefore conclude, that although the evidence is not yet sufficient to enable us precisely to distinguish between the products of the different portions of the Lake Superior region, we still have abundant material to give a tolerably complete conspectus of the character of the entire coleopterous fauna.

The whole country being still almost in a primitive condition, the specimens are equally distributed throughout a large space: the woods will not therefore be found very productive to the collector. In fact nearly all the species were found adjacent to small streams; or else they were driven on shore, particularly on sand beaches, by the winds and waves after being drowned in the lake. So productive was the last method of collecting, that on one occasion more than three hundred specimens of Coleoptera, and many insects of other orders were procured in less than one hour.

There are, however, a few points to which the attention of the future explorer may be directed, as being most likely to reward him for his arduous journey; these are Eagle Harbor on Point Keweenaw, the Hon. Hudson Bay Co.'s fort at the mouth of Pic River, and the islands adjacent to the mouth of Black Bay.

For the sake of making the catalogue as concise as possible, I have used such abbreviations as will render necessary a list of the works cited. Where no authority is appended to a name, it is to be understood that the name is used for the first time in this book. Rarely two references are placed after a name; in this case the latter citation is the more recent, and will be found to give all necessary information respecting synonyms, which are accordingly omitted here.

BOOKS CITED IN THE CATALOGUE.

Am. Tr. Transactions of the American Philosophical Society. New Series.
An. Lyc. Annals of the Lyceum of Natural History of New York.
Aubé. Spécies Général des Coléoptères. (Hydrocanthares.)
B. J. Boston Journal of Natural History.
Beauv. Palisot de Beauvois. Insects d'Afrique, et d'Amerique.
Dej. Spécies Général des Coléoptères de la Collection de M. le Comte Dejean.
Dej. Cat. Catalogue des Coléoptères de sa Collection.
Er. Erichson, Monographia Staphylinorum.
Er. Col. March. Erichson, Die Käfer der Mark Brandenburg.
Er. Germ. Zeit. " in Germar's Zeitschrift für die Entomologie.
Er. Ins. Germ. " Naturgeschichte der Insecten Deutschlands.
Er. Mon. " Entomographien.
Enc. Encyclopédie Méthodique.
Fabr. El. vel *F. El.* Fabricius Systema Eleutheratorum.
Grav. Micr. Gravenhorst, Coleoptera Microptera.
Germ. Ins. Nov. Germar, Insectorum species novæ aut minus cognitæ.
Germ. Zeit. Germar, in Germar's Zeitschrift für die Entomologie.
Gory & Perch. Gory and Percheron, Monographie des Cétoines.
Gyll. Fn. Suec. Gyllenhal, Fauna Suecica.
Hbst. Col. Herbst, Natursystem aller bekannten Insecten: Käfer.
Hd. Haldeman, in locis variis.
J. Ac. Journal of the Academy of Natural Sciences of Philadelphia.
J. Ac. N. S. Ejusd. op. series nova, 1848.
Kb. N. Z. Kirby in Fauna Boreali-Americana. Vol. 4.

Lac. Eroty. Lacordaire Monographie des Erotyliens.
Lac. Chrys. " " des Coléoptères, Subpentamères.
Lap. Bup. Monographie des Buprestides par Laporte et Gory.
Lap. Clyt. " du genre Clytus " " "
Lec. LeConte in Annals of the Lyceum. Vol. 4.
Lin. Fn. Suec. Linnæus Fauna Suecica.
Lin. S. N. " Systema Naturæ, ed. xii.
Mels. Melsheimer, in the Proceedings of the Academy of Nat. Sciences.
N. E. Farmer. New England Farmer.
Nm. Ent. Mag. Newman. The Entomological Magazine.
Ol. Ins. Olivier, Entomologie. Coléoptères.
P. Ac. The Proceedings of the Academy of Nat. Sciences.
Putz. Cliv. Putzeys' Monographie des Clivina, in Memoires de la Société Royale des Sciences de Liége.
Say Exp. Say, in Appendix to Long's Expedition to the St. Peters' River.
Sch. Syn. Schönherr, Synonymia Insectorum.
Sch. Schönherr, Genera et species Curculionidum.
St. Ins. Germ. Sturm's Deutschland's Fauna, Insecten.
Web. Obs. Weber, Observationes Entomologicæ.

CATALOGUE OF INSECTS.

CICINDELA *Lin.*
purpurea *Oliv. Ent.* 2, 83, pl. 14.
marginalis Fabr. El. 1, 240.
longilabris *Say. Exp.* 2, 268.
albilabris Kirby. N. Z. 12.
repanda *Dej.* 1, 74.
hirticollis Say. J. Ac. 1, 20.
hirticollis *Say. Am. Tr.* 1, 411.
albohirta Dej. 2, 425.
12-guttata *Dej.* 1, 73.
Proteus Kirby. N. Z. 9.
vulgaris *Say. Am. Tr.* 1, 409.
obliquata Dej. 1, 72.

CASNONIA *Latr.*
pennsylvanica *Dej.* 1, 172.
LEBIA *Latr.*
divisa.
concinna ‖ *Lec. An. Lyc.* 4, 192.
tricolor *Say. Am. Tr.* 2, 11.
pleuritica *Lec.* 193.
furcata *Lec.* 193.
fuscata *Dej.* 1, 270.
[1] moesta.
viridis *Say. Am. Tr.* 2, 14.
pumila *Dej.* 5, 388.
CYMINDIS *Latr.*
[2] reflexa.

[1] L. moesta.—Nigro-subænea, nitida, thorace capite parum latiore, transverso, antice rotundato, impressione transversa anteriore profunda; anguste marginato, angulis posticis rectis elevatis; elytris tenuissime striatis, striis punctatis, interstitiis planissimis, 3io tripunctato: antennis nigris, concoloribus. Long. .16 unc. Found at Michipicotin on Solidago. Resembles L. viridis (Say) but easily distinguished, apart from color, by the narrower and longer head, and distinctly punctured striæ of the elytra.

[2] C. reflexa.—Piceo-brunnea, pilosa, capite thoraceque grosse confertim punctatis, hoc latitudine breviore, postice angustato, angulis posticis obtusis non rotundatis, margine lato valde reflexo, elytris apice oblique sinuato-truncatis, striato- punctatis, interstitiis planis, disperse punctatis, 3io punctis 3 majusculis; antennis, palpis, pedibusque

DROMIUS *Bon.*
piceus *Dej.* 5, 353.
Cymindis picea Lec. 189.
LIONYCHUS *Schmidt-Goëbel.*
subsulcatus.
Dromius subs. Dej. 2, 451.
latens.
Dromius latens, Lec. 191.
americanus.
Dromius Amer. Dej. 5, 361.
PSYDRUS *Lec.*
piceus *Lec.* 153.
[3] HAPLOCHILE *Lec.*
pygmæa *Lec.* 209.
Morio pygm. Dej. 5, 512.

CLIVINA *Bon.*
americana *Dej.* 5, 503.
DYSCHIRIUS *Bon.*
sphæricollis *Putseys Cliv.* 17.
[4] apicalis.
[5] æneolus.
globulosus *Putz. Cliv.* 20.
[6] parvus.
[7] longulus.
CALATHUS.
gregarius *Dej.* 3, 76.
PRISTODACTYLA *Dej.*
advena *Lec.* 217.

rufo-testaceis. Long. ·4 unc. In sandy places. This species approaches very near to the Rocky Mountain one, which I have considered as cribricollis (Dej.), but the head and thorax are still more coarsely and densely punctured, and the latter more narrowed behind; the elytra are obliquely truncate, in some specimens they are rufous at base, but have no distinct humeral spot, the interstices are flatter, with smaller and more numerous punctures.

[3] By an error of spelling, I formerly wrote Aplochile.

[4] D. apicalis.—Subelongatus, nigro-æneus nitidus, clypeo bidentato, fronte angulatim leviter impressa, thorace ovali, latitudine fere longiore, antice vix angustato, elytris thorace parum latioribus, lateribus vix rotundatis, stria marginali ad humerum abbreviata, tenuiter striatis, striis ante medium punctatis, 2nda 7ma 8va que ad apicem exaratis, interstitiis planis 3io tripunctato, antennarum basi palpisque piceis, vel rufo-piceis. Long. ·12 unc. The anterior tibiæ have the outer spine scarcely longer than the inner, and but slightly curved, on the outer edge is a distinct tooth, and above it two other very obsolete denticles.

[5] D. æneolus.—Æneus, elytris nitidissimis, clypeo valde bidentato, fronte transversim profunde impressa, thorace subgloboso, antice non angustato, lateribus antice leviter rotundatis; elytris fere parallelis, apice rotundatis, striato-punctatis, punctis pone medium externeque obliteratis, stria sutur aliapice distincta, duabusque aliis (exteriore longiore) brevibus exaratis, marginali ad humerum desinente, interstitio 3io tripunctato. Long. ·15 unc. Two specimens. The terminal spines of the anterior tibiæ subequal, scarcely curved; the outer edge with two denticles, the superior scarcely visible.

[6] D. parvus.—This species is only half the size of D. globulosus, but like it has a transverse thorax, narrowed in front. The clypeus is less deeply emarginate, the frontal sulcus not so deep, the elytral striæ and points deeper: the internal terminal spine of anterior tibiæ only one half the length of the outer one; the external margin has but one denticle. Long. ·09.

[7] D. longulus.—This differs from D. globulosus, in having the thorax subglobose, (the length being equal to the breadth,) not narrowed in front; the elytra are more elongate, the striæ are deeper, and can be traced to the apex, although the points vanish at the middle. The 3rd interstice is 3-punctate, the 8th stria profound at apex; antennæ fuscous at apex; internal spine of anterior tibiæ 3-4 as long as the outer one, on the outer margin, the lower denticle acute, the upper one obsolete. Long. ·11.

[8] PLATYNUS *Bon. Brullé*, 1835.
Agonum Bon. Kirby, 1837.
Anchomenus Bon. Er. 1837.
decens.
Feronia decentis Say. Am. Tr. 2, 53.
Anchom. decens Lec. 221.
Anchom. gagates Dej. 3, 107.
depressus. [*Lec.* 221.
Anch. dep. Hd. P. Ac. 1, 299:
marginatus.
Anch. marg. Lec. 221.
angusticollis.
Anch. angus. Lec. 222.
extensicollis.
Feronia extens. Say. Am. Tr. 2, 54.
Anch. extens. Dej. 3, 113.
decorus.
Feronia dec. Say. ib. 2, 53.
Anch. dec. Dej. 3, 115.
subcordatus.
Agonum erythropus Kb. N. Z. 28.
cupripennis.
Feronia cup. Say. Tr. 2, 50.
nitidulus.
Agonum nit. Dej. 3, 143.
chalceus.
Agonum ch. Lec. 224.
cupreus.
Agonum cup. Dej. 5, 735.
[9] atratus.
carbo.
anchomenoides.
Agonum anch. Randall, B. J. 2, 2.
placidus.
Feronia placida Say. Am. Tr. 2, 43.
Ag. luctuosum Dej. 3, 172.
lenis.
Agonum lenum Dej.
picipenne Kb. N. Z. 24.
sordens.
Agonum sord. Kb. N. Z. 25.
[10] ruficornis.
retractus.
Ag. retractum Lec. 228.
nigriceps.
Agonum nig. Lec. 229.

[8] Erichson calls this group Anchomenus, and adds as a reason that Platyna (Wiedeman 1825) is a genus of Diptera. Before that time the three Bonellian genera were considered distinct, and therefore the name was not vacant; Brullé having been the first to unite these genera, had an unquestionable right to select either of the three names for the group. Moreover the name Platynus is suitable for the great majority of the species, and the day has long gone by in science, when a generic name may be changed because its meaning does not accord with the characters of all the species denoted by it.

[9] P. atratus.—Niger nitidus, thorace rotundato, latitudine vix breviore, antice subangustato, basi utrinque late foveato, margine depresso, versus basin anguste reflexo, angulis posticis nullis; impress. basalibus brevibus distinctis; impress. transv. posteriore distincta; elytris thorace latioribus, profunde striatis, interstitio 3io 3-punctato. Long. ·34. Very much like P. melanarius (Ag. melan. Dej.) but distinguished by the smooth basal foveæ and less reflexed margin. The elytral striæ are smooth in one specimen, obsoletely punctured in the other.

P. carbo.—Niger, nitidus, thorace rotundato, latitudine paulo breviore, basi vix rotundato, angulis posticis valde obtusis, rotundatis, basi utrinque late foveato, margine depresso versus basin angustissime reflexo; imp. trans. posteriore profunda, basalibus minutis in foveis sitis; elytris thorace latioribus, tenue striatis, interstitiis planis, 3io 3-punctato. Long. ·35. One specimen. Very like P. (Ag. Dej.), with the basal foveæ deeper and more defined, the reflexed margin narrower and the margin itself thickened. The base of antennæ and palpi have no tendency to become ferruginous.

[10] P. ruficornis.—Elongatus, nigro-piceus nitidus, thorace fere plano, latitudine longiore, postice subangustato, basi cum angulis rotundato, margine versus basin anguste acuteque reflexo, non incrassato, impress. basalibus fere nullis: elytris ellipticis tenue

punctiformis.
Feronia punc. Say. Am. Tr. 2, 58.
Agonum rufipes Dej. 3, 173.
bembidioides.
Sericoda bemb. Kb. N. Z. 15.
Agonum bemb. Lec. 227.
4-punctatus.
St. Fm. Germ. Dej. 3. 170.
POECILUS *Bon.*
lucublandus *Dej.* 3, 212.
Feronia lucub. Say. Am. Tr. 2, 55.
chalcites *Lec.* 231.
Feronia chalc. Say. Am. Tr. 2, 56.
convexicollis *Lec.* 233.
Feronia conv. Say. l. l.
[11] PTEROSTICHUS *Bon. Erichs.*
erythropus.
Feronia ery. Dej. 3, 243.
Platyderus nitidus Kb. N. Z. 29.
Platyderus eryth. Lec. 231.
mandibularis.
Argutor mand. Kb. N. Z. 31.
patruelis.
Feronia patr. Dej. 5, 759.
Argutor patr. Lec. 337.
mutus.
Feronia muta Say. Am. Tr. 2, 44
Adelosia muta Lec. 335.
Luczotii.
Feronia Lucz. Dej. 3, 321.
Fer. oblongonotata Say. Am. Tr. 4, 425.
Adelosia oblong. Lec. 335.
[12] orinomum.
Omaseus orin. Cs. Kb. N. Z. 32.
punctatissimus *Rand. B. J.* 2, 3.
coracinus.
Feronia corac. Nm.
stygicus.
Feronia styg. Say. Am. Tr. 2, 41.

striatis, interstitiis planis, 3io 5-punctato, epipleuris palpis antennisque piceis, his apice rufis, pedibus rufo-testaceis. Long. ·31.

Varies with the 3rd elytral interstice 3-punctate. Twice the size of P. lenis, and distinguished by the thorax narrowed behind, basal impressions indistinct, the reflexed margin broader. P. retractus is much smaller, with a wider thorax and deeper basal impressions.

[11] Under this name, following the example of Erichson, I have grouped all the American species of Dejean's Feronia, excepting the Poecilus, which are sufficiently distinct by the antennæ. In my catalogue of the Carabica, I admitted as distinct genera nearly all the groups proposed by other authors, and attempted to find natural characters for them. What success I have had in finding structural differences, the reader may be able to judge by referring to the work cited: suffice it to say, that the characters therein detailed are entirely too finely drawn for any practical purpose, and by the progressive variation which accompanies the variations of form and sculpture, plainly indicate the existence of one extensive and natural genus: and fortified as I am by the example of Erichson, and the counsel of Zimmerman, I hesitate no longer to merge them into one group, under the name quoted above. An attempt has been made to separate under the name Hypherpes (Chaudoir) all the species without elytral punctures. But the characters of this group will be found as ill-defined as those which have just been suppressed. Feronia lachrymosa (Nm.) can scarcely be told from adoxa but by the superior size, and the presence of elytral punctures; surely it would be the destruction of all natural classification, to separate into different genera, two such closely allied species.

[12] I have had no opportunity of comparing with European specimens, and give the species as identical on the authority of Kirby and Klug, having in my cabinet an Oregon specimen, which has been actually examined by the latter gentleman. Dr. Zimmerman thinks it to be different, and proposes the name *septentrionalis*, which must therefore be adopted if the species prove distinct.

corvinus.
Feronia corv. Dej. 3, 281.
caudicalis.
Feronia caud. Say. Am. Tr. 2, 56.
sodalis.
Feronia sod. Lec. 349.
[13] tenuis.
adoxus.
Feronia adoxa Say. Am. Tr. 2, 46.
Feronia tristis Dej. 3, 324.
fastiditus.
Feronia fastid. Dej. 3, 323.
MYAS *Zieg.*
foveatus *Lec.* 355.
ISOPLEURUS *Kb.*
hyperboreus *Lec.* 357.
Amara hyper. Dej. 5, 800.
septentrionalis *Lec.* 358.
TRIÆNA *Lec.*
angustata *Lec.* 365.
Feronia ang. Say. Am. Tr. 2, 36.
Amara ang. Say. ib. 4.
indistincta *Lec.* 365.
Amara indis. Hd. P. Ac. 1, 300.
depressa *Lec.* 365.
[14] AMARA *Latr.*
inequalis, *Kb. N. Z.* 39.
splendida *Hd. P. Ac.* 1, 300.
gibba.
Celia gibba Lec. 360.
impuncticollis, *Say. Am. Tr.* 4, 428.
Feronia imp. Say. ib. 2, 36.
fallax *Lec.* 362.
convexa *Lec.* 363.
avida.
Zabrus avidus Say. J. Ac. 3, 148.
Pelor avi. Say. Am. Tr. 4, 428.
Bradytus av. Lec. 367.
Amara confinis Dej. 3, 512.
PERCOSIA *Zim.*
obesa *Hd. P. Ac.* 1, 297.
Feronia obesa Say. Am. Tr. 2, 37.
Amara obesa Say. ib. 4, 428.
CURTONOTUS *Steph.* 1828.
Leirus Zim. 1832.
[15] convexiusculus *Steph. Kb. N. Z.* 35.
[16] elongatus.

[13] P. tenuis.—Elongatus, niger nitidus, thorace capite vix latiore, latitudine parum breviore, quadrato, postice leviter angustato, lateribus pone medium sinuatis, angulis posticis rectis prominulis, basi utrinque profunde impresso, bistriato, punctatoque: elytris tenue striato-punctatis, interstitio 3io 3-punctato; palpis pedibusque rufo-piceis. Long. ·32, lat. ·14. Readily known by its narrow form: the head is constricted and punctured behind the eyes: the elytral striæ are fainter towards the apex, which is not at all sinuate.

[14] I have merged into Amara the group *Celia* (Zim.), as it differs from the typical species neither in habitus nor characters, the sole ground for separation being a sexual character of slight import. I have also replaced in the genus, Zabrus avidus (Say) as it has not the characters of Bradytus, (to which I formerly referred it), the tibiæ being alike in both sexes.

[15] I have a specimen which agrees perfectly with Dejean's description, but the thorax is more narrowed behind than in the figure (Icon. Col. Eur. 3, pl. 170, fig. 2.) No opportunity for direct comparison has yet occurred. The species is totally distinct from the two described by me in the 4th vol. of the Annals of the Lyceum.

[16] C. elongatus.—Elongatus, gracilis, rufo-piceus nitidus, thorace quadrato, latitudine non breviore, antice subangustato, lateribus rotundato, angulis posticis subrectis, non rotundatis, basi utrinque bistriato leviterque punctato, elytris thorace latioribus, tenuiter striatis, striis ad basin leviter punctatis. Long. ·4. ♂ with the intermediate tibiæ strongly bidentate, the mentum tooth narrowed in front and deeply impressed.

ACRODON *Zim.*
[17] subænea.
AGONODERUS *Dej.*
pallipes *Dej.* 4, 53.
ANISODACTYLUS *Dej.*
nigerrimus.
Harpalus nig. Dej. 5, 842.
Harp. laticollis, Kb. N. Z. 43.
baltimorensis *Dej.* 4, 152.
EURYTRICHUS *Lec.*
terminatus *Lec.* 387.
Feronia term. Say. Am. Tr. 2, 48.
Harpalus term. Dej. 4, 355.
HARPALUS.
bicolor *Say. Am. Tr.* 2, 26.
erythropus *Dej.* 4, 258.
pleuriticus *Kb. N. Z.* 41.
proximus *Lec.* 398.
herbivagus *Say. Am. Tr.* 2, 29.
megacephalus *Lec.* 397.
[18] laticeps.
rufimanus *Lec.* 402.
varicornis *Lec.* 401.
GEOBÆNUS *Dej. Lec.*
quadricollis *Lec.* 405.
tibialis *Lec.* 405.
Trechus tib. Kb. N. Z. 46.
lugubris *Lec.* 405.
cordicollis *Lec.* 406.
rupestris *Lec.* 406.
Trechus rup. Say.
Trechus flavipes Kb. N. Z. 47.
Acupalpus elongatulus Dej. 4, 457.
STENOLOPHUS *Dej.*
ochropezus *Dej.* 4, 424.
fuliginosus *Dej.* 4, 423.
versicolor Kb. N. Z. 46.
carbonarius *Lec.* 409.
Harpalus carbonarius Dej. 4, 398.
misellus *Lec.* 410.
Acupalpus mis. Dej. 4.
CHLÆNIUS *Bon.*
chlorophanus *Dej.* 5, 662.
sericeus *Say. Am. Tr.* 2, 61.
impunctifrons *Say. ib.* 2, 64.
emarginatus‡ *Kirby N. Z.* 23.
nemoralis *Dej.*
tomentosus *Dej.* 3, 357: *Lec.* 438.
LORICERA *Latr.*
pilicornis *Gyll. F. Suec.* 2, 45: *Dej.* 2, 293.
CYCHRUS *Fabr.*
[19] bilobus *Say.*
SPHÆRODERUS *Dej.*
Brevorti *Lec.* 443.
Lecontei *Dej.* 2, 15: *Lec.* 442.

[17] A. subænea.—This species differs from the smaller and dark colored specimens of A. rubrica (Hd) in being narrower, and more convex. The thorax is scarcely wider than ong, and not nearly so much narrowed in front; the two basal impressions on each side are deeper, the elytral striæ are deeper and more punctured; the color above is dark-piceous, slightly bronzed, antennæ and feet testaceous. Long. ·27.

[18] H. laticeps.—Niger nitidus, palpis solum rufo-piceis, capite magno obtuso, thorace latitudine sesqui breviore, lateribus parum rotundato, basi truncato, angulis posticis subrectis, margine versus basin modice explanato, cum basi obsolete punctato, impressionibus basalibus linearibus, brevibus, linea longitudinali distincta: elytris thorace non latioribus lateribus subrotundatis, tenuiter profunde striatis, interstitiis parum convexis, tibiis posticis et intermediis valde spinulosis. Long. ·8–·5. ♂ Elytris nitidis; ♀ opacis. Like H. *rufimanus*, but three times larger.

[19] C. bilobus.—Purpureo-niger nitidus, thorace subtransverso, postice valde angustato, canaliculato, basi impresso punctatoque; elytris elongato-ovalibus, pone basin subampliatis, apice attenuatis, profunde crenato-striatis æneo-violaceis, antennarum apice palpisque piceis. Long. ·5.

St. Ignace; ♂ has the anterior tarsi scarcely dilated.

CARABUS *Lin.*
serratus *Say. Am. Tr.* 2, 77.
lineatopunctatus Dej.
sylvosus *Say. ib.* 2, 75.
[20] Agassii.
CALOSOMA *Fabr.*
calidum *Fabr.* 1, 211.
frigidum *Kb. N. Z.* 19.
NEBRIA *Latr.*
pallipes *Say. Am. Tr.* 2, 78.
[21] moesta.
[22] suturalis.
OMOPHRON *Latr.*
americanum *Dej.* 5, 583.
Sayi Kb. N. Z. 65.
tesselatum *Say, J. Ac.* 3, 152.
Lecontei Dej. 5, 582.
ELAPHRUS *Fabr.*
[23] politus.

[20] C. Agassii.—Niger, thorace valde rugoso, latitudine paulo breviore, quadrato, postice leviter angustato, margine versus basin anguste reflexo, angulis basalibus retrorsum productis, elytris thorace sesqui latioribus ellipticis, dense seriatim punctatis foveisque parum distinctis 3-plici serie impressis. Long. ·88.

Kakàbeka—Dr. Stout. At first sight seems to be a faded specimen of C. sylvosus (Say), but the thorax is very rugous, the sides more narrowly reflexed, and the basal angles much more produced. The sculpture of the elytra is similar, but more distinct. It is more closely allied to C. tædatus (Fabr.), from Oregon, but the head is less impressed, and the elytra less deeply foveate, with the sides regularly but slightly rounded, not straight and narrowed anteriorly as in C. tædatus. Anything that I can say in praise of the philosopher and gentleman after whom it is named would be quite superfluous.

[21] N. moesta.—Depressiuscula nigra nitida, thorace latitudine duplo fere breviore lateribus marginato, valde rotundato, postice valde angustato, constrictoque, angulis posticis rectis, antice posticeque transversim profunde impresso, punctatoque, impress. basalibus profundis: elytris subparallelis thorace latioribus striis leviter punctatis, 8va fere obliterata, interstitio 3^{io} 5-punctato : antennarum apice tibiis tarsisque rufopiceis. Long. ·41.

This may be Kirby's Helobia castanipes (which I incorrectly cited as N. pallipes Say), as Dr. Schaum writes me it is very like N. Gyllenhalii, to which our insect has the closest resemblance. My specimens have not the striæ between the eyes mentioned by Kirby, nor are the feet castaneous: the margin of the thorax is sometimes obsoletely punctured.

[22] N. suturalis.—Elongata depressa, nigra, thorace latitudine fere duplo breviore, lateribus marginato, margine postice latiore, rotundatoque, basi angustato non constricto, angulis posticis obtusis, basi truncato, cum margine obsolete punctato, antice posticeque profunde transversim impresso, elytris elongatis thorace latioribus obscure rufis, sutura nigricante, striis leviter punctatis interstitiis fere planis, 3^{io} 5-punctato, antennis tibiis tarsisque rufo-piceis vel rufis. Long. ·44.

The 8th stria is less deep than the others, but not obliterated; the punctures in the marginal series are more numerous than in the preceding. Found on the islands at the mouth of Black Bay.

[23] E. politus.—Obscure æneus, politus, capite sparsim punctato, vertice foveato, occipite profunde impresso; thorace capite non latiore antice angulatim valde impresso, dein canaliculato, disco utrinque profunde foveato, ad latera apicem basinque sparsim punctato ; elytris sparsim punctulatis foveis ocellatis purpureis 4-plici serie impressis, pedibus rufo-æneis ; ante-pectore punctato. Long. ·34. One specimen : Maple Island. Dr. Stout.

[24] punctatissimus.
[25] sinuatus.
ruscarius *Say. Am. Tr.* 4, 417.
BLETHISA *Bon.*
quadricollis *Hd. Pr. Ac.* 3. 149.
NOTIOPHILUS *Dumeril.*
[26] punctatus.
porrectus. *Say. Am. Tr.* 2. 4, 418.
PATROBUS *Meg.*
longicornis *Say. Am. Tr.* 4, 421.
Feronia long. Say. ibid, 2, 40.
Patrobus americanus Dej. 3, 34.
EPAPHIUS *Leach.*
micans *Lec.* 414.
fulvus *Lec.* 415.
BEMBIDIUM *Latr.*
sigillare *Say. Am. Tr.* 4, 437.
stigmaticum Dej. 5, 83.
impressum *Gyl. Dej.* 5, 81.
paludosum *St. Ins. Germ.* 6, 179. *Dej.* 5, 79.
lacustre *Lec.* 451.
ODONTIUM *Lec.*
coxendix *Lec.* 452.
Bembidium cox. Say. J. Ac. 3, 151.
nitidulum *Lec.* 452.
Bembidium nit. Dej. 5, 84.
Bemb. coxendix Say. Am. Tr. 4, 436.
OCHTHEDROMUS *Zim. Lec.*
americanus *Lec.* 453.
Bemb. americanum Dej. 5, 84.
salebratus *Lec.* 453.
dilatatus *Lec.* 455.
antiquus *Lec.* 455.
Bemb. antiquum Dej. 5, 88.
planatus *Lec.* 456.

[24] E. punctatissimus.—Læte viridi-æneus, supra et subtus confertissime subtiliter punctatus; thorace subtransverso, capite non angustiore, antice profunde impresso, dein canaliculato, disco utrinque foveato; elytris latitudine sesqui longioribus pone basin leviter sinuatis, foveis ocellatis purpureis 4-plici serie impressis, spatiisque lævigatis 2-plici serie notatis: pectore medio lævi, tibiis femorumque basi testaceis. Long. ·27, lat. ·13.

Sault; common. Punctuation much finer and more dense than in E. ruscarius (Say). The anterior lævigated space is quadrate and extends to the suture: the sides of the abdomen are so finely punctured as to appear granulate.

[25] E. sinuatus.—Læte viridi-æneus, supra et subtus confertissime subtiliter punctatus, thorace latitudine fere longiore, capite parum angustiore, antice profunde transversim impresso, canaliculato, disco utrinque foveato; elytris latitudine duplo longioribus, pone basin profundius sinuatis, dein vix conspicue ampliatis; foveis spatiisque lævigatis sicut in præcedente: pectore medio lævi, tibiis femorumque basi ferrugineis. Long. ·31, lat. ·13. Pic; two specimens. Narrower than the preceding, the punctures of the side of the abdomen are more distinct; but still the pectora are more closely punctured than in E. ruscarius.

[26] N. punctatus.—Nigro-æneus, nitidus capite 7-striato striis externis latis, thorace transverso, postice angustato, angulis posticis rectis, punctato, disco utrinque lævi, basi utrinque foveato: elytris ante medium 1-foveatis, stria scutellari unica notatis, suturali, externisque 8 minus approximatis dense punctatis, stria 7ma mox pone humerum, alteris versus apicem levioribus, 6ta solum integra; tibiis antennarumque art. 2ndo 3io 4to que rufescentibus. Long. ·2.

Size of N. porrectus; but the striæ are more densely punctured, less obliterated, and the feet and antennæ black. It resembles much N. confusus Lec. (semistriatus Say, teste Harris), but the 1st stria is not curved and exarate at tip, the base of the antennæ less decidedly pale, and the scutellar stria is not double.

[30] planipennis.
longulus *Lec.* 456.
patruelis *Lec.* 459.
Bemb. patr. Dej. 5, 69.
variegatus *Lec.* 459.
Bemb. var. Say. Am. Tr. 2, 89.
timidus *Lec.* 460.
versicolor *Lec.* 462.
Notaphus variegatus Kb. N. Z. 58.
affinis *Lec.* 462.
Bemb. affine Say Am. Tr. 2, 86.
Bemb. fallax Dej. 5, 189.
Bemb. decipiens Dej. 5, 159.
4-maculatus *Lec.* 462.
Bemb. oppositum Say. Am. Tr. 2, 86.
[31] axillaris.
frontalis *Lec.* 462.
sulcatus *Lec.* 463.
trepidus *Lec.* 463.
gelidus *Lec.* 464.
nitens.
picipes *Lec.* 465.
Peryphus picipes Kb. N. Z. 54.
tetracolus *Lec.* 465.
B. tetracolum Say.
Peryphus rupicola Kb. N. Z. 53.
substrictus *Lec.* 465.
lucidus *Lec.* 466.
transversalis *Lec.* 466.
Bemb. trans. Dej. 5, 110.
planus *Lec.* 467.
Peryphus planus Hd. P. Ac. 1, 303.
niger *Lec.* 467.
Bemb. nigrum Say Am. Tr. 2, 85.
nitidus *Lec.* 468.
Eudromus nitidus Kb. N. Z. 55.
TACHYS *Knoch.*
xanthopus *Lec.* 469.
Bemb. xanthopus Dej. 5, 60.
incurvus *Lec.* 469.
Bemb. inc. Say Am. Tr. 4, 480.
inornatus *Lec.* 470.
Bemb. inorn. Say ib. 2, 88..
Tachyta picipes Kb. N. Z. 56.
lævus *Lec.* 472.
B. lævum Say. Am. Tr. 2, 87.

[30] O. planipennis.—Depressus, niger pernitidus, thorace quadrato, postice vix angustato, angulis posticis obtusis non rotundatis, impressione posteriore profunda, basi utrinque parum impressa, elytris purpureis, cyaneo-micantibus, profunde striatis, striis antice subpunctatis, punctisque 2 impressis: antennarum basi pedibusque rufis. Long. ·19.

Kaministiquia River below Kakàbeka Falls. This species is very similar to O. purpurascens *Lec.*, but the basal impression of the thorax is single, and less profound; the striæ of the elytra are less punctured; the 8th and 9th striæ are obliterated.

[31] H. longulus.—Elongato-ovalis, rufus, capite thoraceque punctatis, hoc striola utrinque basali, elytris apice oblique subtruncatis, sutura vix acuminata, punctato-striatis, interstitiis uniseriatim sparse punctulatis, maculis utrinque 5 vix conspicue infuscatis. Long. ·11.

Narrower than the others; outline regularly oval: tip of elytra more obliquely sinuate than in H. americanus, but scarcely truncate. The points of the thorax are more distant immediately behind the middle of the disc; the thorax is slightly infuscated at the apex. Varies without any elytral spots.

[31] O. axillaris.—Nigro-æneus, pernitidus, thorace convexo, valde cordato, antice vix impresso, basi utrinque 1-foveato, elytris subtiliter seriatim punctatis, punctis pone medium obliteratis, macula magna axillari, tibiis tarsisque albidis. Long. ·13.

Sault. Very much like O. 4-maculatus, but a little larger; the antennæ, palpi and femora are black, and the punctures of the elytra very small.

HALIPLUS *Latr.*
americanus *Aubé*, 21.
[32] borealis.
longulus.
[33] nitens.
[33] cribrarius.
DYTISCUS *Linné.*
confluentus (confluens) *Say. Am. Tr.* 4, 440.
Ooligbukii Kb. N. Z. 75.
[34] Cordieri *Aubé*, 108.
Harrisii *Kb. N. Z.* 76.
[35] diffinis.
[36] fasciventris *Say. Exp.* 2, 270.
carolinus Aubé, 120.
verticalis *Say. Am. Tr.* 2, 92.
ACILIUS *Leach.*
fraternus *Harris. N. E. Farmer.*
semisulcatus Aubé.
HYDATICUS *Leach.*
liberus.
Dytiscus liberus Say. J. Ac. 5, 160.
H. brunnipennis Aubé, 203.
nigricollis Kb. N. Z. 73.

[32] H. borealis.—Ovalis, rufo-testaceus nitidus, thorace punctato, elytris apice oblique truncatis, sutura acuminata, valde punctato-striata, interstitiis sparsim uniseriatim punctulatis; basi anguste, sutura, apice maculisque utrinque 5 nigris. Long. ·12.

One half larger than H. americanus, and easily known by the want of the basal striola of the thorax; the base of the elytra is blackened along the edge: the spots placed as in H. americanus.

[33] H. nitens.—Ovalis convexus, pallidus pernitidus, capite postice, thorace antice nigro-maculatis, hoc densius punctato (grossius ad basin) ante basin transversim leviter impresso, lævigatoque, elytris valde punctato-striatis, interstitiis uniseriatim punctatis, sutura angustissime, apice, guttisque utrinque 6 minutis nigris. Long. ·15.

Head finely punctured, with a smooth vertical space. Elytra slightly, but not suddenly dilated behind the thorax, then regularly narrowed to the tip, which is obliquely truncate and acuminate: the disc is marked with two spots at the anterior third placed obliquely forward and outwards, just behind the middle 2 or 3 nearly transversely, and 2 or 3 more obliquely backwards and outwards at the posterior fourth. Varies, with the posterior spots wanting. St. Ignace.

[33] H. cribrarius.—Ovalis convexus, pallide testaceus, capite postice, thorace antice nigro-maculatis, hoc apice bisinuato, densius punctato, basi grosse sparse punctato, punctis transversim sub-biseriatim digestis, elytris grosse punctato-striatis, interstitiis uniseriatim punctatis, sutura angustissime, apice guttisque 6 vel 7 parvis nigris. Long. ·17.

Very similar to the preceding, but the points above and beneath are larger. The elytra are less attenuated behind the dilated part, the sides being nearly parallel.

[34] I found the elytra of a ♀, and have seen perfect specimens from Lake Huron. It is smaller than D. Harrisii, the oblique yellow band at the tip of the elytra is very distinct, the sulci terminate at ⅛ from the apex, and are not confluent. In the latter species the ♀ has smooth elytra.

[35] D. diffinis. — Elongato-ellipticus antice vix angustatus, supra nigro-olivaceus nitidus, labro clypeo capitis macula angulata, thoracis margine toto, elytrorum lateribus, corporeque subtus toto testaceis, abdomine utrinque vix infuscato; lobis metasterni postice divergentibus, apice acute rotundatis. Long. 1·15, lat. ·61.

♂ elytris 3-seriatim punctatis, punctis pone medium paucis dispersis.

Eagle Harbor. Mr. Rathvon. Form of confluens, but only one half the size. The sides of the thorax scarcely rounded, the posterior yellow margin scarcely wider in the middle than at the angles.

[36] Varies with the posterior margin of the thorax, narrowly testaceous.

[37] fascicollis *Harris N. E. Farmer?*
zonatus ^teste^ *Aubé*, 214.
COLYMBETES *Clairville.*
sculptilis *Harris l. c.*
triseriatus Kb. N. Z. 73.
[38] binotatus *Harr. l. c.?*
maculicollis Aubé, 245.
agilis *Aubé*, 254.
ILYBIUS *Er.*
[39] pleuriticus.
picipes *Kb. N. Z.* 71.
AGABUS *Er.*
[40] angustus.
erythropterus *Aube*, 305.
Colymbetes ery. Say. Am. Tr. 2, 95.
striatus *Aubé*, 305.
Colymb. stri. Say. Am. Tr. 2, 97.
Agabus arctus Mels. P. Ac. 2, 27.
[41] parallelus.
obtusus.
Colym. obt. Say. Am. Tr. 2, 99.
nitidus? Say. 2, 98.
Agabus gagates Aubé, 306.
stagninus.
Col. stagn. Say. Am. Tr. 2, 100.
Agabus striola Aubé, 308.

[37] This species is more narrowed anteriorly than its European analogue, and wants the narrow rufous line at the base of the thorax; moreover the ♀ has the external basal portion of the elytra more densely and distinctly punctulate.

[38] I know not whether Dr. Harris' name is published. In case it is, Aubé's C. binotatus (p. 247, a West Indian species) must fall. I have the less hesitation in giving our species as identical with the Mexican C. maculicollis, as I found at the Rocky Mountains numerous specimens, which do not differ from those obtained at the north. Mr. Melly, from actual comparison, also informs me that it is identical.

[39] I. pleuriticus.—Angustior oblongo-ovalis convexus, postice suboblique attenuatus, supra æneus, minute reticulatus opacus, capite in vertice binotato, anticeque rufo, elytris pone basin vix dilatatis subparallelis, pone medium gutta oblonga, alteraque versus apicem pallidis, epipleuris pedibusque piceis, vel rufo-piceis. Long. lat.

Narrower than I. biguttulus, less dilated behind, the sides of the elytra being almost parallel for nearly ⅔ of their length, then gradually attenuated to the apex; the irregular series of points are more distinct behind the middle.

I. picipes: What I consider as this species is much smaller, narrower and less convex than I. biguttulus, the thorax less abbreviate, somewhat rounded on the sides; elytra nearly parallel, and less suddenly attenuated at the tip; the confused rows of points are more distinct. My specimen is immature, and the body is rufo-piceous.

[40] A. angustus.—Depressus, anguste ovalis, postice suboblique attenuatus, niger subopacus, capite subtiliter, thorace elytrisque grossius reticulatis, illo margine anguste depresso, lateribus ante medium rotundatis, angulis posticis acutis subproductis, ore antennis, palporumque basi ferrugineis. Long. lat.

Very distinct from its large size and peculiarly shaped thorax. The rows of impressed points on the elytra are distinct, and the reticulations become finer at the apex and margin.

[41] A. parallelus.—♂ ♀ Elongato-ellipticus depressus, niger nitidus subtilissime reticulato-strigosus, capite antice vix ferrugineo, antennis palpisque ferrugineis. Long. ·38, lat. ·2.

Differs from A. striatus in being more elliptical, the two ends being similarly rounded, and the elytra quite parallel for the greater part of their length; the head is wider and the thorax less narrowed in front. The reticulations are a little more evident than in that species.

tæniatus *Aubé*, 311.
Colymbetes tæn. Harris N. E. F.?
ambiguus.
Colymb. amb. Say. Am. Tr. 2, 96.
Ag. infuscatus Aubé, 330.
punctulatus *Aubé*, 332.
semipunctatus.
Col. semip. Kb. N. Z. 69.
fimbriatus.
Ag. reticulatus || *Aubé*, 335.
tristis *Aubé*, 356.
bifarius.
Colymb. bif. Kb. N. Z. 71.
COPELATUS *Er.*
Chevrolatii *Aubé*, 389.
COPTOTOMUS *Say.*
interrogatus *Aubé*, 393.
Colymb. int. Fabr. 1, 267.
Colymb. venustus S. Am. Tr. 2, 98.
Coptot. serripalpis Say. ib. 4, 443.
LACCOPHILUS *Leach.*
maculosus *Say. Am. Tr.* 2, 100.
americanus *Aubé*, 442.
HYDROPORUS *Clairville.*
punctatus *Aubé*, 471.
Laccoph. punct. Say. Exp. 2, 271.
cuspidatus *Germ; Aubé*, 477.
Hygrotus pustulatus Mels. P. Ac. 2, 29.
[42] sericeus.
[43] consimilis.
affinis *Say. Am. Tr.* 2, 104.
nanus Aubé, 504.
[44] 12-lineatus.
similis *Kirby N. Z.* 68.

[42] H. sericeus.—Ovalis convexiusculus, confertissime punctulatus, densius fulvo-pubescens, rufus: clypeo late marginato, thorace lateribus obliquis rectis cum elytris angulum obtusissimum formantibus, antice posticeque anguste nigricante; elytris atro-brunneis, lineis 4 plus minusve interruptis margineque lato ferrugineis, hoc pone medium bimaculato. Long. ·18.

♂ nitidiusculus, pube minus longa, thorace subtiliter punctato.

♀ opaca, pube longiore, tota subtilissime punctata.

The interrupted lines have not a tendency to coalesce into fasciæ, as in H. pubipennis (Aubé), from which it is easily known by the longer pubescence and finer punctuation; the body is less attenuated behind, and a little more convex. The thorax is much more narrowly margined, and, when viewed sideways, forms a very slight angle with the margin of the elytra.

[43] H. consimilis.—Ovalis convexiusculus, postice modice attenuatus, confertissime punctulatus, breviter dense fulvo-pubescens, ferrugineus, clypeo late marginato, thorace lateribus obliquis rotundatis, cum elytris angulum obtusissimum formante, antice posticeque infuscato; elytris atro-brunneis, margine fasciis 2 irregularibus maculaque apicali ferrugineis. Long. ·18.

♂ capite thoraceque nitidulis, hoc distinctius punctato differt.

Spots as in the last, but confluent into bands; from H. pubipennis distinguished by the rounded and more narrowly margined sides of the thorax.

[44] H. 12-lineatus.—Elongato-ovalis minus convexus, omnium subtilissime alutaceus, sparsimque punctulatus, subtus niger, supra testaceus, vertice nigro bimaculato, thorace lateribus subrotundatis, cum elytris angulum formantibus, postice vix transversim depresso nigro bimaculato, elytris versus apicem oblique attenuatis, sutura lineolis utrinque 6 maculisque 2 sub-marginalibus nigris; antennarum basi palpis pedibusque testaceis. Long. ·17.

♂ elytris apice integro vix obliquo.

♀ elytris apice truncato, fere bidentato.

Les Ecrits. Thorax bisinuate at base, external angles not at all rounded or obtuse;

[45] picatus *Kirby ib* 68.
parallelus *Say. J. Ac.* 3, 153.
Kb. N. Z. 67.
interruptus Say. Am. Tr. 4, 445.
niger *Say. ib.* 2, 102.
modestus Aubé, 577.
[46] tenebrosus.
[47] puberulus.
[48] caliginosus.
[49] tartaricus.
[50] varians.
discicollis *Say. Am. Tr.* 4, 446.

in some specimens, besides the basal spots there is an oblique black line towards the margin. The 3rd and 5th elytral lines alone attain the base; at the tip they are gradually shorter externally, and the 4th, 5th, and 6th are united. Seems allied to H. frater Steph. (Conf. Aubé, 528). Were it not for the obsolete punctures and yellow head, it would be H. lævis, Kirby, N. Z., 68.

[45] I must give Kirby's species as distinct, although Dr. Schaum tells me their European analogues are considered identical, lineellus being a ♀ variety of picipes. I have both ♂ ♀ of our species, each agreeing with its opposite sex in sculpture, and differing only in lustre, the ♂ being shining, the ♀ opaque.

[46] H. tenebrosus.—Elliptico-ovalis, minus convexus, niger subtiliter pubescens, minus dense subtiliter punctatus, capite antice posticeque obsolete ferruginco, thorace valde transverso lateribus obliquis leviter rotundatis obsolete ferrugineis, cum elytris angulis non formantibus, disco obsoletius punctato: pedibus obscure ferrugineis. Long. ·17.

Resembles H. americanus, but is darker colored, and less convex; the punctuation of the thorax is less distinct in the middle, that of the elytra less dense; there are traces of a stria ⅓ way between the suture and margin.

♂ pube breviore indistincta puncturaque sparsiore differt.

[47] H. puberulus.—Elongato-ovalis, minus convexus, niger minus dense punctatus pubescens, thorace lateribus rotundatis cum elytris angulum formantibus, disco minus punctato, elytris parallelis, apice oblique attenuatis; antennis palpis pedibusque rufis. Long. ·12.

Resembles the two next, but is narrower, a little more convex, the posterior angles of the thorax are somewhat obtuse, and the sides form an angle with the elytra.

[48] H. caliginosus.—Ovalis minus convexus niger nitidus, minus subtiliter punctatus, sparseque pubescens, thorace lateribus obliquis vix rotundatis, disco obsoletius punctato; elytris basi vix conspicue angustatis, apice oblique attenuatis: antennis palpis pedibusque rufis. Long. ·14.

More convex than the following, less parallel and more acute behind: the punctures of the elytra are much larger and more distant.

[49] H. tartaricus.—Ovalis fere ellipticus, depressiusculus, niger minus dense subtilius punctatus, sparseque pubescens, thorace lateribus obliquis vix rotundatis, disco obsoletius punctato, basi depressa, elytris parallelis, apice subrotundatim attenuatis, antennis palpis pedibusque rufis. Long. ·14.

♂ nitidus: ♀ subtiliter alutacea, opaca.

[50] H. varians.—Ovalis, modice elongatus minus dense punctatus, vix conspicue pubescens, thorace nigro, punctis in disco sparsioribus, lateribus rectis subobliquis, cum elytris angulum obtusum formantibus; elytris lateribus parum rotundatis, apice vix oblique attenuatis, antennis palpis pedibusque testaceis. Long. ·12.

α Capite elytrisque testaceis, his margine, maculaque communi pone medium piceis.

β Capite rufo, elytris nigro-piceis, versus basin indeterminate piceis.

[51] luridipennis.	GYRINUS *Lin.*
[52] notabilis.	affinis *Aubé*, 669.
[53] conoideus.	patruelis.
[54] ovoideus.	conformis *Dej.*
[55] suturalis.	*ventralis Aubé*, 672.
[56] dispar.	ventralis *Kb. N. Z.* 80.

Every intermediate variety occurs: *a* is more common on the south, *β* on the north of the lake.

[51] H. luridipennis.—Elliptico-ovalis, subdepressus, niger dense subtiliter punctatus breviterque pubescens, capite antice posticeque ferrugineo, thorace lateribus obliquis, rectis, anguste ferrugineis, disco sparsius punctulato, elytris apice vix oblique attenuatis fulvis; antennis palpis, pedibusque rufis. Long. ·17. Eagle Harbor.

[52] H. notabilis.—Elongato-ovalis, antice obtusus, postice oblique attenuatus, nigropiceus pubescens, capite punctulato, antice posticeque testaceo, thorace dense punctulato, obsoletius in disco, basi obsolete depressa, lateribus valde obliquis rotundatis, elytris elongatis, confertissime subtiliter punctatis, piceis, margine pallidiore, antennis tenuibus, cum palpis pedibusque rufis. Long. ·21. One specimen, Black Bay.

[53] H. conoideus.—Elongato-obconicus, nitidus, capite rufo, thorace nigro, lateribus rufis obliquis leviter rotundatis, basi utrinque obliquo, non sinuato, obsoletius punctulato, ad latera parce punctato, lineaque punctorum ad apicem; elytris parce punctatis, rufo-testaceis; antennis minus tenuibus cum palpis pedibusque testaceis. Long. ·2.

♂ antennis articulis 3—6 dilatatis, compressis. One specimen. Eagle Harbor.

[54] H. ovoideus.—Convexus, utrinque modice attenuatus, subtus nigro-piceus, supra ochraceus, capite infuscato, macula verticali pallida, thorace brevi lateribus obliquis vix rotundatis, cum elytris angulum obtusum formantibus, basi infuscato, sparsim subtiliter punctulato, punctis majoribus ad basin et latera interjectis, aliisque densioribus ad apicem transversim ordinatis; elytris minus sparsim punctatis, stria suturali vix impressa, sutura antice lævigata: antennis palpis pedibusque ferrugineis. Long. ·13.

♂ femina paulo nitidior. Eagle Harbor.

[55] H. suturalis.—Ovalis modice convexus, postice leviter attenuatus subtus niger, undique densius minus subtiliter punctatus, capite testaceo ad oculos infuscato thorace, lateribus obliquis parum rotundatis cum elytris vix angulatis, testaceo basi apiceque anguste, medioque triangulariter nigro, punctis ad basin et apicem densioribus, transversim ordinatis; elytris lateribus vix rotundatis ad apicem suboblique attenuatis, fuscis, margine basali lateralique cum apice, sutura lineisque 1 vel 2 anticis, antennis pedibusque testaceis. Long. ·13.

At first sight seems to be a variety of the preceding. It is less convex and less narrowed in front. The points of the elytra at the base are unequal, but at the apex they become more dense and equal.

[56] H. dispar.—Regulariter elliptico-ovalis, minus convexus, subtus niger, supra cum antennis pedibusque ferrugineus nitidus, capite thoraceque dense subtiliter punctatis, hoc punctis ad basin et apicem transversim densioribus, lateribus obliquis leviter rotundatis, cum elytris (lateraliter visis) angulum obtusum formantibus; elytris apice rotundatim attenuatis, sparsim subtiliter punctulatis et minus subtiliter sat dense punctatis, præcipue ad apicem. Long. ·15.

Some of the scattered punctures at the base of the elytra have a tendency to form three distant longitudinal bands, the first being near the suture.

lateralis.
minutus *Linné*, *Aubé*, 683.
revolvens.
circumnans.
duplicatus.
longiusculus.
analis *Say. Am. Tr.* 2, 108.
nec Kb. N. Z. 91.
Sayi *Aubé*, 699.
DINEUTES.
assimilis *Aubé*, 778.
Gyr. americanus S. Am. Tr. 2, 107.
Cyclinus assim. Kb. N. Z. 78.
discolor *Aubé*, 778.
Cyclous labratus Mels. P. Ac. 2, 29.
HETEROCERUS *Fabr.*
ventralis *Mels. P. Ac.* 2, 98.
undatus *Mels. ibid.* 2, 98.
angulatus.
apicalis.
cinctus.
ELMIS *Latr.*
bivittatus *Dej. Cat.*
LIMNIUS *Illiger.*
[57] fastiditus.
HYDROCHUS *Germ.*
scabratus *Muls. An. Lugd.* 1, 373.
gibbosus Mels. P. Ac. 2, 99.
rufipes *Mels. ibid.* 100.
HYDRÆNA *Kug.*
tenuis.
OCHTHEBIUS *Leach.*
[58] cribricollis.
[59] nitidus.
HELOPHORUS *Fabr.*
[60] oblongus.
[61] lacustris.
lineatus *Say. J. Ac.* 3, 200.
apicalis.
nitidus.

[57] L. fastiditus.—Fusco-æneus, thorace convexo, pubescente, minus dense punctato, lateribus rectis, marginatis, basi media producto, emarginatoque, angulis posticis acutis, utrinque ad basin impresso; elytris striato-punctatis, interstitiis subtiliter punctulatis breviter flavo-pubescentibus vitta utrinque læte flava ad humerum paulo dilatata. Long. ·11. Maple Island.

[58] O. cribricollis.—Æneo-testaceus margine pedibusque pallidioribus, thorace lateribus rotundato basi bisinuato, grosse punctato, canaliculato, lineaque arcuata utrinque ante medium; elytris punctato-striatis. Long. ·08. Eagle Harbor.

[59] O. nitidus.—Æneo-niger, pernitidus, thorace lateribus rectis basi utrinque obliqua, angulis anticis productis apice rotundatis, profunde canaliculato, antice utrinque bifoveato fovea externa majore, basi utrinque fovea parva, et ad angulos posticos fovea magna exarata, elytris punctis discretis majusculis seriatim positis; antennis pedibusque testaceis. Long. ·07. Eagle Harbor.

[60] H. oblongus.—Elongatus, parallelus, testaceus capite obscure viridi, subtiliter punctato, thorace lateribus rectis basi utrinque obliqua, apice fere truncata, obsolete punctulato, lineis intermediis fere rectis; elytris apice rotundato-subtruncatis, profunde crenato-striatis, gutta parva nigra versus medium utrinque ornatis. Long. ·23. Eagle Harbor.

[61] H. lacustris.—Oblongus, supra obscure testaceus, capite viridi thoraceque granulis minus elevatis dense adspersis, hoc lateribus vix rotundatis, basi utrinque sinuato, angulis anticis prominulis, lineis 5 fortiter impressis, intermediis valde curvatis, elytris pone medium vix oblique attenuatis fortiter crenato striatis, interstitiis 5 to 7 mo que dorso paulo acutis; utrinque versus medium guttis 1 vel 2 fuscis signatis. Long. ·23. Eagle Harbor.

affinis.
[62] scaber.
HYDROPHILUS *Fabr.*
glaber *Hbst. Col.* 7, 298.
lateralis *F. El.* 1, 251.
nimbatus Say. J. Ac. 3, 203.
obtusatus *Say. J. Ac.* 3, 202.
LACCOBIUS *Leach. Er.*
punctatus *Mels. P. Ac.* 2, 100.
HYDROBIUS *Leach.*
(§ *PHILHYDRUS Sol.*)
lacustris.
perplexus.
nebulosus.
Hydrophilus neb. Say. Exp. 2, 277.
CYCLONOTUM *Dej. Muls.*
subcupreum.
Hydrophilus subc. Say. J. Ac. 5, 189.
CERCYON *Leach.*
mundum *Mels. P. Ac.* 2, 102.
ambiguum.
dubium.
vagans (*Crytopleurum Muls.*)
NECROPHORUS *Linné.*
hebes *Kb. N. Z.* 97.
orbicollis *Say.*
var. Hallii Kb. N. Z. 98.
4-maculatus Dej. Cat.
pygmæus *Kb. N. Z.* 98.
velutinus *Fabr. El.* 1, 334.
SILPHA *Linné.*
americana *Linné S. Nat.* 2, 570.
var.? Oiceoptoma affine Kb. N. Z. 103.
inæqualis *F. El.* 1, 340.
lapponica *Hbst. Fabr. El.* 1, 338.
caudata *Say. J. Ac.* 3, 192.
CATOPS *Fabr.*
[63] terminans.
CEPHENNIUM *Müller.*
MEGALODERUS Steph.
[64] n. s.
SCYDMÆNUS *Latr.*
subpunctatus.
pilosicollis.
BRYAXIS *Knoch.*
propinqua.
longula.
FALAGRIA *Leach.*
dissecta *Er.* 49.
var. erythroptera Mels. P. Ac.

[62] H. scaber.—Æneo-niger, capite thoraceque granulis dense scabrosis, hoc basi angustato, lateribus late excavato, dorsoque foveato, lineis 5 impressis, intermediis sinuatis, elytris pone basin sensim ampliatis, versus apicem oblique attenuatis, crenato-striatis, basi bicarinatis, pone basin oblique impressis, interstitiis pone medium alternatim tuberculatis. Long. ·13.

The third and fifth interstices have each three tubercles, the anterior one being small the seventh has two, and the ninth a very slight elevation. The striæ are deeper towards the margin than at the suture.

[63] C. terminans.— Ovatus minus convexus, niger opacus, dense pubescens, ruguloso-punctatus; thorace antice angustato, lateribus rotundato, basi utrinque sinuato, angulis posticis subacutis, elytris stria suturali valde impressa, pedibus fuscis, antennis apice parum incrassatis, apice summo flavo, basi testaceo. Long. ·15. Pic: under old carrion.

♂ tarsi antici, dilatati; tarsi intermedii articulo 1 mo elongato dilatatoque.

[64] This species is the analogue of the European C. minutissimum; it is no larger than a Trichopteryx: I found but a single specimen on St. Joseph's Island, and although it was safely secured in a bottle, it was not there by the time I reached camp. I therefore forbear naming it, merely directing the attention of future explorers to this very interesting species.

depressa.

HOMALOTA *Man. Er.*

pressa.

planata.

pallipes.

flavicans.

polita *Mels. P. Ac.* 2, 31.

attenuata.

dichroa *Er.* 107.

rubricornis.

dubitans.

stricta.

clavifer.

lividipennis *Er.* 129.

OXYPODA *Man.*

sagulata *Er.* 146.

turpis *Mels. Ms.*

moesta.

ALEOCHARA *Grav.*

rubripennis.

nitida *Grav. Mic.* 97; *Er.* 168.

molesta.

GYROPHÆNA *Man.*

amanda.

bellula.

socia *Er.* 189.

corruscula *Er.* 189.

EURYUSA *Er.*

semiflava.

MYLLÆNA *Er.*

terminans.

CONURUS *Steph.*

crassus *Er.* 222.

TACHYPORUS *Grav.*

jocosus *Say. Am. Tr.* 4, 466.

arduus Er. 237.

brunneus *Er. Col. March.* 1, 395.

faber Say. l. l. 468.

punctulatus *Mels. P. Ac.* 2, 32.

TACHINUS *Grav.*

ventriculus *Er.* 920.

gibbulus Er. 252.

luridus *Er.* 920.

hybridus.

puncticollis.

fimbriatus *Grav. Mic.* 191; *Er.* 258.

picipes *Er.* 257.

fumipennis *Er.* 921.

axillaris Er. 261.

obscurus.

conformis.

OLISTHÆRUS *Dej. Er.* 843.

[65] laticeps.

[66] nitidus.

BOLETOBIUS *Leach.*

longiceps.

obsoletus *Er.* 922.

cinctus *Er.* 278.

pygmæus *Man. Brach.* 65; *Er.* 280; 922.

MYCETOPORUS *Man.*

lucidus.

americanus *Er.* 285.

OTHIUS *Leach.*

macrocephalus? *Er.* 297.

lævis.

XANTHOLINUS *Dahl.*

obsidianus *Mels. P. Ac.* 2, 34.

americanus Dej. Cat.

cephalus *Say. Am. Tr.* 4, 452.

consentaneus Er. 326.

hamatus *Say. Am. Tr.* 4, 453.

[65] O. laticeps.—Rufus nitidus, capite nigro postice leviter parcius punctato thorace non angustiore, hoc basi leviter angustato, angulis posticis rectis, paulo impressis, elytris leviter striatis, abdomine fusco, supra sat dense punctulato. Long. ·28. St. Ignace.

[66] O. nitidus.—Rufus nitidus, capite nigro, postice punctulato thorace sesqui angustiore, hoc basi vix angustato, angulis posticis rectis, paulo impressis, elytris nigris striatis, abdomine rufo, supra dense minus subtiliter punctato. Long. ·22. Eagle Harbor.

obscurus *Er.* 330.
var. corvinus Dej. Cat.
STAPHYLINUS *Lin.*
villosus *Grav. Mic.* 160; *Er.* 349.
PHILONTHUS *Leach.*
cyanipennis *Er.* 433.
æneus *Nord. Symb.* 81. *Er.* 437; 928.
Harrisii Mels. P. Ac. 2, 35.
melancholicus.
sparsus.
promtus *Er.* 929.
stygicus.
debilis *Er. Col. March.* 1, 467.
inconspicuus.
morulus.
vapidus.
lomatus *Er.* 482.
consors.
curtatus.
brunneus *Er.* 486.
lugens.
aterrimus *Er.* 492.
egenus.
gratus.
QUEDIUS *Leach. Er.*
obscurus.
corticalis.
morio.
perspicax.
arboricola.
OXYPORUS *Fabr.*
vittatus *Grav. Micr.* 195; *Er.* 558.
LATHROBIUM *Grav.*
Zimmermani.
simile.
concolor.
nigrum.
longiusculum *Gr. Micr.* 181; *Er.* 597.
LITHOCHARIS *Dej.; Boisd.*
confluens *Er.* 615.
SUNIUS *Leach. Er.*
longiusculus *Er.* 643.
PÆDERUS *Grav.*
littorarius *Grav. Mon.* 142; *Er.* 656.
STENUS *Latr.*
Juno *Fabr. El.* 2, 602; *Er.* 694.
stygicus *Er.* 698.
lugens.
longicollis.
planifrons.
bisulcatus.
egenus *Er.* 698.
simplex.
terricola.
strumosus.
punctatus *Er.* 744.
EVÆSTHETUS *Grav.*
americanus *Er.* 747.
BLEDIUS *Leach.*
ruficornis.
annularis.
divisus.
PLATYSTETHUS *Man.*
americanus *Er.* 784.
OXYTELUS *Grav.*
misellus.
TROGOPHLŒUS *Man.*
planus.
[67] Argus.
pumilus.
ANTHOPHAGUS *Grav.*
verticalis *Say. Am. Tr.* 4, 463.
memnonius.
LESTEVA *Latr.*
biguttula.
ACIDOTA *Leach.*
subcarinata *Er.* 863.
patruelis.
tenuis.

[67] This species is remarkable for possessing two ocelli: but the structure of the abdomen proves it to belong to the Oxytelini, and in no part of the body does it show any difference from Trogophlœus: it and the preceding species belong to the division possessing a visible scutellum.

LATHRIMÆUM *Er.*
sordidum *Er.* 871.
DELIPHRUM *Er.*
seriatum.
[68] LATHRIUM.
convexicolle.
OMALIUM *Grav.*
longulum.
complanatum.
protectum.
ANTHOBIUM *Leach.*
simplex.
ventrale.
[69] dimidiatum *Mels. P. Ac.* 2, 43.
confusum.
PROTEINUS *Latr.*
parvulus.
MEGARTHRUS *Kirby.*
excisus.
MICROPEPLUS *Latr.*
[70] costatus.
TRICHOPTERYX *Kirby.*
discolor *Hald. J. Ac. N. S.* 1, 108.
aspera *Hald. ib.* 109.
ANISOTOMA *Illiger.*
[71] assimilis.
[72] indistincta.
[73] collaris.
[74] strigata.

[68] Mandibulæ edentatæ. Maxillæ mala exteriore cornea (interiore invisa.) Palpi maxillares tenues, art. 2ndo 4to que elongatis. Tibiæ omnino muticæ. Tarsi breves, tenues, articulis 4 primis æqualibus, posticis art. 4to subtus producto, breviter calceato.

Frons inimpressus, ocellis supra oculos sitis, minus distinctis. Proximus videtur Olophro, at tarsorum structura abhorret. Discedit porro statura longiore, elytrisque abdominis segmentum 1 mum solum tegentibus. Victus riparius.

L. convexicolle.—Elongatum nigrum, thorace convexo, lateribus rectis submarginatis, angulis anticis rotundatis, basi cum angulis posticis rotundata, sat dense punctato, obsolete canaliculato, ante basin leviter foveato, elytris grossius punctatis sutura leviter elevata, abdomine subtilissime alutaceo, ano pedibus antennisque rufopiceis. Long. ·19. Eagle Harbor.

[69] Mas abdomine nigro; femina sesqui major, abdomine concolore testaceo.

[70] M. costatus.—Niger thorace celluloso, elytris versus apicem transversim impressis, tricostatis interstitio externo punctulato, abdomine late marginato, segmentis 3 primis utrinque carinula brevi instructis, 1mo ad basin subtiliter canaliculato. Long. $\frac{2}{3}$ lin.

The feet are piceous: seems allied to M. tesserula Curtis. Er. 913.

[71] A. assimilis.—Ovalis nigro-picea, subtiliter dense punctata, thorace antice angustato, lateribus rotundato, basi utrinque punctis seriatim transversim positis, elytris punctato-striatis, interstitiis alternatim punctis majusculis uniseriatim positis. Long. ·16. Eagle Harbor.

♂ Tibiis posticis elongatis curvatis.

[72] A. indistincta.—Fere hemispherica, piceo-rufa, obsolete sparsim punctulata, thorace lateribus minus rotundato, basi subsinuata, punctis utrinque notata, elytris punctato-striata, interstitiis alternatim punctis 3 vel 4 majusculis. Long. ·11.

[73] A. collaris.—Ovalis, convexa, rufo-testacea, antennis capite thoraceque piceis, hoc lateribus valde rotundato, dense punctato, basi truncata punctis majoribus utrinque notata; elytris profunde punctato-striatis, interstitiis vix subtilissime punctulatis, alternatim punctis 5 vel 6 majusculis. Long. ·12. Eagle Harbor.

♂ tibiis posticis curvatis.

[74] A. strigata.—Hemispherica rufa, thorace lateribus rotundato, basi truncato, lævissimo, elytris tenuiter punctato-striatis, interstitiis transversim subtiliter rugulosis. Long. ·08.

CYRTUSA *Er.*
[75] globosa.
[76] STERNUCHUS.
gibbulus.
AGATHIDIUM *Illiger.*
[77] ruficorne.
[78] revolvens.
PHALACRUS *Payk.*
[79] difformis.
OLIBRUS *Er.*
[80] apicalis.
BRACHYPTERUS *Kugellan.*
urticæ *Kug. Er. Ins. Germ.* 3, 132.
COLASTUS *Er.*
semitectus *Er. Germ. Z.* 4, 243.
truncatus.
Nitidala truncata Rand. B. J. 2, 18.
tantillus.
CARPOPHILUS *Leach.*
niger *Er. Germ. Z.* 4, 263.
Cercus niger Say. J. Ac. 3, 195.
EPURÆA *Er.*
flavicans.
vicina.
parvula.
longula.
parallela.
retracta.
rufa *Er. Germ. Z.* 4, 273.
Nitid. rufa Say. J. Ac. 5, 180.

[75] C. globosa.— Hemispherica, nigro-picea, nitida, thorace subtiliter dense punctulato basi truncato, angulis posticis vix rotundatis, margine diaphano: elytris dense punctulatis, punctisque vix majoribus seriatim positis, antennarum basi, tarsis tibiisque piceis, his anticis non dilatatis. Long. ·13.

[76] STERNUCHUS. Antennæ capillares, articulo 1mo crassiore majore; 3 ultimis parum dilatatis, omnibus setis 2 longis apicalibus. Metathorax subtus permagnus, prominens, planus, antice declivus, pedibus intermediis in declivitate profunde sitis, approximatis. Coxæ anticæ, exsertæ, conicæ, posticæ permagnæ laminatæ, abdominis partem anteriorem obtegentes. Abdomen parvum, 5-articulatum, (articulis 2 primis consolidatis ?) Tarsi filiformes consolidati, unguibus simplicibus.

Head large, semicircularly rounded anteriorly, acutely angulated on the sides behind, labrum very short, almost concealed by the margin of the clypeus. Thorax very short, not emarginate in front, base rounded, angles none. Elytra covering the abdomen, declivous, scarcely convex behind. Palpi filiform. I should have considered this insect a Cybocephalus, but for the filiform tarsi. The structure of the antennæ differs from Cyllidium, but I am by no means certain that I have placed it in a proper position: it seems to have some relation to Clambus, but the great size of the metasternum and posterior coxæ prevents a complete examination of the lower surface.

S. gibbulus. Globatilis, gibbus, niger lævissimus, antennis ore pedibusque flavis. Long. ⅓ lin.

[77] A. ruficorne. Globatile supra nigrum, elytris vix punctulatis, stria suturali postice profunda, antennis pedibusque rufis, abdomine sæpius ferrugineo. Long. ·08. Hab. ubique.

[78] A. revolvens.— Globatile at minus convexum, nigrum, elytris dense subtiliter punctatis, obsoletissime striatis, stria suturali, profunda. Long. ·14.

[79] P. difformis.— Hemisphericus, rufescenti-piceus, thorace vix obsolete punctulato, lateribus subrectis, basi cum angulis posticis rotundata, elytris sat dense punctulatis, stria suturali profunda. Long. ·08.

♂ Mandibula sinistra cornu erecto curvato longitudine caput æquante.

[80] O. apicalis.— Breviter ovalis, postice vix angustatus, convexus, piceus nitidus, thorace basi truncato, elytris impunctatis, stria suturali impressa, aliisque 1 vel 2 obsoletissimis, apice corporeque subtus rufo, antennis pedibusque flavis. Long. ·08.

N. B. The maxillary palpi have the last joint somewhat securiform.

PHENOLIA *Er.*
grossa *Er. Z.* 4, 300.
Nitid. grossa Fabr. El. 1, 347.
OMOSITA *Er.*
colon *Er. Germ. Z.* 4, 299.
MELIGETHES *Leach.*
obsoleta.
AMPHICROSSUS *Er.*
[81] concolor.
IPS *Fabr.*
sepulchralis *Rand. B. J.* 2, 19.
Dejeanii Kb. N. Z. 107.
filiformis.
bipunctatus.
PELTIS *Geoff.*
fraterna *Rand. B. J.* 2, 17.
ferruginea‡ *Kb. N. Z.* 104.
septentrionalis *Rand. l. l.* 17.
THYMALUS *Latr.*
fulgidus *Er. Z.* 5, 458.
CICONES *Curtis.*
fuliginosus.
Synchita ful. Mels. P. Ac. 2, 111.
CERYLON *Latr.*
affine.
unicolor.
Latridius uni. Zieg. P. Ac. 2, 270.
CUCUJUS *Fabr.*
clavipes.
LÆMOPHLŒUS *Dej. Er.*
biguttatus.
Cucujus big. Say. J. Ac. 5, 267.
DENDROPHAGUS *Fabr.*
[82] glaber.
BRONTES *Fabr.*
dubius *Fabr. El.* 2, 97.
SILVANUS *Latr.*
[83] planus.
PARATENETUS *Spin.*
[84] fuscus.
PARAMECOSOMA *Curtis.*
[85] denticulatum.
inconspicuum.
ATOMARIA *Kb.*
similis.
longula.
cingulata.
CORTICARIA *Marsham.*
serricollis.
denticulata *Kb. N. Z.* 110.
similis.
affinis.
convexa.
reticulata.
cavicollis *Man. Germ. Z.* 5, 50.
LATHRIDIUS *Illiger.*
reflexus.

[81] A. concolor.—Ellipticus convexus, ferrugineus, punctatus, pubescens, thorace tenuiter marginato, lateribus modice rotundatis. Long. 15. Pic.

[82] D. glaber.—Elongatus piceus, glaber, capite thoraceque punctatis, hoc longitudinaliter biimpresso, lateribus sinuato, elytris punctato-striatis, margine cum antennis pedibusque rufo. Long. ·27.

[83] S. planus.—Valde depressus, rufus, capite thoraceque dense punctatis, hoc angulis posticis, late emarginatis, denteque vix conspicuo ante medium armato, angulis anticis rotundatis, elytris subtilissime punctulatis pubescentibus, stria suturali tenui impressa. Long. ·12.

[84] P. fuscus.—Oblongo-ovatus, antice angustatus, convexus, ferrugineo-fuscus, grossius punctatus, sparse pubescens, thorace lateribus subangulatis, pone medium 4-dentatis, ante medium crenatis, basi truncata elytris thorace latioribus sutura nigra. Long. ·12.

[85] P. denticulatum.—Elongato-oblongum, ferrugineum, punctatum minus subtiliter flavo-pubescens, thorace transverso lateribus paulo rotundatis crenulatis, basi media marginata, utrinque impressa, elytris stria suturali parum profunda. Long. ·08.

MYCETOPHAGUS *Fabr.*
pictus.
TRIPHYLLUS *Latr.*
didesmus. [261.
Mycetophagus did. Say. J. Ac. 5,
DERMESTES *Lin.*
murinus *Lin. Er. Ins. Germ.* 3, 429.
BYRRHUS *Linné.*
[86] americanus.
cyclophorus *Kb. N. Z.* 117.
picipes *Kb. N. Z.* 116.
varius *Fabr. El.* 1, 105.
[87] eximius.
[88] tesselatus.
SYNCALYPTA *Dillwyn.*
[89] echinata.
PLATYSOMA *Leach.*
depressum *Er. Jahr.* 111.
HISTER *Lin.*
abbreviatus *Fabr. El.* 1, 89.
depurator *Say. J. Ac.* 5, 33.
americanus *Payk.* 31.
subrotundus *Say. J. Ac.* 5, 39.
PAROMALUS *Er.*
bistriatus *Er. Jahr.*
SAPRINUS *Leach.*
pensylvanicus *Er.* 184.
assimilis *Er.* 184.
distinguendus.
proximus.
mancus.
Hister m. Say. J. Ac. 15, 41.
fraternus *Lec. B. J.* 5, 77.
Hister f. Say. J. Ac. 5, 40.
PLATYCERUS *Latr.*
[90] depressus.
quercus *Schön.*
Pl. securidens Say. J. Ac. 3, 249.
Lucanus querc. Weber Obs. 1, 85.
GEOTRUPES *Latr.*
miarophagus *Say. J. Ac.* 3, 211.

[86] B. americanus.—Oblongo-ovatus, antice acutus, convexus, niger dense breviter fusco-pubescens, thorace nigro cinereoque variegata, elytris sutura vittisque 4 nigris, guttis albis interruptis, quæ spatium transversum antice dentatum, postice lateribus obliquis, medio recte truncatum, formant; guttisque nonnullis aliis versus apicem oblique retrorsum positis; tenuiter striatis. Long. ·4. Twice the size of B. cyclophorus; found from Niagara to Lake Superior.

[87] B. eximius.—Oblongus antice acutus, lateribus parallelus, niger fusco-pubescens, nigro flavoque variegatus; thorace nigro, cinereo flavoque variegato, elytris striatis, sutura vittisque 4 nigris, his guttis interruptis, lineam ante medium transversam dentatam, figuram semicircularem antice dentatam, lineamque versus marginem antrorsum obliquam formantibus, his omnibus postice flavo tomentosis. Long. ·2. Pic. The middle part of the anterior margin of the semicircular figure forms a broad common cinereous spot.

[88] B. tesselatus.—Elongatulus, utrinque subacutus, virescente-niger nigro-pubescens, elytris striatis cinereo tesselatis, subtus niger, pedibus piceis. Long. ·12. Pic.

[89] S. echinata.—Breviter ovata, utrinque attenuata, nigra parce cinereo-pubescens, setis erectis clavatis nigris adspersa, in elytra longioribus uniseriatim in striarum interstitiis positis; striis tenuibus, marginali sola profunda. Long. ⅜ lin. Eagle Harbor.

[90] P. depressus.—Depressus, niger vix æneus, thorace lateribus pone medium angulato, angulis posticis obtusis minime rotundatis, elytris profundius punctatis, striatisque. Long. ·62. Twice the size of P. quercus, the elytral stria are alternately a little approximated; the mandibles of the ♂ are much dilated at the apex, but less curved than in P. quercus. A very small specimen has the sides of the thorax behind the angle emarginate, so that the basal angles become still more prominent and scarcely obtuse.

APHODIUS *Ill.*
[91] hyperboreus.
omissus. [*nec Say.*
concavus Hd. J. Ac. N. S. 1, 103,
pinguis *Hd. l. c.* 103.
[92] angularis.
[93] consentaneus.
4-tuberculatus *Fabr.*
curtus *Hd. l l.* 105.
OXYOMUS *Latr.*
strigatus.
Aph. strigatus Say. J. Ac. 3, 212.
RHYSSEMUS *Muls.*
[94] cribrosus.
ÆGIALIA *Encycl.*
[95] lacustris.

91 A. hyperboreus.—Oblongus rufo-piceus nitidus, capite nigro, thorace lateribus punctato, disco lævissimo, angulis posticis obtusis rotundatis, basi vix marginata utrinque obliqua, nigro lateribus obsolete rufis, elytris crenato-striatis, interstitiis fere planis, lævissimis. Long. ·3. Pic. The clypeus is smooth, with only a few points at the side, the margin reflexed, and slightly emarginate. Belongs to Erichson's division D, as well as the three following species. It is very similar to *A. omissus*, but distinguished (apart from color) by the smooth clypeus and impunctured elytra. I have changed the name of the next species, as it cannot be Mr. Say's A. concavus: that author makes no mention of the large scutellum, which he would not have failed to observe in comparison with other species. I know not how Mr. Haldeman omitted this character which would serve at once to distinguish the species in question, and A. pinguis from all the other American species seen by him.

92 A. angularis.—Oblongus niger nitidus supra undique sparse subtiliter punctulatus, thorace lateribus rotundato, angulis posticis obtusis non rotundatis, basi vix marginata, utrinque oblique vix sinuata, lateribus punctatis, disco parce punctato, elytris profundius crenato-striatis. Long. ·26. Pic. Variat elytris piceis, pedibus rufo-piceis.

Agrees with A. pinguis in being covered with a fine punctuation; the clypeus is more broadly emarginate, and the posterior angles of the thorax not at all rounded; the basal margin of the thorax is interrupted and indistinct. The ♂ has the thorax a little wider than the elytra. Belongs also to Erichson's division D.

93 A. consentaneus.—Elongatus, rufo-testaceus, elytris pallidioribus, capite thoraceque subtiliter sat dense punctatis, hoc lateribus parum rotundato angulis posticis obtusis valde rotundatis, basi tenuiter marginata, elytris thorace non latioribus profunde crenato-striatis. Long. ·2.

Clypeus margined, scarcely emarginate, frontal suture straight: the punctures of the thorax are intermixed with a few very minute points. Belongs to division E, of Erichson.

94 R. cribrosus.—Piceus, opacus, thorace lateribus rectis, angulis posticis late emarginatis, basi vix rotundata, grosse confertim cribrato, canaliculato, elytris antice subangustatis, basi emarginatis, acute 10-costatis, sulcis uniseriatim leviter punctatis. Long. ·16.

Head convex, punctured, clypeus scarcely margined, oblique each side. I should refer this species to Euparia, were not the posterior tibiæ destitute of the rows of bristles, and the external spur which distinguish that genus; they have two scarcely discernible rudiments of teeth on the outer edge. The podex is entirely concealed by the elytra.

95 Æ. lacustris.—Oblonga, convexa, postice subdilatata, nigra nitida, thorace transverso, antice angustato, basi marginata, utrinque oblique subsinuata, angulis posticis rotundatis, lateribus marginatis, anticeque impressis, sat dense grossius punctato, elytris valde crenato-striatis, interstitiis convexis lævibus. Long. ·18.

Head convex, rough anteriorly with elevated granules, clypeus finely margined, widely emarginate. There are also two species found on the Atlantic coast.

LACHNOSTERNA *Hope.*
[96] quercina.
Mel. querc. Kn. N. Beit. 74.
[97] anxius.
[98] consimilis.
[99] futilis.
SERICA *M'Leay.*
vespertina *Dej. Cat.*
Mel. vespertina Say. J. Ac. 3, 244.
[1] tristis.
DIPLOTAXIS *Kirby.*
tristis *Kb. N. Z.* 130.

DICHELONYCHA *Har. Kb.*
hexagona.
Melol. hex. Germ. Ins. Nov. 124.
elongata *Harris.*
Melolontha elongata F. El. 2, 174.
virescens *Kb. N. Z.* 134.
testacea *Kb. N. Z.* 135.
Backii *Kb. N. Z.* 134.
OSMODERMA *Lepell.*
scabrum. *Gory & P. Cet. tab.* 8, *fig.* 2.
♂ *Gymnodus foveatus Kb. N. Z.* 140.
♀ ——— *rugosus Kb. N. Z.* 140.

[96] L. quercina.—Castanea nitida, supra glabra, antennis pedibusque testaceis, thorace minus subtiliter punctato, antice angustato, lateribus parum dilatatis, angulis posticis rectis, basi media late minus extensa, elytris obsolete 3-costatis sat dense punctatis rugosisque, umbone humerali minus elevata, angulo suturali obtuso, pygidio parce punctato. Long. ·93.

♂ antennarum clava parte reliqua longiore, corpore cylindrico.

♀ antennarum clava brevi, corpore postice leviter dilatato.

This is one of a group of very closely allied species, which I have divided according to the form and punctuation of the thorax, and the form of the sutural angle of the elytra. It is the common species everywhere, and is probably Mel. quercina Knoch.

[97] L. anxia.—Nigro-castanea nitida, supra glabra, antennis pedibusque rufo-testaceis, thorace sat dense distinctius punctato, antice angustato, lateribus parum dilatatis, angulis posticis rectis, basi media late minus extensa, elytris leviter 3-costatis distinctius sat dense punctatis, umbone humerali prominulo, angulo suturali obtuso, pygidio parce punctato, basi longitudinaliter rugoso. Long. ·92.

♀ Corpore postice modice dilatato. More dilated behind than the preceding, with larger punctures on the thorax and elytra.

[98] L. consimilis.—Postice non dilatata, castanea, nitida supra glabra, antennis pedibusque testaceis thorace subtilius parce punctato, antice angustato, lateribus modice dilatatis, angulis posticis rectis, basi media late extensa, elytris dense subtiliter punctatis rugosisque, angulo suturali valde obtuso, pygidio parce punctato, basi subruguloso. Long. ·93.

♂ clava antennarum parte reliqua vix longiore.

Differs from the large eastern species (Mel. brunnea Kn.) in having the thorax less dilated on the sides, the posterior angles not acute, and the sutural angle of the elytra very obtuse.

[99] L. futilis.—Dilute castanea supra glabra nitida, antennis testaceis, thorace sat dense minus subtiliter punctato, latitudine triplo breviore, antice angustato, lateribus modice dilatato, angulis posticis obtusis, basi late rotundato, elytris sat dense punctatis sub umbone humerali modice elevata late impressis, angulo suturali subobtuso, pygidio punctato, abdomine densius subtiliter punctulato. Long. ·6.

[1] S. tristis.—Oblongus convexus, piceus punctatus, capite pone oculos lævi, thorace latitudine duplo breviore, antice angustato, lateribus ante medium rotundatis, angulis posticis rectis paulo rotundatis, basi bisinuata, margine tenui basali lævi: elytris obsolete cyaneo micantibus, lateribus parallelis, leviter sulcatis, in sulcis punctatis, interstitiis lævibus, pedibus rufo-piceis, antennis testaceis 9-articulatis. Long. ·32. Clypeus flat, densely punctured, margin scarcely elevated, broadly emarginate, marked anteriorly with a fine transverse line.

DICERCA *Esch.*
divaricata.
Stenuris divaricata Kb. N. Z. 154.
Buprestis div.? Say. J. Ac. 3, 163.
Dicerca dubia Mels. P. Ac. 2, 142.
aurichalcea *Mels. P. Ac.* 2, 142.
parumpunctata Mels. ibid.
tenebrosa.
Stenuris teneb. Kb. N. Z. 155.
lacustris.
lugubris.
bifoveata.
ANCYLOCHEIRA *Esch.*
lineata *Dej. Cat.*
Buprestis lineata Fabr. El. 2, 192.
Nuttalli.
consularis Dej. Cat.
Anoplis Nuttalli Kb. N. Z. 152.
maculiventris.
Bup. maculiv. Say. Exp. 2.
Bup. 6-*notata Lap. Bup. pl.* 32.
Anoplis rusticorum Kb. N. Z. 151.
striata *Dej. Cat.*
Bup. striata Fabr. El. 2, 192.
PHÆNOPS *Esch.*
assimilis.
[2] longipes.
Bup. longipes Say. J. Ac. 3, 164.
CHRYSOBOTHRIS *Esch.*
dentipes.
Buprestis den. Germ. Ins. Nov. 38.
femorata *Dej. Cat.*
Bup. femorata F. El. 2, 208.
scabripennis *Lap. Bup. pl.* 9, *fig.* 71.
Odontomus trinervia Kb. N. Z. 157.
AGRILUS *Meg.*
lacustris.
advena.
FORNAX *Lap.*
spretus.

CRATONYCHUS *Dej.*
puncticollis.
recticollis.
decumanus *Er. Germ. Z.* 3, 104.
communis *Er. ibid.* 3, 102.
ADELOCERA *Latr.*
[3] brevicornis.
LIMONIUS *Esch.*
confusus *Dej. Cat.*
quercinus *Dej. Cat.*
Elater quer. Say. An. Lyc. 1, 262.
CAMPYLUS *Fisch.*
denticornis *Kb. N. Z.* 145.
flavinasus Mels. P. Ac. 2, 219.
productus ? *Rand. B. J.* 2, 8.
CARDIOPHORUS *Esch.*
vagus.
CRYPTOHYPNUS *Esch.*
insignis.
silaceipes *Germ. Z.* 5, 139.
lacustris.
tumescens.
simplex.
misellus.
dorsalis *Germ.* 5, 147.
renifer.
AMPEDUS *Meg.*
lugubris *Germ.* 5, 165.
semicinctus.
El. semicinctus Rand. B. J. 2, 10.
apicalis.
El. apicatus Say. Am. Tr. 4.
Amp. melanopygus Germ. 5, 161.
phoenicopterus *Germ.* 5, 161.
luctuosus.
ferripes.
sparsus.
lutosus.
PRISTILOPHUS *Latr.*
fusiformis.

[2] Kirby gives this as identical with the European P. appendiculata: the characters in this group are rather obscure, and I prefer continuing it as distinct until I have an opportunity for comparison.

[3] This species is very near to A. conspersa, (Germ. Zeit. 2,257.)

CORYMBITES *Latr.*
anchorago.
Ctenicerus Kendalli Kb. N. Z. 149.
Elater anchorago Rand. B. J. 2, 5.
resplendens *Germ. Z.* 4, 60.
Ludius resp. Esch. Thon. Arch. 2, 34.
Elater œrarius Rand. B. J. 2, 7.
cylindriformis *Germ.* 4, 64.
[4] mirificus.
DIACANTHUS *Latr.*
medianus *Germ.* 4, 71.
submetallicus *Germ.* 4, 72.
æneolus.
bicinctus.
Ludius bic. Dej. Cat.
curiatus.
El. curiatus Say.
Ludius propola Dej. Cat.
appropinquans.
Elater appro. Rand. B. J. 2, 5.
El. œripennis Kb. N. Z. 150.
splendens *Zieg. P. Ac.* 2, 44.
furcifer.
triundulatus.
Elater 3-und. Rand. B. J. 2, 12.
spinosus.
dubius.
suturellus.
DOLOPIUS *Meg.*
fucosus.
indentatus.
mixtus.
incongruus.
stabilis.
umbraticus *Dej. Cat.*
pauperatus *Dej. Cat.*
filiformis.
pulcher.
obesulus.
CYPHON *Fabr.*
obscura *Guérin. Mon.* 4.
variabilis *Guérin. ib.*
PYRACTOMENA *Dej.*
borealis.
Lampyris bor. Rand. B. J. 2, 16.
falsa.
PYGOLAMPIS *Dej.*
ardens.
tædifer.
ELLYCHNIA *Dej.*
neglecta *Dej.*
corrusca *Dej.*
Lamp. corrusca Fabr. El. 2, 100.
lacustris.
CÆNIA *Nm.*
dimidiata *Lec. J. Ac. N. S.* 1, 76.
CELETES *Nm.*
mystacina *Lec. ib.* 77.
tabida *Lec. ib.*
EROS *Nm.*
coccinatus *Lec. l. l.* 77.
modestus *Lec.* 80.
æger *Lec.* 80.
socius *Lec.* 81.
PODABRUS *Fisch.*
diadema *Dej.*
Canth. diad. Fabr. El. 1, 298.
[5] modestus.
Canth. mod. Say. J. Ac. 3, 179.

[4] C. mirificus.—Niger, thorace transverso antice augustato, lateribus rectis, angulis posticis productis carinatis, rufo, macula magna rotundata nigra fere ad apicem extendente, densius subtiliter punctato, elytris flavis, sutura usque ad dodrantem, macula humerali, lineaque submarginali a medio postico tendente nigris; tenuiter striatis interstitiis punctatis, tibiis tarsisque testaceis, illis apice fuscis. Long. ·41.

The prosternal spine is more deflexed than in any other species I have yet seen.

[5] P. modestus. —Niger tenuiter pubescens, capite antice flavo, postice dense punctato, thorace subtransverso, lateribus rotundato, obsolete punctato, postice canaliculato margine flavo, angulis posticis rectis, elytris scabris, sutura margine antennis palpisque basi flavis. Long. ·45.

[6] rugosulus.
[7] punctatus.
[8] marginellus.
lævicollis.
Malthacus lævi. Kb. N. Z. 248.
[9] puberulus.
[10] curtus.

Telephorus *Geof.*
bilineatus.
Cantharis bilin. Say. J. Ac. 3, 182.
Curtisii *Kb. N. Z.* 247.
[11] nigrita.
scitula.
Cantharis sc. Say. J. Ac. 5, 169.

The feet are either yellow or fuscous, the 2nd and 3rd joints of antennæ equal, and each $\frac{2}{3}$ the length of the 4th; claws with a large tooth.

[6] P. rugosulus.—Niger tenuiter pubescens, capite antice flavo, (clypeo apice fusco) postice dense punctato, thorace subtransverso, apice angustato, basi truncato, antice transversim impresso, canaliculato, punctato, lateribus flavis; elytris dense scabris, lineolis 3 obsoletis; coxis, antennarum articulo 1mo, palporum basi pedibusque anticis flavis. Long. ·32.

The 3rd joint of the antennæ is longer than the 2nd, but shorter than the 4th. The anterior thighs are commonly dusky beneath, sometimes all the feet are black: the claws are bifid.

[7] P. punctatus.—Niger densius cinereo-pubescens, capite antice obscure rufo, postice dense punctato, thorace quadrato, lateribus versus basin sinuatis, angulis posticis prominulis, rufo, dense punctato, disco utrinque pone medium elevato, elytris minute scabris, lateribus margine antice pallido. Long. ·28.

The 2nd joint of the antennæ is $\frac{2}{3}$ the length of the 3rd, which is equal to the 4th; the palpi are longer than in the preceding: claws with a broad tooth.

[8] P. marginellus.—Niger, cinereo-pubescens, mandibulis, antennarum articulis 2, palpisque testaceis, capite postice dense punctato, thorace quadrato, lateribus fere rectis anguste testaceis angulis posticis prominulis, minus dense punctato, disco utrinque modice elevato, elytris, subtiliter scabris, sutura margineque tenui pallidis. Long. ·31.

Like the last in form; the palpi are shorter and more dilated, the thorax less punctured, the 3rd joint of the antennæ is but little longer than the 2nd.

[9] P. puberulus.—Ater, undique subtiliter cinereo-pubescens, thorace quadrato, angulis posticis prominulis, late canaliculato, disco subtiliter alutaceo utrinque pone medium elevato, lævique, elytris subtilissime scabris, pedibus antennisque fuscis basi testaceis, his articulis æqualibus. Long. ·25.

Like *lævicollis*, but the thorax is pubescent. The palpi are filiform, claws with a broad tooth.

[10] P. curtus.—Latiusculus niger subtiliter pubescens, thorace brevissimo, utrinque truncato, lateribus obliquis rectis, lævi, læte flavo apice nigra; elytris subtilissime punctatis, abdominis segmento singulo testaceo-marginato, antennarum basi mandibulisque testaceis. Long. ·17.

The eyes are scarcely prominent, the 3rd and 4th joints of the antennæ are equal, each being twice the length of the 2nd; claws dilated at the base.

[11] T. nigrita.—Niger, undique cinereo-pubescens, thorace subquadrato, latitudine sesqui breviore, antice vix angustato, undique marginato, angulis posticis vix rotundatis, disco lævi, utrinque pone medium modice elevato, elytris distinctius punctatis, margine antice testaceo. Long. ·22.

The 3rd and 4th joints of the antennæ are equal, each twice as long as the 2nd; claws with a tooth: palpi moderately dilated. Varies with the mouth, base of antennæ, margin of thorax and anterior feet testaceous.

[12] nigriceps.
SILIS *Meg.*
[13] longicornis.
[14] difficilis.
MALTHINUS *Latr.*
fragilis.
parvulus.
niger.
COLLOPS *Er.*
tricolor *Er. Monog.* 57.
Malachius tric. Say. J. Ac. 3, 182.
CLERUS *Fabr. Klug.*
undatulus *Say. B. J.* 1, 163.
nubilus *Kl. Mon. Cl.* 386.
Thanasimus abdominalis Kb. 244.
thoracicus *Ol.* 4, 18, *pl.* 2, 22.
HYDNOCERA *Nm.*
difficilis *Lec. An. Lyc.* 5, 27.
XYLETINUS *Latr.*
fucatus *Dej. Cat.*
DORCATOMA *Fabr.*
ocellatum *Say Exp.* 2, 273.
ANOBIUM *Fabr.*
foveatum *Kb. N. Z.* 190.
errans *Mels. P. Ac.* 2, 309.
OCHINA *Zieg.*
nigra *Mels. P. Ac.* 2, 308.
ANTHICUS *Fabr.*
4-guttatus *Hald. P. Ac.* 1, 304.
[15] terminalis.
[16] difficilis.
[17] scabriceps.

[12] T. nigriceps. — Pallidus, sparse longius cinereo-pubescens, capite postice nigro, thorace latitudine vix breviore, subquadrato, margine undique elevato, disco modice elevato, medio late impresso, elytris distinctius punctatis, medio leviter infuscatis. Long. ·17.

The 3rd and 4th joints of the antennæ are equal, each being one third longer than 2nd; palpi a little dilated, postpectus fuscous; claws bifid.

[13] S. longicornis. — Nigra, sparse pubescens, thorace latitudine triplo breviore, læte flavo, margine antice posticeque nigro elevato, angulis posticis acute incisis, lævi; elytris minus subtiliter punctatis mandibulis flavis. Long. ·2.

The antennæ are very long; the 2nd joint very short and the 3rd equal to the 4th; the lateral margin of thorax is very narrow.

[14] S. difficilis. — Nigra, cinereo-pubescens, thorace latitudine triplo breviore, antice angustato, læte rufo, margine nigro, antice posticeque elevato, angulis posticis acute incisis, vix subtilissme punctulato, elytris punctatis, mandibulis flavis. Long. ·2.

Lake Superior and Sta. Fe, more densely pubescent than the former, with broad lateral margin to the thorax; the antennæ longer than the body, 3rd joint hardly equal to the 4th.

[15] A. terminalis. — Elongatus subdepressus dense punctatus breviter pubescens; capite nigro, linea angusta lævi, thorace latitudine longiore, rufo, basi subangustato marginatoque, elytris parallelis fuscis basi late indeterminate testaceis, maculaque rotundata ad trientem secundum testacea; subtus niger, pedibus antennisque testaceis. Long. ·10. Lake Superior and New York.

[16] A. difficilis. — Elongatus fuscus, albido pubescens, capite thoraceque rufo-testaceis illo disperse punctato, spatio indistincto lævi, hoc capite vix angustiore, campanulato, antice rotundato, pone medium angustato, versus basin cylindrico, dense minus subtiliter punctato, basi marginata, elytris elongatis parallelis, minus subtiliter punctatis, cum antennis pedibusque testaceis. Long. ·1.

Variat, fascia fusca transversa ad elytrorum medium.

[17] A. scabriceps. — Elongatus niger densius albido pubescens capite thoraceque confertissime rugoso-punctatis, hoc capite vix angustiore, campanulato, antice rotundato, pone medium angustato, basi cylindrico marginatoque, elytris punctatis, apice obsolete rufescente, antennis piceo-testaceis. Long. ·1.

[18] granularis.
[19] pallens.
SCHIZOTUS *Nm.*
cervicalis *Nm. Ent. Mag.* 5, 374.
POGONOCERUS *Fisch.*
concolor *Nm. l. c.* 5, 375.
PEDILUS *Fisch.*
CORPHYRA Say. B. J. 1, 189.
lugubris.
imus Nm. l. c. 375.
Anthicus lug. Say. J. Ac. 5, 246.

MORDELLA *Fabr.*
atrata *Mels. P. Ac.* 2, 313.
biguttula.
[20] pectoralis.
[21] ANASPIS *Latr.*
nigra. [1, 99.
Hallomenus nig. Hd. J. Ac. N. S.
ventralis *Mels. P. Ac.* 2, 312.
[22] filiformis.
flavipennis *Hd. l. l.* 100.

Variat; *α* elytrorum basi rufescente; *β* antennis pedibus elytrisque testaceis, hoc fascia lata ad medium fusca; capite thoraceque fuscis: *γ* capite thoraceque testaceis, elytris fascia indistincta.

A very variable species, distinguished from the preceding only by its scabrous head and thorax and longer pubescence. The varieties did not occur mixed together; the type and *α* were very abundant near Pt. Porphyry: *β* and *γ* are found along the entire coast of the lake.

[18] A. granularis. — Subelongatus, convexus, niger breviter albo-pubescens; capite thoraceque dense minus subtiliter granulosis, illo basi subemarginato, angulis acutis, linea longitudinali tenui lævi; hoc capite non angustiore, latitudine breviore, obovato basi truncata, obsolete marginata; elytris parallelis dense minus subtiliter punctatis, apice rufescente, antennis tibiis tarsisque testaceis. Long. ·13.

Variat, *α* capite thoraceque fuscis, elytris testaceis fascia lata ad medium nigra.

β testaceus, elytris fascia fusca indistincta.

[19] A. pallens. — Pallide testaceus, convexus, albido-pubescens, oculis nigris, capite triangulari basi emarginato, angulis acutis, minus dense punctato, linea longitudinali lævi, thorace capite non angustiore, latitudine breviore, obovato, obsolete canaliculato, sat dense punctato, elytris subtilius punctatis, apice subtruncatis, abdomine nigro-fusco. Long. ·11.

[20] M. pectoralis. — Angusta, nigra dense pubescens, thorace latitudine sesqui breviore, lateribus rectis, macula parva flava utrinque versus apicem, elytris postice paulo attenuatis macula magna basali ad suturam fere extendente, sutura margineque pone medium anguste flavis, abdominis segmento singulo flavo-marginato, antennis pedibus pectoribusque flavis, his macula magna utrinque nigra. Long. ·13. Kakabeka.

[21] I have found it necessary to divide this genus, and therefore give the characters of my two groups.

ANTHOBATES.

Tarsi anteriores articulo 3to subcalceato, emarginato que; 4to minuto vix conspicuo. Ungues simplices, basi dilatati. Abdomen conicum, stylo anali nullo: coxis anticis permagnis, conicis, prosternum obtegentibus. This genus contains Anaspis 3-fasciata Mels. P. Ac. 2, 313, and two other similarly colored species.

ANASPIS.

Tarsi anteriores articulis decrescentibus, 4to perbrevi, bilobato. Ungues basi late vix dentati. Abdomen et coxæ ut supra.

[22] A. filiformis. — Linearis, rufo-testacea, dense flavo-pubescens, thorace capite parum latiore, latitudine sesqui breviore, angulis posticis rectis, elytris subtilissime transversim rugosis, abdomine fusco, antennis nigris, basi testaceis. Long. ·1.

EPICAUTA *Dej.*
cinerea *Dej. Cat.*
Lytta cin. Fabr. El. 2, 80.
[23] fissilabris.
SPHÆRIESTES.
[24] virescens.
CEPHALOON *Nm.*
lepturides *Nm. Ent. Mag.* 5, 377.
varians *Hd. J. Ac. N. S.* 1, 95.
DITYLUS *Fischer.*
cœruleus *Hd. ib.* 1, 96.
ASCLERA *Dej. Schmidt.*
puncticollis *Hd. ib.* 96.
PYTHO *Latr.*
nigra *Kb. N. Z.* 164.
MELANDRYA *Fabr.*
[25] maculata.
ORCHESIA *Latr.*
gracilis *Mels. P. Ac.* 3, 57.
XYLITA *Payk.*
[26] buprestoides *Pk: teste Kb. N. Z.* 240.
SERROPALPUS *Hell.*
substriatus *Hd. J. Ac. N. S.* 1, 98.
obsoletus *Hd. l. c.* 98.
SCRAPTIA *Latr.*
biimpressa *Hd. l. c.* 100.
CISTELA *Fabr.*
sericea *Say. J. Ac.* 3, 270.
PLATYDEMA *Lap.*
clypeata *Hd. J. Ac. N. S.* 1, 102.
[27] NELITES.
æneolus.
UPIS *Fabr.*
ceramboides *Fabr. El.* 2, 584.
Tenebrio reticu. Say. Exp. 2, 279.
variolosus *Beauv.*
[28] CRYMODES.
discicollis.

[23] E. fissilabris. — Nigra opaca, confertissme subtiliter punctata, breviter pubescens, fronte macula parva rufa, labro sparse punctato, brevi, profunde emarginato. Long. ·68. Kakábeka.

Very different from E. atrata in the form of the labrum.

[24] S. virescens.—Elongatus, niger, supra obscure virescens, nitidus, capite thoraceque dense punctatis, hoc capite non latiore, lateribus rotundatis, basi angustato, elytris thorace sesqui latioribus, subtiliter punctato-striatis, sutura interstitiisque alternis punctis paucis seriatis, antennis capite thoraceque longioribus, basi rufo-piceis. Long. ·12.

[25] M. maculata. — Fusca, nitida, punctata, breviter vix conspicue pubescens, thorace fere semicirculari, basi media late lobata, angulis posticis acutis, impressione magna utrinque a medio ad basin extendente, elytris fascia lata ad medium apiceque cum pedibus palporumque basi testaceo-pallidis. Long. ·35.

An Emmesa connectens Nm. Ent. Mag. perperam descripta ?

[26] I have not been able to compare this with European specimens.

[27] NELITES. Clypeus antice prolongatus, non marginatus. Palpi maxillares cylindrici, articulo 4 to longiore truncato. Tarsi postici articulo 1 mo elongato. Antennæ apice sensim leviter incrassatæ. Differs from Hoplocephala in having the clypeus not margined anteriorly; the antennæ are less incrassated, the penultimate joints being scarcely transverse; I know not how it differs from Phyletes (Meg.), having had no opportunity of examining the latter.

N. æneolus.— Supra obscure viridi-æneus, nitidus, ovalis convexiusculus, capite thoraceque punctatis, hoc transverso, lateribus rectis, margino anguste reflexo diaphano, elytris thorace latioribus punctato-striatis, interstitiis uniseriatim subtilissime punctulatis, subtus niger, antennis, ore pedibusque rufo-piceis. Long. ·15. Pic, to Fort William.

[28] CRYMODES. Corpus alatum elongatum. Antennæ capite sesqui longiores, granosæ, articulis 3 ultimis subabrupte majoribus. Clypeus antice truncatus, vix marginatus, labro brevi. Mandibulæ prominulæ, apice acute incisæ. Palpi maxillares apice trun-

[29] PRIOGNATHUS.
monilicornis.
Ditylus mon. Rand. B. I. 2, 22.
ATTELABUS *Lin.*
pubescens *Say. J. Ac.* 5, 252.
ARRHENODES *Stev.*
maxillosus *Sch.* 1, 326.
CLEONUS *Sch.*
obliquus.
LISTRODERES *Sch.*
humilis *Sch.* 2, 284.
ALOPHUS *Sch.*
subguttatus.
HYLOBIUS *Germ.*
heros.
assimilis *Sch.* 2, 345.
confusus *Kb. N. Z.* 196.
OTIORHYNCHUS *Germ.*
subcinctus.
PISSODES *Germ.*
nemorensis *Germ: Sch.* 3, 262.
affinis *Rand. B. J.* 2, 24.
GRYPIDIUS *Sch.*
gibbifer.
ERIRHINUS *Sch.*
sparsus.
ANTHONOMUS *Germ.*
signatus *Sch.* 7, 221.
PHYTOBIUS *Schmidt.*
inæqualis.
Ceutorynchus inæ. Say. Curc. 22.
ORCHESTES *Illiger.*
pallicornis *Sch.* 3, 505.
CEUTORHYNCUS *Schüppel.*
nigrita *Dej.*
nodicollis *Dej.*
COSSONUS *Fabr.*
platalea *Say. Curc.* 24.
RHYNCOLUS *Creutzer.*
pulvereus.
HYLURGUS *Latr.*
americanus *Dej. Cat.*
BOSTRICHUS.
conformis *Dej. Cat.*
cum duobus alteris.
CIS *Latr.*
obesus.
rugosus.
Triphyllus rug. Rand. B. J. 2, 26.
SPONDYLIS *Fabr.*
[30] laticeps.

cati, articulis subæqualibus. Mentum quadratum, latitudine fere duplo brevius, antice subrotundatum. Pedes tenues, tarsi articulo 1 mo longiore. Approaches Boros, but the antennæ are very different: the clypeus is not prolonged in front of the antennæ, and the lateral margin bends downwards before reaching the eyes.

C. discicollis. — Elongatus, piceus, punctatus, capite lateribus parallelis antice acuto, thorace capite plus sesquilatiore, transverso rotundato, basi angustato truncatoque; planiusculo, ad latera, et in disco leviter bi-impresso; elytris thorace non latioribus parallelis, versus suturam indistincte striatis. Long. ·62.

[29] PRIOGNATHUS. Corpus alatum elongatum. Caput elongatum antice acutum oculis parvis integerrimis, clypeo impresso, marginato labro valde transverso. Antennæ longe ante oculos sitæ, capite thoraceque longiores, articulo 3 io leviter elongato, 3 ultimis subrotundatis. Mandibulæ apice incisæ, intus serratæ. Palpi maxillares articulo ultimo leviter inflato, truncato. Mentum transversum, antice truncatum. Pedes tenues, tarsi articulo 1 mo longiore.

I know not what induced Mr. Randall to class this insect with the Œdemeridæ, it is plainly a Tenebrionite, although the position of the antennæ with reference to the eyes is peculiar. Only the last three joints of the antennæ are moniliform.

[30] S. laticeps. — Niger, punctatus, thorace cordato, capite non latiore, obsolete carinato, elytris costis 3 vel 4 minus distinctis, antennis capite thoraceque fere longioribus. Long. ·75.

CRIOCEPHALUS *Muls.*
agrestis *Hald. Am. Tr.* 10, 35.
TETROPIUM *Kirby.*
ISARTHRON Dej. Muls.
cinnamopterum *Kb. N. Z.* 174.
CALLIDIUM *Fabr.*
dimidiatum *Kb. N. Z.*
Clytus palliatus Hd. l. l. 41.
proteus *Kb. N. Z.* 172.
collare *Kb. N. Z.* 171.
CLYTUS *Fabr.*
speciosus *Say. Am. Ent. pl.* 53.
undulatus *Say. ib.*
undatus Kb. N. Z. 175.
Sayi Lap. Clyt.
var. lunulatus Kb. N. Z. 175.
[31] gibbulus.
ÆDILIS *Serv.*
[32] despectus.
AMNISCUS *Dej.*
macula *Hald. l. c.* 48.
Lamia macula Say. J. Ac. 5, 268.
POGONOCHERUS *Meg.*
[33] penicillatus.
TETRAOPES *Dalm.*
5-maculatus *Lec. Hd. l. c.* 53.
SAPERDA *Fabr.*
[34] adspersa.
[35] moesta.

Eagle Harbor, Mr. Rathvon. Very similar in appearance to S. buprestoides, but the form of the thorax is different. The posterior tibiæ are scarcely dilated at the end.

[31] C. gibbulus. — Niger pubescens, thorace oblongo, modice elevato, confertissime punctato, elytris confertim subtiliter punctatis, basi subgibbosis, gibberis minus elevatis, ante medium rufis, lineis 2 obliquis apiceque densius cinereo-villosis: antennarum articulo 1mo tarsisque rufis. Long. ·27.

Very similar to C. verrucosus, but the thorax is less elevated and the elytra more distinctly punctured and much less gibbous: the 3rd joint of the antennæ is not armed with a spine: in the markings there is no difference, except that the cinereous lines are less oblique.

[32] Æ. despectus. — Niger cinereo-pubescens, supra punctis pluribus nigro-pubescentibus variegatus, thorace transverso lævi, basi abrupte constricto, serieque transversa, punctorum notato; elytris apice truncatis macula oblonga sublaterali versus medium fasciaque angulata pone medium nigris. Long. ·41.

The antennæ and posterior tibiæ are annulated, the former in both sexes but little longer than the body: the femora have one or two black spots. It is found everywhere; the ♀ has the anal segment elongate and truncate.

[33] P. penicillatus. — Cylindricus, niger dense cinereo-pubescens, thorace lateribus spinoso, disco valde tuberculato, calloque parvo pone medium elevato, confertissime punctulato, elytris apice truncatis, cinereo, fuscoque variegatis, fascia lata ante medium albida; 3-carinatis, carina 1ma pilis longis nigris fasciculata, interstitiis minus dense punctatis. Long. ·27. Pic. The antennæ and feet are annulate.

I am doubtful if the next species (Tetraopes) is found at Lake Superior.

[34] S. adspersa. — Nigra, ochraceo dense pubescens, thorace latitudine sesquibreviore, grossius ocellatim punctato, punctis nigris; spatio utrinque fere ad apicem extendente minus dense pubescente; elytris postice subangustatis, sutura acuminata, sparsim grosse nigro-punctatis, huc illuc spatiis densius pubescentibus, quorum unum mox pone medium oblique versus suturam ascendit. Long ·9.

Very close to S. calcarata, but the color differs, and the thorax is shorter.

[35] S. moesta. — Nigra cinereo-pubescens, grosse confertim punctata, thoraçe latitudine vix breviore, basi leviter angustato, cinereo-bivittato, elytris apice rotundatis, antennis corpore brevioribus annulatis, basi nigris. Long. ·5. Pic.

The claws are entire, although at first view it would seem to be a Phytoecia; the head has a black, finely impressed frontal line. The eyes are almost divided.

MONOCHAMUS *Dej. Kb.*
scutellatus *Hd. l. c.* 51.
resutor Kb. N. Z. 167.
[36] mutator.
RHAGIUM *Fabr.*
lineatum *Sch. Syn.* 3, 414.
[37] ARGALEUS.
attenuatus.
Pachyta atten. Hd. Am. Tr. 10, 59.
[38] nitens.

[39] EVODINUS.
monticola.
Leptura mont. Rand. B. J. 2, 27.
[40] ACMÆOPS.
discoidea.
Pachyta disc. Hald. l. l. 60.
proteus.
Leptura Proteus Kb. N. Z. 186.
Pachyta sublineata Hd. 60.
[41] strigilata.

[36] M. mutator. — Niger, pube cinereo variegatus, thorace confertim rugoso-punctato, spinis horizontalibus dense albido-pubescentibus, scutello albo, elytris dense punctatis, punctis antice elevatis; rufo-piceis, pube cinereo, fuscoque variegatis, antennis nigris, cinereo annulatis. Long. ·98.

This is very similar to M. confusor Kb. (maculosus Hd.), but the thorax, which is smooth in that species with a few small punctures, is rugosely punctured, and the suture of the elytra is slightly prolonged. The ♂ has very long black antennæ.

[37] ARGALEUS. Caput mox pone oculos non constrictum ore attenuato, palpis labialibus modice dilatatis. Antennæ ante oculos insertæ, longæ; oculi antice emarginati, postice truncati. Tibiæ posticæ apice truncatæ, calcaribus terminaliter sitis. Thorax spinosus. Elytra triangularia.

Differs from Toxotus cylindricollis, &c., in the form of the eyes, as well as the situation of the terminal spurs of the tibiæ. The spinous thorax gives an appearance like Rhagium. To this genus belongs the European Toxotus cursor.

A. nitens. — Minus elongatus, niger subtiliter dense punctatus longe cinereo-pubescens, thorace canaliculato, antice angustato, basi apiceque profunde constricto, lateribus acute tuberculatis, elytris postice angustatis, apice subtruncata, glabris punctatis luteis, disco sæpius infuscato; antennis articulo 4to abbreviato. Long. ·6. Pic.

[39] EVODINUS. Caput mox pone oculos angustatum, ore attenuato; palpis apice oblique truncatis. Antennæ ante oculos insertæ, longæ: oculi magni vix emarginati. Thorax lateribus acute tuberculatus. Mesosternum angustum, parallelum, coxis magnis. Elytra triangularia, apice truncata.

E. monticola.—Niger, fulvo-pubescens, thorace canaliculato, utrinque constricto, elytris subtilissime rugose punctatis, flavis, utrinque maculis 2 parvis ante medium transversim sitis, alteris 2 majoribus lateralibus, apiceque nigris: antennis rufescentibus, corpore vix brevioribus. Long. ·4. On the flowers of Cornus.

[40] ACMÆOPS. Caput mox pone oculos angustatum, palpis apice recte truncatis. Antennæ ante oculos insertæ. Thorax apice constrictus, vel tuberculatus, vel gibbus, vel simplex. Mesosternum triangulare.

A numerous group, which may be divided into two sections.

A. Body thick, mouth short.—Pachyta thoracica Hd. some new species, with the European P. virginea and collaris.

B. Body more slender, mouth elongated: the species cited above, with 4-vittata.

[41] A. strigilata.—Niger, punctatus, flavo-pubescens, capite elongato, subrostrato, thorace convexo antice angustato, tenuiter canaliculato, minus dense punctato, elytris latiusculis, postice non angustatis, apice truncata, luteis humeris apiceque infuscatis. Long. ·28.

Lept. strig. Payk. Fn. Suec. 3, 112.
Pachyta strig. Muls. Long. 246.
Lept. semimarginata? Rand. B. J. 2, 30.
[42] ANTHOPHILAX.
viridis.
malachitica.
Leptura mal. Hd. l. c. 64.
Stenura cyanea Hd. P. Ac. 3, 151.
STRANGALIA *Latr. Serv. Muls.*
§*STENURA Serv.*
nigrella.
Lept. nigrella Say. J. Ac. 5, 279.
[43] plebeja.
Leptura pl. Rand. B. J. 2, 28.
cordifera.
Leptura cord. Ol. Ins. 4, 73.
6-maculata.
Leptura 6-mac Lin. Kb. N. Z. 182.
subargentata.
Leptura subarg. Kb. N. Z. 184.

LEPTURA *Lin.*
canadensis *Fabr. El.* 2, 357.
♂ *tenuicornis Hd. l. c.* 64.
proxima *Say. J. Ac.* 3, 420.
chrysocoma *Kb. N. Z.* 182.
rufula.
Pachyta ruf. Hd. l. c. 60.
pubera *Say. J. Ac.* 5, 279.
[44] tibialis.
mutabilis *Nm. Ent. Mag.*
luridipennis Hd. l. c. 63.
sphæricollis *Say. J. Ac.* 5, 280.
DONACIA *Fabr.*
proxima *Kb. N. Z.* 225.
episcopalis *Lac.* 1.
magnifica.
hirticollis *Kb. N. Z.* 226.
rudicollis Lac. Chrys. 1, 108.
porosicollis *Lac. ib.* 1, 150.
fulgens.
distincta.

The 3rd and 4th joints of the antennæ are equal, and a little shorter than the 5th. Varies with the elytra fuscous. I have diligently compared this with European specimens, without finding any difference.

[42] ANTHOPHILAX. Caput mox pone oculos constrictum, palpis dilatatis, labialibus multo latioribus. Antennæ 11-articulatæ, inter oculos insertæ; oculi emarginati. Thorax angulis posticis rectis, utrinque modice constrictus, lateribus acute tuberculatis.

To this group belongs Pachyta 4-maculata of Europe. Differs from Strangalia and Leptura by the dilated labial palpi.

A. viridis. — Nigra, capite thoraceque virescentibus, punctatis, cinereo-pubescentibus, hoc antice angustato, utrinque constricto, leviter canaliculato, lateribus subacute spinoso, elytris grosse confluenter punctatis, substriatis, splendide viridi-æneis, apice rotundata, antennis apice, tibiis basi rufescentibus. Long. ·6. Eagle Harbor.

The 3rd joint of the antennæ is longer than the 4th.

[43] S. plebeja.—Elongata, nigra, confertim punctata, thorace longe flavo-pubescente lateribus parum rotundatis, utrinque tenuiter profunde constricto, angulis posticis laminatim productis; elytris testaceis, postice sensim angustatis, paulo dehiscentibus, apice intus incisa. Long. ·55.

Precisely similar to S. nigrella, except in the color of the elytra: the pubescence of the thorax is long and prostrate; while in S. nigrella it is short and erect.

[44] L. tibialis. — Nigra, breviter flavo-pubescens, capite thoraceque confertissime punctatis, hoc convexo, antice parum angustato, apice, basique constricto, lateribus vix rotundato, elytris confertim punctatis, subparallelis, apice paulo dehiscentibus, introrsum oblique leviter truncatis, flavo-testaceis, macula laterali ad medium, altera majore pone medium, apiceque nigris, tibiis tarsisque flavis, illis apice fuscis. Long. ·43.

pusilla *Say. J. Ac.* 5, 293.
fulvipes *Lac. Chrys.* 1, 192.
cuprea *Kb. N. Z.* 225.
gracilis.
aurifer.
gentilis.
emarginata *Kb. N. Z.* 224.
flavipes *Kb. N. Z.* 223.
jucunda.
confusa.

ORSODACNA *Latr.*
tibialis *Kb. N. Z.* 221.
testacea.

SYNETA *Esch. Lac.*
rubicunda *Lac. l. c.* 1, 230.

[45] TARAXIS.
abnormis.

CRYPTOCEPHALUS *Fabr.*
4-maculatus *Say. J. Ac.* 3, 441.
tridens. *Mels. P. Ac.* 3, 172.

PACHYBRACHYS. *Dej.*
M-nigrum *Hd. J. Ac. N. S.* 1, 261.
abdominalis *Hd. ibid.* 263.

HETERASPIS *Dej.*
pumilus *Dej. Cat.*

PACHNEPHORUS *Dej.*
10-notatus.
Colaspis 10-not. Say. J. Ac. 3, 445.
Pach. variegatus Dej. Cat.

METACHROMA *Dej.*
gilvipes *Dej. Cat.*
canella *Dej. Cat.*
Crypt. canellus F. El. 2, 52.
4-notata.
Colaspis 4-not. Say J. Ac. 3, 446.

NODA *Dej.*
puncticollis *Dej. Cat.*
parvula *Dej.*

FIDIA *Dej.*
lurida *Dej.*

COLASPIS *Fabr.*
lineata.

PHYLLODECTA *Kb.*
[46] vitellinæ *teste Kb. N. Z.* 216.

HELODES *Fabr.*
trivittata *Say. J. Ac.* 5, 298.

PHYTODECTA *Kb.*
[47] rufipes *teste Kb. N. Z.* 213.

LINA *Meg.*
discicollis.
consanguinea.

CHRYSOMELA *Lin.*
scalaris *Lec. An. Lyc.* 1.
spirææ *Say.*
confinis Kb. N. Z 211.
elegans *Oliv.* 91, 94. *fig.* 92.

PLECTROSCELIS *Chevr.*
chalcea *Dej.*

[45] TARAXIS.— Antennæ basi distantes, breviusculæ articulo 1 mo majore crassiore, 3 io secundo sesqui longiore 5 to 4 to que paulo brevioribus, reliquis longitudine crassioribus. Oculi emarginati. Coxæ anticæ parvæ globosæ, approximatæ, prosterno non prominulo. Abdomen articulo 5 to majore inferne emarginato, segmentulo anali aucto. Tarsi articulo 3 io lato, parum emarginato, unguibus late appendiculatis. Palpi apice acuminati. Thorax elytris angustior a medio ad basin valde angustato constrictoque, apice iterum leviter constricto, elytris cylindricis apice rotundatis.

T. abnormis. — Testacea, nitida, grosse punctata, thorace linea minus distincta lævi, elytris ad scutellum et pone medium, cum vertice, pectoribusque rufescentibus. Long. ·15. Pic. Looks like a minute Syneta, but at once distinguished by the abdomen, tarsi and antennæ.

[46] I give this as identical with the European on Kirby's authority. I have not been able to compare specimens.

[47] I have had no opportunity of comparing with European specimens.

confinis *Dej.*

DISONYCHA *Chevr.*

5-vittata.

Altica 5-vit. Say. J. Ac. 4, 85.

GRAPTODERA *Chevr.*

cuprea.

ignita.

Alt. ignita Ill. Mag. 6, 117.

GALLERUCA *Fabr.*

canadensis *Kb. N. Z.* 219.

cribrata *Dej.*

gelatinariæ *Fabr. El.* 1, 490.

[48]sagittariæ *Gyll. teste Kb.* 219.

notulata *Fabr. El.* 1, 489.

Olivieri *Kb. N. Z.* 218.

HIPPODAMIA *Chevr. Redt.*

abbreviata *Dej. Cat.*

Coccinella abb. Fabr. E. 1, 360.

parenthesis.

Coccinella par. Say. J. Ac. 4, 93.

C —— *tridens Kb. N. Z.* 229.

5-signata.

Coccinella 5-sig. Kb. N. Z. 230.

13-punctata *Dej. Cat.*

Cocc. 13-punc. Lin. Fn. Su. 481.

C — *tibialis Say. J. Ac.* 4, 94.

COCCINELLA *Lin.*

9-notata *F. El.* 1, 366.

5-notata *Kb. N. Z.* 230.

3-fasciata *F. El.* 1, 363.

tricuspis *Kb. N. Z.* 231.

incarnata *Kb. ib.*

venusta *Mels. P. Ac.* 3, 178.

notulata Dej. Cat.

15-punctata *oliv.*

mali Say. J. Ac. 4, 93.

pullata *Say. J. Ac.* 5, 302.

notans Rand. B. J. 2, 49.

confuse-signata.

picta *Rand. B. J.* 2, 51.

concinnata Mels. P. Ac. 3, 177.

immaculata *Fabr. El.* 1, 357.

PSYLLOBORA *Chevr.*

20-maculata.

nana Dej. Cat.

Cocc. 20-mac. Say. J. Ac. 4, 98.

BRACHIACANTHA *Chevr.*

bis-5-pustulata *Fabr. El.* 1 384.

ursina F. ib. 386.

var. minor.

disconotata.

consimilis.

[49] OXYNYCHUS.

moerens.

SCYMNUS.

[50]caudalis.

[48] Nor have I compared this species.

[49] Corpus alatum breviter oblongum antice subangustatum glabrum. Antennæ capite breviores articulo 2 ndo majusculo, ultimo ovali majore. Ligula emarginata. Scutellum distinctum. Ungues simplices. Epipleuræ impressæ. Abdomen articulo 1 mo laminarum margine externo curvato.

O. moerens. — Niger nitidus, punctulatus, thoracis margine, elytrorumque gutta minuta pone medium testaceis, antennis tarsisque rufis. Long. ·1. St. Ignace.

Variat, α elytris gutta altera parva humerali testacea, margine pone medium rufescente. β Niger immaculatus.

[50] S. caudalis, — Breviter ovalis, convexus, punctatus, niger, thorace lateribus, antennis, palpis pedibus, abdominisque segmentis 2 ultimis rufis, mesosterno lato, fere truncato, abdominis laminis integris, basi punctatis, ad marginem segmenti 1 mi fere extendentibus. Long. ·09. ♂ articulo ultimo abdominis late profunde emarginato, pedibus capiteque rufis.

Pl. 8.

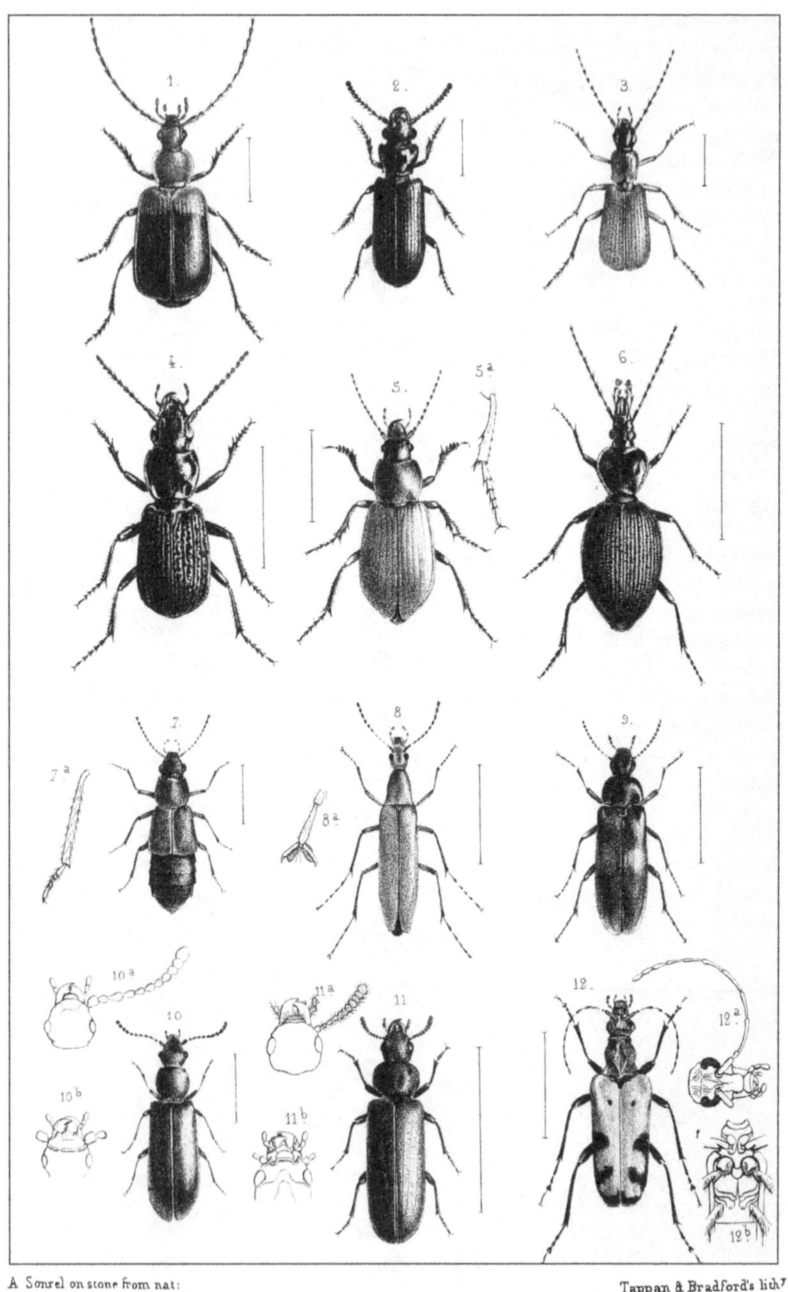

A Sonrel on stone from nat:

Tappan & Bradford's lith.

1. Lebia concinna Lec
3. Platynus nigriceps Lec
5. Amara elongata Lec
7. Latrium convexicolle Lec.
9. Melandrya maculata Lec.
11. Cryphæus disicollis Lec.

2. Haplochile pygmæa Lec.
4. Pterostichus punctatissimus Randall.
6. Cychrus bilobus Say.
8. Cephaloon lepturides Newmann.
10. Priognathus monilicornis Lec.
12. Evodinus monticola Randall.

[51]lacustris.
punctum.
[52]ornatus.

ORTHOPERUS.
flavidus.
CORYLOPHUS.
lugubris.

Plate 8th represents twelve new species of the Coleoptera, described in the preceding Catalogue.

On glancing over the catalogue which is just ended, the entomologist cannot fail to be struck with two very remarkable characters displayed by the insect fauna of these northern regions. First, the entire absence of all those groups which are peculiar to the American continent. Thus, there is no Dicælus, no Pasimachus among the Carabica; the Brachelytra are represented only by forms common to both continents. Among the Buprestidæ is no Brachys; in the Scarabæidæ, the American groups (except Dichelonycha) are completely unrepresented; in brief, there is scarcely a genus enumerated which has not its representative in the Old World. The few new genera which I have ventured to establish, are not to be regarded as exceptions, they are all closely allied to European forms, and by no means members of groups exclusively American.

Secondly, the deficiency caused by the disappearance of characteristic forms, is obviated by a large increase of the members of genera feebly represented in the more temperate regions, and also by the introduction of many genera heretofore regarded as confined to the northern part of Europe and Asia. Among these latter are many species which can be distinguished from their foreign

[51] S. lacustris.—Breviter ovalis, convexus, punctatus, niger, mesosterno lato fere truncato, abdominis laminis integris, basi punctatis, ad marginem segmenti 1mi fere extendentibus. Long. ·09. ♂ articulo ultimo abdominis profunde triangulariter impresso; basi minus dense punctata; pedibus vel rufis, vel piceis, rufo-marginatis.

♀ abdomine integro, æqualiter dense punctato, antennis pedibusque nigris, posticis nonnunquam rufo

[52] S. ornatus.—Ellipticus, convexus dense subtiliter punctatus, niger elytris utrinque macula magna obliqua ante medium, alteraque magna orbiculata pone medium læte rufa, antennarum basi tibiis tarsisque fusco-rufis, abdominis laminis extrorsum omnino obliteratis, mesosterno lato, parum emarginato. Long. ·08.

analogues only by the most careful examination. This parallelism is sometimes most exact, running not merely through the genera, but even through the respective species of which they are composed; thus of the two species of Olisthærus, each is most closely related to its European analogue, *O. laticeps* being similar to *O. megacephalus*, while *O. nitidus* can scarcely be known from *O. substriatus*.

While upon this subject, we may take occasion to distinguish the different kinds of replacement of species, which are observed in passing from one zoölogical district to another more or less distant. There appear to be four distinct modifications by which faunas are characterized.

1st. When the same species, or organic forms, so similar as to present no appreciable difference, appear at points so situated as to preclude the possibility of any intercommunication. These are most rare, and are only observed when the physical circumstances under which the species exists are nearly identical.

2d. When a species in one district is paralleled by another in a different region so closely allied that upon a superficial glance they would be regarded as the same. These are called *analogous species;* e. g., the Olisthæri, Spondyli, Bembidia, Helophori, &c., &c., of the preceding catalogue, as compared with European species.

3d. Where several species in one region are represented by several others of the same genus, which perform a similar part in the economy of nature, without, however, displaying any farther affinity to each other. These are called *equivalent species;* e. g., most of the species of Cicindela, Brachinus, Clytus, Donacia, &c., of America, as compared with those of the eastern world.

4th. Where the members of a group are represented collectively by kindred species in another district, which however display such differences of structure that each may at once be referred to its proper locality; e. g., most of the Melolonthæ among Coleoptera, and the entire group of Quadrumana among mammalia.

Now it will be observed, that in proceeding from the Arctic circle to the tropics, the prominent character of the fauna is successively modified by these peculiarities. We pass from a region where the fauna is the same at remote points, through one where the productions are similar, but not identical, to one finally, where the equilibrium of forms is still preserved, but where the general arrangement is totally different, the prominent groups of one continent being either feebly represented on the other, or else entirely wanting.

It does not become us, in the present imperfect state of tropical exploration, to determine what groups are peculiar to each continent; we can merely say that particular forms are more abundant in certain regions. For by a strange fatality, (at least in Coleoptera,) no sooner is any group admitted by a common consent to be exclusively American, than suddenly, as if produced by the well-known jugglery of those countries, a species starts up in Central Asia, or Africa, (e. g., Galerita, Agra, Sandalus.) Still, enough remains to show us that the prevailing character of tropical fauna is individuality; the production of peculiar forms within limited regions: while the distinguishing feature of temperate and arctic fauna is the repetition of similar or identical forms through extensive localities.

On proceeding now to illustrate these deductions by special examples from the catalogue before us, it will be seen that the parallelism of species in temperate and frigid climates can be demonstrated more particularly in the genera which are more universally diffused over the earth, or in those which are especially confined to temperate regions, than in such as receive their principal development within the tropics. Thus for instance, among the great group of Carnivorous Coleoptera, the terrestial species, (although well represented in cold climates,) contain an immense number of genera, each of which (with few exceptions) seems to have a particular locus, external to which it is feebly represented. Accordingly in this group, the parallelism of species is by no means clear, and the forms are rather to be considered equivalent than analogous. On the other hand, among the aquatic Predaceous Coleoptera, the genera are but few, and the tribe is more abundant in cold regions; and in these the parallelism is most exact, so that there are but few mentioned in the preceding pages, that have not their exact counterparts in Europe. The characters appended to the new species will render this sufficiently obvious to the student, while the relations of those previously described by Kirby and Aubé have afready been clearly pointed out by those authors.

Passing on to the other water-beetles, the species of Helophorus and Ochthebius will afford other striking examples of this parallelism. Among the Brachelytra are numerous other instances, the most remarkable being the genus Olisthærus, already alluded to. Proteinus and Megarthrus also for the first time appear on this continent. The Aphodii with large scutellum, the Ditylus, Pytho, Sphæriestes, and Spondylus are also good illustrations. Among the Elateridæ are numerous instances, but having not yet submitted this group to philosophical study, I have not ventured to describe the new species, but have merely indicated them by names. For the present therefore, any remarks on the parallelism of the forms in this group must be postponed. Notwithstanding this approximation to a uniform, subarctic standard, we still find in these boreal regions, a prevailing character of North American fauna—the extreme paucity of Curculionidæ. The Donaciæ too, although numerous, do not afford any prominent parallelism. The American species can only be regarded as equivalent to the European.

On concluding this short essay on the geographical distribution of Coleoptera in the northern part of our continent, I feel that some cause must be assigned for the brief manner in which such extensive material has been disposed of. Enough has been given to point the laws of distribution, and to show that they accord most perfectly with those derived from other branches of natural history, while during the yet imperfect condition of entomological science in this country, a minute analysis of the components of the entire fauna would be a work of immense labor, and would in fact be rendered nugatory, until all the species are described, and all the groups submitted to a philosophical revision. My complete success in tracing the parallelism between the Pselaphidæ of Europe and North America (in an unpublished monograph of this family) leads me to believe that a rich store of material is herein presented to such minds as are satisfied with statistical comparisons between the inhabitants of different zoölogical districts; and that nothing but industry and a free access to the most common European insects is required to produce a most formidable list of analogous species. I shall rest satisfied with having shown that this parallelism exists even more accurately than in the vertebrate class, and with having pointed out examples far more numerous than those furnished by the higher animals: the more so, since I feel that one already conversant with entomological names will find no difficulty in extending the already long list of parallel species, while to the general reader, who desires only the deductions of science, without entering upon the tedious processes by which they are obtained, a catalogue of mere technicalities, which fail to convey a single idea to his mind, will be equally useless and uninteresting.

I purposed in the present essay to trace, as far as possible, the mechanism of the agency by which the present distribution of species has been effected, and to reduce its most obvious results to some fixed principles. Fearful, however, lest my views should be considered as derived exclusively from a consideration of insects, and their phenomena of distribution, I prefer waiting until a sufficient familiarity with other sciences will enable me to be less partial in my choice of illustrations. I do this with the less regret as I find some of my deductions are at variance with many of the most ancient, and most firmly established prejudices of our nature, and before venturing any assertion, which even in appearance deviates from "general impressions," it is at least prudent to be supported by facts drawn from more extended observation than is furnished by one or two limited departments of knowledge.

In the rapids at Niagara have been observed large numbers of the singular animal described by Dekay (in the Zoölogy of New York) as a new genus of Crustacea, under the name of Fluvicola Herricki. They were attached to stones just below the surface of the water, and crawled but slowly; when seized, they endeavor to contract themselves into a ball.

These animals have a marvellous resemblance to the extinct group of Trilobites, although, as will be seen in the sequel, they are the larvæ of an insect. Mr. Agassiz informs me that a similar form has long been known to the zoölogists of Continental Europe as Scutellaria amerlandica, but I have not been able to find any published account of it.

On turning over some stones near the river bank, I was agreeably surprised to find many specimens which had left the water for the purpose of changing into pupæ. The elliptical shield of the superior surface, which gives the animal its Crustacean appearance, was firmly adherent to the stone by its ciliated margin, and formed an excellent protection under which the later transformations could take place with safety. In fact, the superior shield being cast off with the larva skin, served in place of the cocoon or nest constructed by many larvæ, before transforming.

I regret that in the short account given by me at the recent meeting of naturalists in Cambridge, I was induced to speak of this discovery, without having access at the time to specimens. Those which I expected to find at Boston had been lost, and my former examination of the pupæ collected by myself was very slight. I referred the insect to the order of Neuroptera, and I must here return my sincere thanks to my friend Dr. Harris, for a hint towards its true nature.

For the opportunity of examining some very large and well developed larvæ, I am indebted to my friend I. C. Brevoort, who procured them at Niagara in July of the previous year.

The body proper of the larvæ is elongate, the head being free, (i. e. not retractile,) but concealed under the large shield, like a prolongation of the dorsal epidermis of the prothorax. On each side are six small, approximate ocelli, anterior to which is the antenna, a little longer than the head, and two-jointed; each joint having a tendency to become divided at its middle, so that on a superficial inspection there would appear to be four joints. These organs are inserted at the outer extremity of the clypeo-cranial suture; the labrum is large, and a little emarginate in the middle. The lower part of the head is covered by a large mentum, which prevents the mandibles and maxillæ from being seen. The maxillary palpi are half the length of the antennæ, filiform, rather stout, and three-articulated, the joints being equal. The labial palpi are bent down and covered by the epidermis. In the very young larvæ the palpi are still shorter in proportion to the antennæ. A more full description of the parts of the mouth must be reserved for a separate treatise, when their structure can be illustrated by plates. The abdomen is furnished on each side with six bunches of long branchial filaments, which proceed from the interstices between the articulations; there is a larger bunch of filaments connected with the anal aperture, which may be retracted, and is ordinarily not visible in dead specimens; exterior to these filaments on each articulation is a small fovea. The articulation itself is prolonged each side, for a short distance between the laminæ of the expanded epidermis, so that the outline of the proper fleshy portion is serrate. The legs are slender, the tarsus inarticulate, and furnished with a single claw.

The pupa is broadly oval, and depressed. The head is concealed under a hood formed by the prolongation of the epidermis of the prothorax. This hood is produced at the posterior angles, so that it becomes exactly similar to the thorax of a Lampyris.

The front between the antennæ is transversely elevated, so that the mouth is situated on its inferior surface. The antennæ are three times longer than the head, and inside of the pupa skin (in much developed specimens) are seen to be serrate, and eleven-jointed; the palpi are two-thirds the length of the antennæ, and are somewhat dilated at the extremity. The labial palpi are very short. The labrum is transversely cordate. The wings are bent under the body. The superior ones exhibit the structure of elytra, and have four slight longitudinal ribs: the inferior are membranous, and show a slight transverse nervure near the middle. The abdomen is six-jointed and serrate at the sides, owing to the angular prolongation of each joint, and is entirely free from branchial appendages. The last joint is rounded. The feet are slender, and not armed with a claw. The mesopectus is deeply channeled.

After the description just given of the pupa, no one will doubt that the insect belongs to the Coleoptera; and from the serrate outline of the abdomen, one would be inclined to refer it to the groups possessing larvæ like the Lampyris, Lycus, &c. The separation of the prothorax and its great development, as well as the structure of the superior wings, absolutely exclude it from the Neuroptera, to which I at first referred it.

The peculiar structure of the head of the pupa, and the great length of the palpi, point clearly to Eurypalpus, a curious genus, which is placed by authors near Cyphon, which, as is well known, is closely allied to the Lampyridæ. Eurypalpus differs very much from all the allied genera, in being aquatic. It is furnished with slender legs, but the tarsi are long, especially the last joint, which has two very strong claws, (as in Macronychus) to fit it for clinging to stones in a rapid current. The mesopectus of Eurypalpus is likewise deeply channeled. The elytra are also furnished with three or four very obtuse elevated lines. As yet there is but a single species of the genus known. E. Lecontei, (Dej. Cat.) I am not aware that any description has been published of it.

Thus is settled the history of the transformations of an anomalous form, which has much perplexed naturalists for many years. Its history shows the care with which our investigation should be made, when we are upon unknown ground. But where the homologies of the animal with other aquatic larvæ provided with branchia are so exact, it is a little remarkable that its larval character should remain so long unnoticed. The *only difference* between it and an ordinary larva (either of Coleoptera or Neuroptera) is the prolongation of the dorsal epidermis, to form a shield under which the true body is concealed. Similar prolongations are found in nearly all orders of insects.

I know not how Dr. Dekay fell into the mistake of considering the elongate palpi as a second pair of antennæ: and surely such an anomalous form as a Crustacean with *six legs*, and a head *separate* from the thorax, deserved a more careful examination, before receiving a definite place in the system.

The figure in the New York Zoölogy, (as the animal is very peculiar in its form,) bears a certain vague resemblance to what it was intended to represent; but for all systematic purposes, it is, like nearly all the plates in that part of the State Survey, perfectly worthless.

V.

CATALOGUE OF SHELLS, WITH DESCRIPTIONS OF NEW SPECIES.

BY DR. A. A. GOULD.

Helix albolabris, *Say*. Northern shore, Michipicotin.
" tridentata, *Say*. Niagara, Mackinaw.
" thyroidus, *Say*. Niagara, Mackinaw.
" alternata, *Say*. Niagara, Mackinaw.
" palliata, *Say*. Niagara.
" monodon, *Rackett*. Niagara, Mackinaw.
" perspectiva, *Say*. Niagara, Mackinaw.
" striatella, *Anthony*. Fort William, Cape Gourganne, N. E. of St. Ignace.
" concava, *Say*. Niagara.
" arborea, *Say*. Mackinaw, Fort William, Cape Gourganne, St. Ignace.
" electrina, *Gould*. Cape Gourganne.
" chersina, *Say*. Michipicotin, Cape Gourganne.
[1] Vitrina limpida, *Gould*. Cape Gourganne.
Succinea ovalis, *Gould*. Fort William.
" obliqua, *Say*. Niagara, Northern Coast.
" avara, *Say*. Niagara.
Physa heterostropha, *Say*. Black River, Pie Island, Fort William.

[1] VITRINA LIMPIDA, *Gould* (*V. pellucida*, Say, in Long's Expedition. II. 258.) Having made a critical comparison of our Vitrina with the V. *pellucida* of Europe, with which species it has hitherto been regarded as identical, I am induced to believe that they are different species. The American shell is more globose ; the plane of the aperture is more oblique, and the basal portion of the lip sweeps round from the columella in a rapidly curving arc, instead of stretching off almost horizontally; indeed the whole aperture is more nearly circular. These differences become quite obvious when the shell is greatly magnified. The color of the European shell is always more or less green or yellow, whereas the American specimens are colorless, and decidedly more fragile. In size, they are about one fourth smaller than the foreign ones, and have, least, half a whorl less in the spire. It is indeed more nearly like V. *subglobosa*, Mich. which, however, has a much more elevated spire, and its basal face much more inflated.

[2]Physa vinosa, *Gould*. Northern coast, Michipicotin.
" ancillaria, *Say*. Niagara, Sault St. Marie, Michipicotin.
Limnea jugularis, *Say*. Northern Coast.
" caperata, *Say*. Niagara, Black River.
" humilis, *Say*. Michipicotin, Cape Gourganne.
[3] " catascopium, *Say*. Northern shore, Fort William.
" desidiosa, *Say*. Northern shore.
[4] " lanceata, *Gould*. Pic, Gourganne.

[2] PHYSA VINOSA, *Gould*, T. tenui, ovato-globosâ, badiâ, spiraliter minutissimè striatâ, epidermide tenui indutâ; spirâ obtusâ, anfr. 4, ultimo permagno; apertura ovato-lunatâ, ⅘ longitud. testæ adequante, hepaticâ; columellâ rectâ, tenui. Long. ⅘, lat. ½ poll. *Proceed. Bost. Soc. Nat. Hist.*, II., 263, *Dec.* 1847.

I quote the above description of a species first brought from Lake Superior by Dr. C. T. Jackson, and hitherto found only in the region of that lake. Prof. A. found it on the north shore, at Michipicotin. It is well characterized by its inflated form, delicate structure, striated surface, its wine-red color externally, and its liver-brown color within. It resembles, somewhat, P. *ancillaria*, which differs in form by having shouldered whorls, and its greatest diameter behind the middle. Unfortunately, the figure has been drawn from a very small specimen, and does not exhibit the characters of a full-grown specimen. (See pl. 7, figs. 10 and 11.)

[3] LIMNEA CATASCOPIUM. There is no slight difficulty in defining the limits of allied species in this genus. While real specific characters are very few and ill defined, the variations of species are very numerous and wide in their range; nevertheless, by a certain facies, or by collecting large numbers at a given locality, we are able to pronounce shells which are very different in their aspect to be specifically identical. These remarks apply with special force to the species above named. Some of the specimens are elongated and slender, while others are short and ventricose; some are thin and fragile, others dense and firm; some are smooth or with a delicately corrugated epidermis; others are indented and broken into numerous facets; some have a very largely developed fold on the pillar, while others present a simple column; in some the columella is curved and flexuous, in others it is direct; some have regular and symmetrical outlines made up of cylindrical whorls, while others have a very acute angle and a broad shoulder at the posterior part of the body whorl; and the color may be amber, brownish, livid or cinereous. There can be little doubt that these wide variations have been regarded as different species, as indeed they could not fail to be, were only isolated specimens examined; but when we come to compare large numbers collected in company, we see the connecting links and the necessity of retaining them under one name. Among them we find L. *pinguis*, Say, which Mr. Haldeman has already referred to this species; and also L. *emarginata*, Say, which, from the few specimens he had seen, Mr. Haldeman deemed to be a well marked species. The numerous specimens since brought from the Lake Superior region render it sufficiently certain that it is only a variety of L. *castascopium*, with the last whorl more or less angular posteriorly, and with a straight pillar which gives to the base of the aperture a peculiarly broad and distorted form.

Amid all the variations, however, there is a certain aspect of the aperture which is characteristic. It is large when compared with that of L. *umbrosa*, or L. *elodes*; it is nearly semicircular, while in large specimens of L. *desidiosa*, where the proportional size of the aperture is more nearly the same, its posterior outline is broad and nearly transverse.

[4] LIMNEA LANCEATA, *Gould*. Testa mediocri, fragili, diaphanâ, corneâ, attenuatâ, striis incrementi et striis volventibus argutè reticulatâ; spiræ anfr. 6 planiusculis, per-

Planorbis bicarinatus, *Say*. Sault St. Marie, Black River.
" parvus, *Say*. Sault St. Marie.
Valvata tricarinata, *Say*. Black River.
Amnicola grana, *Say*. Fort William, Cape Gourganne.
Paludina ponderosa, *Say*. Niagara.
Melania livescens, *Menke*. (niagarensis, *Lea*.) Niagara.
" subulata. Niagara.
Cyclas similis, *Say*. Sault St. Marie.
" partumeia (young) ? *Say*. Fort William.
[5] Pisidium dubium, *Say*. Fort William, Michipicotin.
Unio radiatus, *Gmel*. Northern shore.
Anodonta Pepiniana, *Lea*. Northern shore, Cape Gourganne.

The number of bivalve shells seems to diminish very abruptly at the chain of the great lakes; so that of the great number of species, so profuse also in the number of individuals, in the States bordering on the south, scarcely ten species, and those not abundant, are found to the north; and all these are meagre in development, and of the simplest form and color.

obliquis, ultimo $\frac{3}{4}$ testæ æquante; aperturâ angustâ, dimidiam longitudinis fere adequante, posticè acuta, plica columellari conspicuâ, acutâ, vix spirali; labro fasciâ castaneâ submarginali picto. Long. $\frac{4}{5}$, lat. $\frac{1}{4}$, poll. *Proceed. Bost. Soc. Nat. History*, III. 64. Oct. 1848. (See pl. 7, figs. 8 and 9.)

A medium sized species, with an elongated, delicate, minutely reticulated shell, composed of about six very oblique flattish whorls, the last of which constitutes three fourths of the whole shell. The aperture is narrow, having a sharp, slightly winding fold on the pillar, and a submarginal brown stripe just within the lip.

Next to L. *gracilis*, this is the most delicate species we have. It may be compared with L. *attenuata* and L. *reflexa*, from both of which it differs in the flatness of its whorls, in its narrow, elongated aperture, and in being only half their size. It is much like *Physa hypnorum* reversed.

[5] Pisidium dubium. The separation from Cyclas of some species under the name of *Pisidium* being regarded as legitimate, I place this shell under that genus. The shells brought from Lake Superior seem, however, to differ somewhat from specimens from the Atlantic region. They are smaller, more elevated, less sulcated, and the hinge is less robust. I had designed to apply to them the specific name P. *tenellum*, but unfortunately the specimens were mislaid before I had examined them with sufficient care to give the characters with the requisite precision.

VI.

FISHES OF LAKE SUPERIOR COMPARED WITH THOSE OF THE OTHER GREAT CANADIAN LAKES.

Besides the interest there is everywhere in studying the living animals of a new country, there is a particular interest to a naturalist in ascertaining their peculiar geographical distribution, and their true affinities with those of other countries. It is only by following such a course, that we can hope to arrive at any exact results as to their origin. In this respect the freshwater animals have a peculiar interest, as from the element they inhabit, they are placed under exceptional circumstances.

Marine animals, as well as those inhabiting dry land, seem to have a boundless opportunity before them to spread over large parts of the earth's surface, and their locomotive powers would generally be sufficient to carry them almost anywhere; but they do not avail themselves of the possibility; notwithstanding their facilities for locomotion, they for the most part remain within very narrow limits, using their liberty rather to keep within certain definite bounds. This tendency of the higher animals especially, to keep within well-ascertained limits, is perhaps the strongest evidence that there is a natural connection between the external world, and the organized beings living upon the present surface of our globe. The laws which regulate these relations, and those of geographical distribution in particular, have already been ascertained to a certain extent, and will receive additional evidence from the facts recorded during our journey.

The freshwater animals are placed in somewhat different circumstances. Their abode being circumscribed by dry land within limits which are often reduced to a narrow current of water, and being farther, for the most part, prevented by structural peculiarities from passing from the rivers into the ocean, they are confined within narrower limits than either terrestrial or marine types. Within these limits again they are still farther restricted; the shells and fishes of the head-waters of large rivers, for instance, being scarcely ever the same as those of their middle or lower course, few species extending all over any freshwater basin from one extreme of its boundary to the other; thus forming at various heights above the level of the sea, isolated groups of freshwater animals in the midst of those which inhabit the dry land. These groups are very similar in their circumscription to the islands and coral reefs of the ocean; like them they are either large or small, isolated and far apart, or close together in various modes of association. In every respect they form upon the continents as it were a counterpart of the archipelagoes.

From their circumscription, these groups of lakes present at once a peculiar feature in the animal kingdom, their inhabitants being entirely unconnected with any of the other living beings which swarm around them. What, for instance, is there apparently in common between the fishes of our lakes and rivers, and the quadrupeds which inhabit their shores, or the birds perching on the branches which overshadow their waters; or what connection is there between the few hermit-like terrestrial animals that live upon the low islands of the Pacific, and the fishes which play among the corals, or in the sand and mud of their shores? And nevertheless there is but one plan in the creation; freshwater animals under similar latitudes are as uniform as the corresponding vegetation, and however isolated and apparently unconnected the tropical islands may seem, their inhabitants agree in their most important traits.

The best evidence that in the plan of creation animals are intended to be located within circumscribed boundaries, is farther derived from their regular migrations. Although the Arctic birds wander during winter into temperate countries, and some reach even the warmer zones; although there are many which, from the colder temperate climates, extend quite into the tropics, there is nevertheless not one of these species which passes from the northern to the southern hemisphere; not one which does not return at regular epochs to the countries whence it came from. And the more minutely we trace this geographical distribution, the more we are impressed with the conviction that it must be primitive, that is to say, that animals must have originated where they live, and have remained almost precisely within the same limits ever since they were created, except in a few cases, where, under the influence of man, those limits have been extended over large areas. To express this view still more distinctly, I should say that the question to be settled is, whether for instance the wild animals which live in America originated in this continent, or migrated into it from other parts of the world; whether the black bear was created in the forests of New England and the Northern States, or whether it is derived from some European bear, which by some means found its way to this continent, and being under the influence of a new climate, produced a new race; whether the many peculiar birds of North America which live in forests composed of trees different from those which occur either in Europe or Asia, whether these birds, which themselves are not identical with those of any other country, were or were not created where they live; whether the snapping turtle, the alligator, the rattlesnake, and other reptiles which are found only in America, have become extinct in the Old World after migrating over the Atlantic, to be preserved in this continent; whether the fishes of the great Canadian lakes made their appearance first in those waters, or migrated thither from somewhere else? These are the questions which such an inquiry into the geographical distribution of animals involves; it is the great question of the unity or plurality of creations; it is not less the question of the origin of animals from single pairs or in large numbers; and, strange to say, a thorough examination of the fishes of Lake Superior, compared with those of the adjacent waters, is likely to throw more light upon such questions, than all traditions, however ancient, however near in point of time to the epoch of creation itself.

In order to proceed methodically in this investigation, our first step must be to examine minutely, whether the fishes of Lake Superior are the same as those of other lakes in this or any other country, and if not, how they differ. To satisfy ourselves in this respect, we shall successively examine all the families of fishes which have representatives in those great freshwater seas.

PETROMYZONTIDÆ (Lamprey-eels.)

There are families in all departments of nature, whose peculiarities call for an investigation of their more general relations rather than of their structural details. The Petromyzons are in this case. Closely allied together and circumscribed in a most natural family, it is a question whether they should be entirely separated from all other fishes to form a great group by themselves, or whether they belong to one of those great divisions in which the individual members differ widely from each other. In other words, should the Petromyzons stand by themselves in a natural classification of fishes, as Prince Canino and Joh. Müller have placed them, or shall we combine them with skates and sharks, as Cuvier has done? To answer such a question, it is necessary to discuss beforehand principles of the utmost importance in the study of natural history, and above all to settle the following difficulty:—Is the study of anatomical structure an absolutely safe guide in the estimation of the relations of animals to each other? Cuvier, who made the study of comparative anatomy the foundation of classification, carried out this principle in a most remarkable manner, and improved the natural arrangement of animals most surprisingly; indeed, he made zoölogy truly a science by it; but with a tact that characterizes genius, he limited the absolute consequences of this law by a true appreciation of the relative value of characters; introducing at the same time with the principle of classification according to the structure of animals, that of subordination of characters, without which the first great principle might mislead us, instead of helping to ascertain the true relations of organized beings. Now it seems to me as if zoölogists and anatomists had of late insisted too strictly upon the absolute differences which exist between animals, instead of attempting to appreciate the relative value of the differences noticed. Of course, as this latter point rests almost within the limits of individual appreciation, it is more difficult to find the right path here, than in almost any other department of zoölogical investigations; but I hope to be able to introduce another great principle of zoölogical classification, which shall afford a safe guide to settle such doubts; I mean the study of embryonic development.

Let me now show, in the present instance, how I consider it possible

17

to be led by anatomical evidence considered in its absolute results, to combinations strictly opposed to those which an additional acquaintance with embryonic development might indicate.

Guided by his admirable natural feeling of affinities, Cuvier placed in one and the same great division, sharks, skates, and lamprey-eels. Influenced by anatomical investigation, and indeed by the most minute and admirable knowledge of their anatomical structure, derived from unparalleled investigations, Joh. Müller concluded, on the contrary, that the Cyclostomata were to be separated from the other cartilaginous fishes, and placed by themselves at the other end of the class. Who is right in this case cannot be ascertained by any farther anatomical investigation; it has thenceforth become a matter of individual appreciation, unless we introduce another principle, by which we can weigh the real value of these remarkable differences. Such a principle, I think, we have in the metamorphosis of embryonic life. Indeed, if it can be shown, that besides the differences which exist in all fishes between their earliest forms and their full-grown state, there are peculiarities in sharks, skates, and lamprey-eels common to all of them, from an early period of development, which remain characteristic throughout life, it must be acknowledged that these families belong to one and the same great group, notwithstanding their extreme differences in their full-grown condition. Now, such facts exist. In the first place, it is impossible, without disturbing their true affinities, to consider an extraordinary development of pectoral and ventral fins as a standard to appreciate fundamental relations between fishes, *as in all fishes, without exception, they are both wanting in earlier life*, and as there is scarcely a family in which ventrals at least, are not wanting in some genus or other. We might just as well place Petromyzons among the eels, as their common English name purports, on the ground of the deficiency of their abdominal and thoracic organs of locomotion, as separate them from the other Placoids. Again, the peculiarities in the development of the dorsal, caudal, and anal fins in sharks and skates, and the differences which exist between them and the Petromyzons, indicate in no way their affinity or their difference; in Petromyzon we have the embryonic condition of vertical fins, where a continuous fold in the skin of the middle line extends, as in all embryo fishes, from the back round the tail, towards the abdominal region. In the sharks we have distinct vertical fins, as they generally grow out of the continuous, embryonic odd fin; whilst in skates these fins disappear almost entirely, or are considerably reduced. That animals in their embryonic condition are neither so elongated as many of cylindrical form in their full-grown state, nor so short as some others, is ascertained by the embryology of snakes and toads. Thus, all the great external differences which exist between skates and sharks on one side, and Petromyzon on the other, do not show that these animals do not belong to the same natural group, as we have even among the full-grown ones, what we may call transitions between the extreme forms; for instance, sharks with more elongated body than others, with more extensive vertical fins, even with two dorsals and some without ventrals. Again, the remarkable form of skates arises solely from an extraordinary development of the pectorals; they are nevertheless closely allied to sharks, notwithstanding the striking difference in the position of the gill-openings.

As for the anatomical differences which exist among these fishes, and upon which so much stress is placed as to make the want of a heart, in Amphioxus, the foundation for a peculiar *class* to include that single fish, let us not forget, that there is an epoch in embryonic life, when no vertebrated animal has yet a heart; when the vertebral column is a mere soft continuous cord; when the brain is scarcely subdivided into lobes; when the head, as such, is not yet distinct from the trunk; when the mouth is a mere circular opening at the anterior extremity of the body; when the gills are simple fissures on the sides of the head, or at what is to be a head, without branchiostegal rays or operculum, or protecting covering of any kind.

Whoever is familiar with the anatomy of fishes must perceive, after these remarks, that the peculiarities which characterize Petromyzon, have a bearing upon the embryonic condition of their structure even in their full-grown state, and do not by any means mark a difference between them and the sharks and skates, any more than between them and any other family of fishes. On the contrary, should it be possible, after these statements, to show that there are important characters, common to Petromyzon, sharks and skates, notwithstanding their extreme external differences, it should be acknowledged that Cyclostomata and Plagiostomata are only different degrees of one and the same great type. Now, such characters we have; in the first place, in the structure of the mouth, which differs so widely from that of the other fishes, and agrees so closely in all Placoids, as Müller himself has shown in his Anatomy of Myxinoids. Next, the teeth also agree, in being arranged in several concentric series, and also in their microscopical structure, as well as in their mode of attachment to the skin lining the jaw, and not to the bone itself. We have other hints of the relation between Cyclostomes and Plagiostomes in their spiracles, and also in their numerous respiratory apertures, so that, after due consideration, I come to the conclusion that the Myxinoids and Petromyzons, far from being the types of peculiar subclasses, are simply embryonic forms of the great type to which sharks and skates belong, bearing to these powerful animals, in a physiological point of view, the same relation which exists between Ichthyodes and the tailless batrachians.

Of Cyclostomata, two species have been mentioned as occurring in the colder parts of North America, both referred by Dr. Richardson to the genus Petromyzon proper, but of which I have seen no trace myself in the great lake region, though I know Petromyzons to occur below Niagara Falls. However, I am able to add a new species of this family to the fauna of those waters, which belongs to the genus Ammocœtes, and was found in the mud in Michipicotin River, at the landing place of the Factory, the first specimens of which were picked up by the students when dragging their canoes along the shore.

Ammocœtes borealis, Agass.

This pretty little species differs from all those already known, by easily appreciable characters. It is at first sight plainly distinguished from the *Ammocœtes bicolor*, Les. and *A. branchialis*, Dum. whose dorsal fin is, as it were, divided into two lobes by a very low emargination; but it resembles the *Am. concolor*, Kirt. and *unicolor*, Dekay, in its dorsal fin, being uniformly continuous. It differs, however, from this latter, whose form is much more elongated, by the extent of its dorsal fin, which equals one half of the whole length of the body, whilst in the *Am. unicolor* it extends scarcely before the anus. In the individual which has served for this description, the whole length exceeds a little five inches.

The general form of the body is compressed, differing still in that respect from *A. unicolor*, which is subcylindrical, whilst the *concolor* is cylindrical at its anterior, and compressed at its posterior part. Our species is, on the contrary, in some manner ribbon-like, and its length goes on diminishing regularly from the neck towards the tail, where it ends in an attenuated and obtuse caudal lobe. The neck is prominent, but the skull is declivous. The upper lobe of the mouth, which terminates the anterior extremity, is concave, the opening of the cavity which it circumscribes being turned downwards. The anterior margin of the lip is concave, the lateral margins describe a convex lobe to the angles of the mouth. The lower lip is completely distinct from the upper, small and fixed upon the anterior of the lateral margins of the upper; it is slightly concave about the middle of its circumference. The convex lateral lobes are elliptical. The mouth, placed in the centre of the funnel formed by the two lips, is proportional to the size of the fish. When it is shut it seems to be cleft vertically, though in reality it is circular. The branched fringes which surround the mouth, are especially developed on the lower lip and at the angles of the mouth; they lengthen, but are reduced in thickness, on the inner side of the upper lip, under the form of an isosceles triangle, whose interior is equally furnished with them. The opening of the nose is situate in a circular depression between the anterior extremity of the skull and the inner margin of the upper lip. This depression is continued upwards, and terminates about the middle of the skull. The eyes are very small and placed on the sides of the head, at the height of the angles of the mouth, in a slight furrow of the face. The branchial openings are subcircular or convex in front, truncated behind, and open in a wrinkled furrow half an inch long, in form of a very elliptical curved line. The first branchial opening is at a distance of $\frac{3}{16}$ of an inch behind the angles of the mouth. The anus opens in a depression at a distance of $\frac{3}{8}$ of an inch from the extremity of the caudal fin; it is cleft longitudinally, and bordered by two thinned lips. The anal fin, very low at its origin immediately behind the anus, widens a little as it advances towards the caudal, with which it unites after having produced a more marked lobe. The dorsal fin is higher, but like the anal grows in height towards the posterior extremity, and forms like it a more dilated lobe before it unites with the caudal. This latter extends over an equal length above and below the tail. It is separated from the dorsal and anal fins by a notch, beyond which the fin arises to the height of the terminal lobes of the two anterior fins, and preserves the same height along the whole circumference of the tail, under the form of an elongated oval. Undulated, annular, transverse lines, distinct enough on the sides of the body, corresponding with the lateral muscles of the trunk, are very marked.

This species is from Michipicotin, where we have picked up a rather large number of specimens.

Lepidosteus.

This genus of fishes is known throughout the United States under the name of gar-pike. It is a very singular animal, and its history is closely connected with the most important progress which has recently been made in ichthyology.

The first knowledge naturalists had of this remarkable fish was derived from Catesby, who published a figure and a short account of it in his Natural History of South Carolina.

Linnæus, who received specimens of the same species from Dr. Garden of South Carolina, introduced it into his *Systema Naturæ* under the name of *Esox osseus*, supposing it allied to the common pickerel, because its dorsal and anal fins are opposite to each other and far back, near the end of the tail.*

Lacépède, who first noticed some of its peculiarities, removed it from the genus Esox, and established a distinct genus for it, under the name of *Lepisosteus*, which name, however, not being quite grammatically correct, I afterwards modified to *Lepidosteus*, which is now generally received.

The French naturalist knew a second species of that genus, from the Mississippi, which he called *Lepidosteus Spatula*. Afterwards

* For some zoölogical particulars respecting this fish, see preceding Narrative, page 33.

Rafinesque described several more, which, however, can scarcely be identified, as his descriptions are so very short and imperfect as to give little information upon their structure. In his Animal Kingdom, Cuvier characterized the genus Lepidosteus more correctly than his predecessors, without, however, noticing the great difference which exists between this genus and the common *Abdominales* among which he places it.

It was my good fortune early in the course of my scientific studies to perceive the striking differences which exist between these Lepidostei and all the other fishes now living upon our globe; and at the same time to call the attention of naturalists to the close relationship which exists between them and the fossil fishes of the earlier geological ages. So that, after an extensive study of the remains of these ancient inhabitants of olden time, Lepidosteus has become notable as the only living representative of the large group of fishes which peopled, almost exclusively, the waters during the early ages of the earth's history, and which has gradually decreased in number, until, at last, he was left almost alone to remind the observers of the present age, of a once powerful and widely spread dynasty among the watery tribes.

These facts call for a close examination of this singular fish. In the first place, let me say, that all the species of *Lepidosteus*, of which I now know ten distinct species, inhabit exclusively the fresh waters of North America. This is, in itself, a remarkable fact, most important in the history of nature, as it shows that far from deriving its inhabitants from other parts of the world, America has had, and has now, animals which are entirely peculiar to it, and which have nowhere any near relatives.

I am well aware that the Bichir of the Nile is remotely allied to the gar-pikes, and that another species of Polypterus occurs also in the Senegal; but this genus constitutes also by itself a peculiar group, and can only be considered as distantly related to the Lepidostei.

Another remarkable peculiarity in the geographical distribution of these fishes consists in the fact that different species are limited to different water basins, as the species of the Middle and Southern Atlantic States are as different from those of the Western waters as they are from the species which occur in the Northern lakes; so that, not only is the genus located in a peculiar continent, but the individual species are also confined to special regions of this country, from the great Canadian lakes to the freshwaters of Florida, and from the Atlantic rivers to the numerous affluents of the Mississippi. New England, however, has no species, and this is the more surprising as they occur further north in the St. Lawrence, and further south in the Delaware.

The question now arises, how this genus of fishes stands in its class; and whether, notwithstanding their peculiarity, they may not be associated with some other families.

Before answering this question, let me insist upon another fact, that, even if we take into account the nominal species of Rafinesque and that beautiful species of the Northern lakes first described by Dr. Richardson, the Lepidostei are only ten in number. And if we introduce into the same general division, the Polypteri, we shall have a natural group of fishes containing in the present creation not more than a dozen species. And even should we suppose that some more relatives of that group may be discovered in the course of time, we can by no means suppose that this family would ever contain as large a number of species as most of the other families of the class. We need only remember the innumerable species of suckers, or of cat-fishes, which occur every where in our fresh waters, or the various kind of perch, mackerel, codfish, &c., which swarm in the ocean, and among which the new discoveries to be expected can hardly be fewer than among our Lepidostei, to be satisfied that there is here a remarkable contrast between these families. It is therefore a fact plainly shown by this evidence, that the most natural groups of animals which we discover in nature, differ widely among themselves in the number of their representatives.

It is not less obvious, that these groups differ from each other in a very unequal degree, taken as general groups or considered in the isolated members of their families.

The amount of difference which distinguishes the gar-pikes from the common pickerels, or from the trouts, or from the herrings, or from the suckers, is far greater, for instance, than that which distinguishes the pickerels from the trouts, or the trouts from the herrings; and again, the generic differences which occur among the trouts, the graylings and white-fishes, and distinguish them from true salmon, are far greater than that which exists between the chubs, gudgeons, barbels or carps; and the specific distinctions which may be noticed in these different genera are again of an unequal value. So that we arrive at once to this important conclusion, that natural groups in the animal kingdom show naturally differences of unequal value, and that all attempts on the part of naturalists to equalize the divisions which they acknowledge in their researches, must, as a matter of course, result in failure; and I have not the slightest doubt that our classifications have not been more improved, and that we have made less extensive progress in the knowledge of the true relationship between the various groups of the animal kingdom, for the very reason that we have too often aimed at an arrangement which the most familiar facts in nature plainly contradict. Instead of this desired uniformity, we sometimes observe a numerous group of closely allied species corresponding to another group with few, but more distinct and more widely different species, and even isolated types, the relation of which seems to branch in all directions, without ever coming very close to any other group. Now, unless our classifications admit, as a natural limit, this diversity, it will be impossible ever to form a system which will answer to the natural affinities really existing in nature. As I have said on another occasion,* classification should be a picture from nature, and not an artificial frame of our own invention, into which natural objects are more or less conveniently brought together.

Another important point of view, of which naturalists should never lose sight, is the relation which exists between animals now found alive on various parts of the surface of our globe, and those known to us only from fossil remains discovered in strata of a different geological age.

The Lepidosteus, however isolated in the present creation, had once many and very diversified representatives all over the globe. Fossils of the same family of which the gar-pike is the type, have been found all over Europe in the oldest fossiliferous beds, in the strata of the age of the coal; in the new red sandstone; in the oölitic deposits,

* See *Principles of Zoölogy*, by L. Agassiz and A. A. Gould, Vol. II.

and even in the chalk and tertiary beds. They existed in the same wide range upon the continent of North America, and have been found in Asia as well as in New Holland; so that this family, now limited to the continent of North America, and, if we include in it the Bichir also, to two river basins of Africa, — was once cosmopolite in its geographical distribution.

The natural consequence from such evidence is, that we cannot arrive at a true insight into the relations of the animal creation, unless we study, at the same time, the living animals, and those which have become extinct; and that a natural classification must associate the fossils promiscuously in their natural relationship with the living types. The separation of palæontology from zoölogy, for the sake of convenience in the study of geological phenomena, has been very injurious to the real progress of zoölogy, and is so entirely unscientific, that until they are again combined under the same head, even in our elementary text books, we can hardly expect that zoölogy will make the progress which extensive investigations carried on singly, in the study of living and fossil animals, would lead us to expect.

Moreover, the identification of fossils requires a close investigation of such characters as are shown in the only remains of extinct species which have been preserved, and which are, almost exclusively, their solid parts. It is therefore very important that, in zoölogical investigations, more attention should be paid to the characters derived from such parts as are the only ones accessible in the study of fossils.

The mutual advantages to be derived from such a course cannot but be strikingly felt by those who have devoted their attention to the study of fossils. It may even be said that the condition of fossil remains, as they generally occur in rocks, has led naturalists to study more carefully the living species, than they did before. I need only mention the minuteness with which the skeletons of living animals have been described since it has been necessary to identify extinct species from isolated bones.

The skeletons of fishes, which were neither correctly figured in zoölogical drawings of these animals, nor minutely examined in their structure, are no longer considered as unworthy of the attention of minute observers. Even our knowledge of the structure of the shells in mollusca and of the wings of insects, has been improved with reference to the identification of fossil remains. It is therefore plain that comparative anatomy should be more extensively and intimately combined with zoölogy than is generally the case. The classification of the animal kingdom should no longer be based simply upon the structure of the animals, but form and structure should everywhere and always be considered in their intimate connections.

I have already alluded to the narrow circumscription of the genus Lepidosteus, within the limits of the temperate zone of North America. In like manner, also, the Marsupialia, for instance, are almost wholly confined to New Holland, and the Edentata to Brazil. All this goes to show that there is an important connection between a given country and its inhabitants, which rests with the primitive plan of the creation.

The limited existence of Lepidosteus in North America in the present creation has, no doubt, reference to the fact that North America was an extensive continent long before other parts of the globe had undergone their most extensive physical changes. Or in other words, that the present character of this continent has not been much altered from what it was when the ancient representatives of Lepidosteus lived; while in other parts of the world, the physical changes have been so extensive as to exclude such forms from among the animals suited for them.

We have therefore here a hint towards a more natural and deeper understanding of the laws regulating the geographical distribution of animals in general.

There are animals and plants whose detailed history is, as it were, at the same time, the history of that branch of science to which they belong. This is particularly the case with those animals, which, from particular circumstances, have thrown unusual light upon the relations which exist between them and their allied types. There are even a few such animals, the study of which has actually marked the advance of science. I cannot notice on this occasion the gar-pike without being strongly reminded how strikingly this has been the fact with Lepidosteus. The first sight I had of a stuffed skin of that fish in the Museum of Carlsruhe, when a medical student in the University of Heidelberg, in 1826, convinced me that this genus stood alone in the class of fishes; and that we could not, by any possibility, associate it with any of the types of living fishes, nor succeed in finding, among living types, any one to associate fairly with it. It was a fact, at once deeply impressed upon my mind, that it stands isolated among all living beings; and this early impression has gradually led me to the views respecting classification which I have expressed above, and which have frequently guided me in appreciating both the various degrees of relationship, and also the differences which I have noticed among different families; and, I may say, has also kept me free from fanciful attempts at symmetrical classifications.

Somewhat later, my investigations of the fossil fishes led me to the distinct appreciation of the great difference there is between the characters of the class of fishes in early geological ages; I also noticed that all the bony fishes of former ages are more or less allied to the gar-pike, and widely different from the types of fishes now prevailing. But the real nature of this difference was only gradually understood. I had not yet perceived that the fishes of older times had peculiar characters of their own, not to be found either among the more recent fossils or among the living representatives of that class. But the opportunity of study ing the skeleton of Lepidosteus, which was afforded me in Paris by Cuvier, showed at once, that these fishes have reptilian characters.*

The articulation of their vertebræ differs from that of the vertebræ of all other fishes no less than the structure of their scales. Their extremities, especially the pectoral limbs, assume a higher development than in fishes generally. Their jaws also, and the structure of their teeth, are equally peculiar. Hence, it is plain that, before the class of reptiles was introduced upon our globe, the fishes, being then the only representatives of the type of vertebrata, were invested with the characters of a higher order, embodying, as it were, a prospective view of a higher development in another class, which was introduced as a distinct type only at a later period; and from that time the reptilian character, which had been so prominent in the oldest fishes, was gradually reduced, till, in more recent periods, and in the present creation, the fishes lost in

* For further details, see my Recherches sur les Poissons Fossiles, Vol. II. part 2, p. 1—73.

the successive creations all this herpetological relationship, and were, at last, endowed with characters which contrast as much, when compared with those of reptiles, as they agreed closely in the beginning. Lepidosteus alone reminds us, in our time, of these old-fashioned characters of the class of fishes, as it was in former days.

An opportunity afforded me by John Edward Gray, Esq., of the British Museum, of examining a specimen of this genus, preserved in alcohol, furnished another evidence that the reptilian character of Lepidosteus was not only shown in its solid parts, but was even exemplified in the peculiar structure of its respiratory apparatus and its cellular air bladder, as I have pointed out in the Proceedings of the Zoölogical Society of London.*

One step further was made during this excursion, when, at Niagara, a living specimen of Lepidosteus was caught for me, and to my great delight, as well as to my utter astonishment, I saw this fish moving its head upon the neck freely, right and left and upwards, as a Saurian, and as no other fish in creation does.

This reptilian character of the older fishes is not the only striking character which distinguishes them. Investigations into the embryonic growth of recent fishes have led me to the discovery that the changes which they undergo agree, in many respects, in a very remarkable manner, with the differences which we notice between the fossils of different ages; so much so, that the peculiar form of the vertebral column, and especially its odd termination in very young embryos, where the upper lobe of the caudal fin is prolonged beyond the lower lobe, and forms an unequal, unsymmetrical appendage upwards and backwards, agrees precisely with the form of the tail of the bony fishes of the oldest geological deposits; so that these ancient fishes may be said to have embryonic peculiarities in addition to their reptilian character. This fact, so simple in itself, and apparently so natural, is of the utmost importance in the history of animal life. It has gradually led me to more extensive views, and to the conviction that embryonic investigations might throw as much light upon the successive development of the animal kingdom during the successive geological periods, as upon the physiological develop-

* Proceed. Zoöl. Soc. of London, Vol. II. page 119.

ment of individual animals; and, indeed, I can now show, through all classes of the animal kingdom, that the oldest representatives of any family agree closely with the embryonic stages of the higher types of the living representatives of the same families; or, in other words, that the order of succession of animals, through all classes and families, agrees, in a most astonishing measure, with the degrees of development of young animals of the present age.

This being the case, it is obvious that a minute investigation of the embryology of Lepidosteus would throw a vast amount of light upon the history of the succession of fishes, of all geological periods; and also would probably give the first indication of the manner in which the separation of true ichthyological characters from reptilian characters, was gradually introduced; as it is more than probable, from all we know otherwise of the embryology of animals, that the young gar-pike, in its earliest condition, will have characters truly ichthyological, and only assume, gradually, the peculiar reptilian characters which distinguish it. But notwithstanding all my efforts to secure the Lepidosteus in the breeding season, I have failed up to this day to gain the desired information. It only remains for me, therefore, to urge naturalists living near the waters inhabited by Lepidosteus to take up the subject as early as an opportunity is afforded them.

Although Lepidosteus does not occur in Lake Superior, I have deemed it sufficiently important to introduce these remarks here, as this fish occurs in all the northern lakes except Lake Superior, as far north even as Mud Lake, below Sault St. Marie. Its presence in these waters is another of the striking differences which exist between the ichthyological fauna of Lake Superior, and that of the other lakes; and shows once more, within what narrow limits animals may be circumscribed, even when endowed with the most powerful means of locomotion, and left untrammeled by natural barriers.

This Lepidosteus is one of the swiftest fishes I know. He darts like an arrow through the waters, and the facility with which he overcomes rapids, even the rapids of the Niagara, shows that the falls of St. Mary would be no natural barrier to him, if there were no natural causes to keep him within the limits in which he is found, and which extend from Lake Michigan, Lake St. Clair, and Mud

Lake, through Lake Erie, and Ontario, down to the St. Lawrence and its outlet into the sea, into which this fish never ventures far, though he does not altogether avoid brackish and salt water.

Dr. Richardson was the first naturalist who described the northern Lepidosteus. He mentions it in his Fauna Boreali-Americana, under the name of *Lepidosteus Huronensis*, and gives a correct and detailed description of it. Nevertheless, it has been since mistaken, and referred to the southern species first described by Catesby and Linnæus, from which it is however very distinct, both by the proportions of its parts, its scales, its fins, and especially by the form of its frontal bones, in which the supra-orbital emargination is much lower and more elongated. Again, notwithstanding the description of Dr. Richardson, Dr. Dekay has redescribed it under the name of *Lepidosteus Bison;* and Zadock Thompson has described a young specimen under the name of *Lepidosteus lineatus.* At first, his description would seem to indicate a really distinct species; but I have ascertained, by a series of specimens, that the differences pointed out are really the characters of the young, and have no value as specific characters; the detached lobe formed by the upper raylets of the caudal fin is gradually united with the lower rays,* and the longitudinal stripe, which is well marked in young specimens of a few inches in length, gradually vanishes, to leave only a few spots upon the sides, which even disappear entirely in the oldest individuals. The vertical fins alone remain spotted in the adult. The natural color of this fish is a light greenish gray, passing downwards into a dull white.

Acipenseridæ (*Sturgeons.*)

The family of Sturgeons is well characterized and easily distinguished from any other in the class. These fishes have generally been placed in the order of Chondropterygians, near the sharks, until I objected to this association, and attempted to show that, not-

* It is a very remarkable fact that several fishes of the old Red Sandstone period have, in their full-grown state, a peculiar form of their caudal fin, which is nearly identical with the form of the caudal fin of the young Lepidosteus; a form which is otherwise unknown to me at present in the whole class of fishes.

withstanding their extraordinary peculiarities, they are more closely related to the gar-pikes, than to any other group of fishes. This view, though at first strongly opposed, is now generally admitted, having been sustained both by anatomical and palæontological evidence.

The sturgeons are generally large fishes, which live at the bottom of the water, feeding with their toothless mouths upon decomposed organized substances. Their movements are rather sluggish, resembling somewhat those of the codfish tribe.

Their geographical distribution is quite peculiar, and constitutes one of their prominent peculiarities. Located as they are, in the colder portions of the temperate zone, they inhabit either the fresh waters or the seas exclusively, or alternately both these elements, remaining during the larger part of the year in the sea, and ascending the rivers in the spawning season. Although adapted to the cold regions of the temperate, they do not seem to extend into the arctic zone, and I am not aware that they have been observed in any of the waters of the warmer half of the temperate zone. The great basin of salt water lakes or seas which extends east of the Mediterranean, seems to be their principal abode in the Old World, or at least the region in which the greater number of species occur; and each species takes a wide range, extending up the Danube and its tributaries, and all the Russian rivers emptying into the Black Sea. From the Caspian they ascend the Wolga in immense shoals, and are found farther east in the lakes of Central Asia, even as far as the borders of China. The great Canadian lakes constitute another centre of distribution of these fishes in the New World, but here they are neither so numerous, nor do they ever occur in contact with salt water in this basin.

Northwards, there is another great zone of distribution of sturgeons, which inhabit all the great northern rivers emptying into the Arctic Sea, in Asia as well as in America. They occur equally in the intervening seas, being found on the shores of Norway and Sweden, in the Baltic and North Sea, as well as in the Atlantic Ocean, from which they ascend the northern rivers of Germany, as well as those of Holland, France, and Great Britain. Even the Mediterranean and the Adriatic have their sturgeons, though few in number. There are also some on the Atlantic shores of North America, along the British Possessions as well as the Northern and Middle United States. They seem to be exceedingly numerous in the northern Pacific, being found everywhere from Behring's Straits and Japan to the northern shores of China, and on the north-west coast of America, as far south as the Columbia River. Again, the so called western waters of the United States have their own species, from the Ohio down to the lower portion of the Mississippi, but it does not appear that these species ascend the rivers from the Gulf of Mexico. I suppose them to be rather entirely fluviatile, like those of the great Canadian lakes.

Beyond the above limits southwards there are nowhere sturgeons to be found, not even in the Nile, though emptying into a sea in which they occur; and as for the great rivers of Southern Asia and of tropical Africa, not only the sturgeons, but another family is wanting there, I mean the family of Goniodonts which in Central and Southern America takes the place of the sturgeons of the North. Again, all the species in different parts of the world are different.

It is a most extraordinary fact, which will hereafter throw much light upon the laws of geographical distribution of animals and their mode of association, viz., that certain families are entirely circumscribed within comparatively narrow limits, and that their special location has an unquestionable reference to the location of other animals; or in other words, that natural families, apparently little related to each other, are confined to different parts of the world, but are linked together by some intermediate form, which itself is located in the intermediate track between the two extremes. In the case now before us, we have the sturgeons extending all around the world in the northern temperate hemisphere, in its seas as well as in its fresh waters, all closely related to each other. Neither in Asia nor in Africa is there an aberrant form of that type, or any representative type in the warmer zones; but in North America we have the genus Scaphirhynchus, which occurs in the Ohio and Mississippi, and which forms a most natural link with the family of Goniodonts, all the species of which are confined exclusively to the fresh waters of Central and South America. The closeness of this connection will be at once perceived by attempting to compare the species of true Loricariæ with the Scaphirhynchus. I know very well, that the affinities of Goniodonts and Siluroids with sturgeons are denied, but I still strongly insist upon their close relationship, which I hope to establish satisfactorily in a special paper, as I continued to insist upon the relation between sturgeons and gar-pikes, at one time positively contradicted, and even ridiculed. I trust then to be able to show, that the remarkable form of the brains of Siluridæ comes nearer to that of sturgeons and Lepidostei, than to that of any other family of fishes. This being the case, it is obvious, that there must be in the physical condition of the continent of America some inducement not yet understood, for adaptations so special and so different from what we observe in the Old World. Indeed, such analogies between the organized beings almost from one pole to another, occur from man down to the plants in America only, among its native products; while in the Old World plants as well as animals have more circumscribed homes, and more closely characterized features in the various continents at different latitudes.

As for the species of sturgeons which occur in the Canadian lakes, I know only three from personal examination, one of which was obtained in Lake Superior, at Michipicotin, another at the Pic, and the third at the Sault: though I know that they occur in all other Canadian lakes, yet it remains to be ascertained how the species said to be so common in Lake Huron, compare with those of Lake Superior, and with those in the other great lakes and the St. Lawrence itself. As for the Atlantic species, ascending the rivers of the United States west and south of Cape Cod, I know them to differ from those of the lakes, at least from those which I possess from Lake Superior. The number of species of this interesting family which occur in the United States is at all events far greater than would be supposed from an examination of the published records. Upon close comparison of the specimens in my collection from different parts of the country, and in different museums, as those of the Natural History Society of Boston, of Salem, of the Lyceum of New York, my assistant, Mr. Charles Girard, and myself have discovered several species not yet described. For this comparison I was the better prepared as I had an opportunity in former years of studying almost all the European species in a fresh condition, during a prolonged visit in Vienna.

Acipenser lævis, Agass.

This species, one of the largest of the genus, is from the Pic. The length of the specimen, of which I possess the head and the fins, and which was in fresh condition when I examined it, was four feet six inches. The head, which is contained two and a half times in the whole length, is subconical and a little flattened below; the upper surface forms an uniformly descending line from the occiput to the extremity of the snout, somewhat elliptical beyond the eyes, thus giving to the latter a slightly recurved appearance. From the level of the eyes to the centre of the skull, on the middle line of the head, there exists an equally elongated surface, more flattened, being the rudiment of a longitudinal dimple; finally, on the occipital part of the skull we observe a small keel, where the two bones of this region begin to become convex, in order to pass to the cutting plates of the back.

The surface of the bones which form the exterior covering of the head, is invested with small tubercles of enamel, of a circular form with obtuse summits. At first without apparent order, at the very centre of the bone they become linear, radiating to the circumference. Their greatest development occurs in the occipital region and on the transverse line level with the nostrils. On the middle part of the head these tubercles become thinner, and on the extremity of the snout they are reduced to a fine reticulation. The sides of the head have only a very few asperities. The only bone on which they are developed is the operculum, and it is only in its posterior half that they radiate from the centre towards the margin. A few rows only are directed towards the upper part of the head. The other bones constituting the opercular apparatus are covered with a membrane finely roughed at the surface. The bones placed at a small distance behind the eye and limiting the anterior margin of the branchial cavity, bear a few blunt tubercles irregularly distributed on their surface. The branchiostegal membrane is naked and smooth, attached by a thin shred to the posterior part of the operculum, and passes before the pectoral fin, to which it is contiguous; beyond this it dilates, in order to shut the branchial cavity at the lower part of the head, forming a very open curved line; finally it terminates at a small distance from the mouth. The eyes are at a distance of three and six-eighths inches from the end of the snout. Their form is subcircular, their pupil transversely cleft. Their immediate covering is a smooth membrane, which continues below to the anterior extremity, where it becomes reticulated, but without any appearance of the smallest plate on its surface. At the anterior and upper part of the eye is a small protuberance projecting over the depression in which the nostrils are situated. These latter open at the surface by two orifices on each side. The one of an elliptical form with a free opening, occupies a prominent position, so that it would be observed from both sides of the head, looking at it from above. The other, a larger one, has the form of a crescent, with its convexity turned towards the eye, and placed a little obliquely on the vertical line, extending below the lower line of the eye for two-thirds of its length.

The lower portion of the head appears as a flat surface rising insensibly from the anterior margin of the mouth to the extremity of the snout. This latter rises gradually in an oblique line, which begins in front of the barbels. The middle line is convex, the margins are inclined. The barbels, four in number, are situated in pairs on both sides. The two pairs are a little more distant from each other than the two barbels of the right and left side. Their length is nearly the same, of about two inches; their form subconical, growing thinner at their extremity. Behind the barbels we notice a subquadrangular depression in which their base is concealed when they bend backwards. The mouth is situated on the anterior half of the lower part, in a transversal notch; it extends from one side of the head to the other, the posterior margin being almost straight, the anterior having an elliptical outline on the middle line. A thick membrane, with a glandular and undulating surface, surrounds the jaws, leaving the symphysis of the lower jaw free. Both extremities are attached to the anterior third part of both lower maxillary bones, sending a small membranous expansion towards the symphysis, taking afterwards the direction towards the angles of the mouth.

Here the membrane is thickened considerably, and continues so on the whole circumference of the upper jaw, following its outlines.

The mouth is protractile, and when projected outwards carries with it the surrounding membrane. The jaws are weak, both maxillary branches of the upper and lower jaw uniting by means of a tendinous membrane. The extremity of the tongue is round, covered with a thick membrane, with a wrinkled surface perforated with small mucous holes.

A thick layer of mucosity covers the surface of the head. This mucosity is secreted by the crypts of the skin; these are especially very conspicuous on the space situated between the mouth and the snout, and on the upper side of the latter. They have the appearance of irregular meshes excavated in the skin, at the bottom of which we distinguish, by means of a magnifying glass, the crypts which line its surface.

The body is of a regular form, diminishing insensibly from the anterior side backwards to the dorsal and anal fins, behind which it decreases rapidly towards the tail. This latter goes on tapering, then turns up obliquely, arching itself slightly over the lower lobe of the caudal. The surface of this caudal prolongation is covered with small elongated escutcheons, which become the more slender the more they rise along the caudal arch. They begin above the last escutcheon of the lateral row, much resembling the scales of the tail in Lepidosteus.

The five rows of escutcheons on the sides of the body and along the back are scarcely visible, for they are hidden in the thickness of the body.

The upper lobe of the caudal fin is composed in its whole extent of spinous rays, generally short and much inclined backwards, diminishing in length the more they recede, and becoming rudimentary at their termination. The lower lobe, which gives to the caudal fin its general form, is exclusively composed of articulated and dichotomous rays. Those of the lower margin, much the largest and longest, remain undivided for two-thirds of their length; they seem even to follow a direction peculiar to them by a slightly concave line. The other rays grow more and more slender the more they rise above the lobe. They bifurcate first in the middle, and subsequently several times at a distance which varies for every ray. The lower lobe of the caudal extends not so far backwards as the upper. This latter has the form of a very open arch; the lower is convex below. The line which joins both extremities is oblique within the upper half; on the middle line it becomes concave, giving to the posterior margin of this fin the form of an irregular crescent.

The dorsal fin is equally notched, forming a crescent on its terminal margin. All the rays which compose it are articulated. Those of the anterior margin, four times longer than those of the posterior, are arched backwards, undivided through their whole extent. The other rays dichotomize in the same manner as those of the caudal.

The anal, longer than broad, is placed opposite and somewhat behind the dorsal. Its form is oblong, the inner and outer margins are rounded; the posterior margin is straight, bending slightly inwards at the middle. The rays are similar to those of the dorsal. Those of the lower margin being the longest and remaining undivided through the whole extent; those of the outer margin dichotomize like those of the dorsal.

The ventral fins, as broad as they are long, are placed half way between the pectorals and the anal. Their posterior margin is almost square, the inner slightly sinuous, the outer rounded. The rays of the former dichotomize from their basis, those of the latter are undivided, like those of the other fins.

The pectorals are of all fins the most developed. Their greatest length is seven inches and a half, and their breadth nearly four inches. Their form is a rather regular oval, setting aside their margin of insertion, which for two-thirds of its extent, from the outer margin, forms a straight line, directing itself obliquely towards the interior of the fin, whilst on the other third we observe a curve which brings the inner margin of the fin back upon itself. The rays of this margin become excessively slender, and remain undivided, like those of the outer margin. Those of the centre dichotomize according to the common rule.

The number of rays in the fins is as follows: P. 39 to 40; D. 34; V. 26; A. 23. We may count as many as fifty to sixty on the lower lobe of the caudal, but they become indistinct beyond this number.

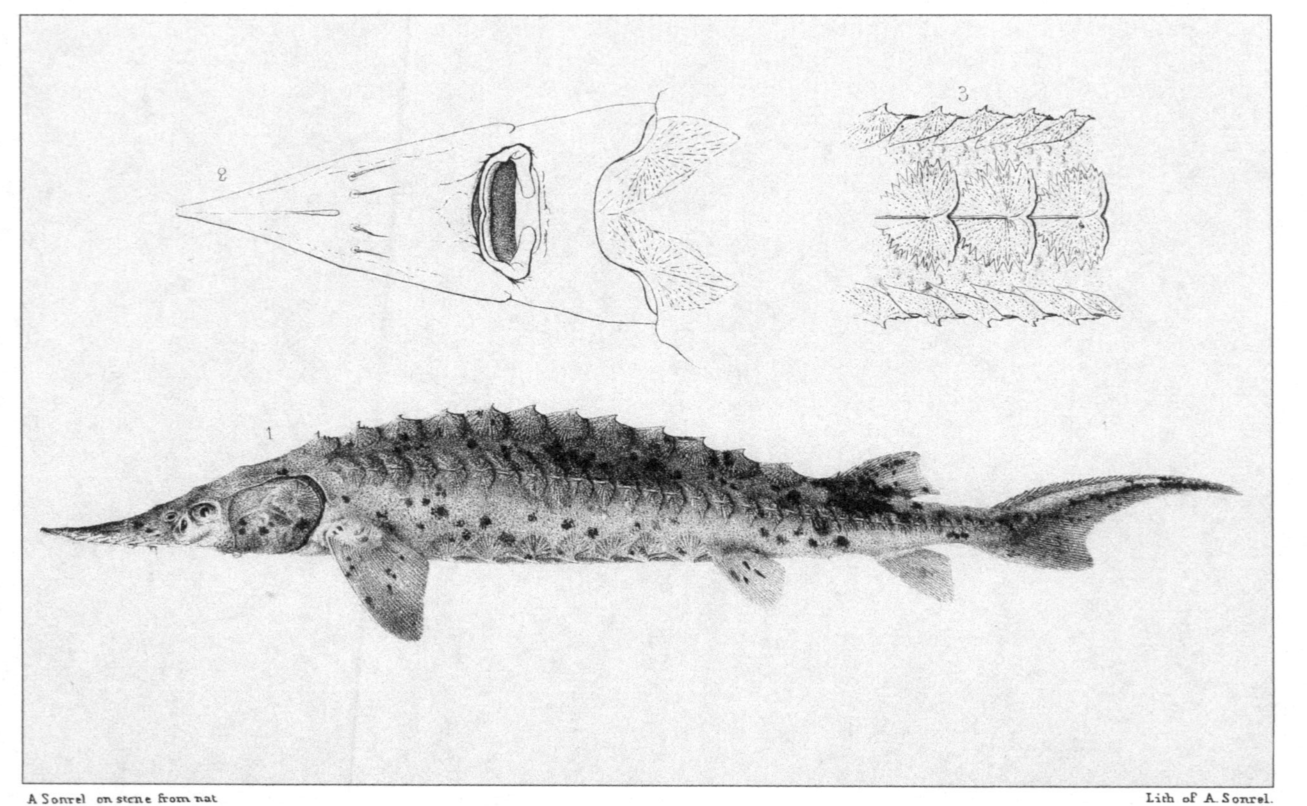

A Sonrel on stone from nat

Lith of A. Sonrel.

ACIPENSER CARBONARIUS Ag

A character common to all fins is to have the outer margin sensibly thicker than the inner, which becomes thin and membranous. It is also in this outer margin that are found the largest rays, arched from within outwards, undivided in the greatest part of their extent, thus giving them a peculiar aspect. Small tubercles are observed in the outer third of the rays where they are most dichotomized.

The color is of an uniform blackish brown, which extends to the fins; it is a little less intense on the head, on the lower half of the sides below the middle line it has a yellowish reflection. A pale white exists over the lower part of the head and the abdomen, as far as the under surface of the tail.

This species resembles the *A. rubicundus* of Lesueur, who describes two varieties of it, one found with the true *rubicundus* in Lakes Erie and Ontario; the other inhabiting the River Ohio. The descriptions which he has given of them do not enable us to recognize our species in either of these varieties.

Acipenser carbonarius, Agass.

The general form of this species is rather thick and short than slender. The back is proportionally very elevated and very convex from the occiput to the anterior margin of the dorsal fin, from whence the body begins to grow considerably slender towards the tail, which last rises obliquely in order to form the higher arch of the caudal fin. (Plate 5, fig. 1.)

The total length is one foot two inches and a half. The head is contained three times and a half in this length. The face, from the anterior margin of the branchial cavity to the extremity of the snout, equals the fourth part of the length of the trunk. The snout, from the orifices of the nostrils is contained seven times in this length.

The head itself is depressed, flattened, uniformly inclined from the occiput beyond the nostrils, where the snout rises considerably, growing thinner on its margins, which circumstance gives it a convex form. Seen from above, its shape is that of an elongated triangle. The upper surface is quite uniform, having only one slight depression on the middle line, bordered by two small carinæ of the frontal and parietal bones. Small plates continue on the snout to its extremity, and are prolonged on the sides before the nostrils, but do not reach the lower circumference of the eye. All these bones are covered with fine granules, disposed in linear rows in the direction of the head. The eyes occupy the upper region of the face. They are oval and have their largest diameter longitudinal. They are surrounded with a smooth zone on their lower circumference, limited above by the bones of the skull, and behind by a bone which separates them from the opercular apparatus and the branchial cavity. Another bone, which is triangular, being the continuation of the preceding, limits the posterior margin of the face and completes the anterior margin of the branchial cavity. The nostrils, situated in a depression which is reserved for them before the eyes, open, as is common, at the surface, by two holes pierced laterally, of which the upper, the smallest, is subcircular and free, the lower oblong, vertical and protected by a small membrane at its anterior margin. The small plates which cover the snout reach not so far as the bone of the lower angle of the face. The opercular bone is covered with these fine granules disposed in striæ radiating from the centre. The membrane which invests it and which shuts the respiratory opening in front, is covered with a fine rasp, which continues on the sides of the head to the angle of the mouth. The branchiostegal membrane proper is naked and very thin. It surrounds the opercular bone from the upper margin of the branchial cavity, and is prolonged and becomes wider a little above the branchial opening behind the pectoral fins and beneath the head.

The inferior surface of the head is level, with the snout a little raised. The mouth opens in a depression behind the eyes. Its general form is the same as in the *A. lævis*, (see pl. 5, f. 2.;) it is protractile as in this latter, but the membranous fold which surrounds the jaws, is smooth on its whole anterior circumference, where it appears only as a wrinkle surrounding the jaw. It thickens at the angles of the mouth and terminates in a flattened flap, of glandular appearance, on the third quarter of the extent of the lower jaw, leaving the symphysis bare. The palate and the tongue have sinuous and transverse wrinkles on their surface.

Four thread-like barbels half an inch long, are placed mid-way between the mouth and the termination of the snout, a little nearer however to the mouth. On this face, though generally flattened, we may observe a median longitudinal swelling, having on each side a depression with widened margins. This skin is bare, although covered upon its surface with a net of irregular meshes in which we observe small holes which secrete the mucosity, as in *A. lœvis*.

The escutcheons of the dorsal row are twelve in number, well developed, and a rudimentary thirteenth applied to the anterior margin of the dorsal. They are so near to one another that some are even slightly imbricated. Their general form is heart-shaped, broader than long, the two sides limited by a regular denticulated curved line, rising abruptly so as to form a very sharp median carina, terminated at the two posterior thirds in a hook, whose point is turned backwards. Their surface is covered with radiating lines, owing to the linear arrangement of their tubercles, which are excessively small, and acute. On the space between the posterior margin of the dorsal and the origin of the caudal we observe three small plates. The largest is situated on the side of the dorsal, the two smaller follow immediately and are arranged in pairs. Their surface is equally covered with small acute tubercles, but the centre is scarcely indicated by a larger tubercle, whence the others radiate. (See pl. 5, fig. 3.)

The lateral escutcheons are from thirty-two to thirty-three in number, of irregular oblong form, with the two sides retracted. The anterior margin is concave, the posterior convex, slightly notched in the middle. The median carina is but slightly prominent, the sides of course but little inclined; the hook which rises above it is slightly curved backwards; sometimes it is bifurcated at its point. The surface, as usual, is covered with small granules in radiating rows. Their position in relation to the body is oblique from before backwards. They are less serrated than those of the back, and diminish gradually as they approach the tail.

The escutcheons of the abdominal region, from seven to eight in number, extending over the space contained between the posterior margin of the pectoral and the anterior margin of the ventral fins, resemble much in their general outlines those of the back. Their form is perhaps more rounded, though they do not form a regular circle. They are quite as much inclined, and their hooks are stronger, and more arched at the point. The radiating striæ are also more visible.

In front of this double row of escutcheons and as if forming their immediate continuation on the inner side of the pectoral fins, and in front of them, we observe a subtriangular bone, the anterior side of which is concave, bordering the branchial opening beneath. These two bones are contiguous on their anterior angle, and form by their reunion a convex curved line along the sides of the mouth, to which the branchiostegal membrane is attached. A prominent carina, but unprovided with hooks, extends along the median line from the posterior angle. A single wrinkle indicates on the middle of the anterior angle the rudiment of a carina. The striæ radiate from those two centres. The bone of the anterior part of the pectorals and upon which these fins articulate, is small and hidden under the skin.

An odd elliptical escutcheon with regular outlines is situated in the middle of the space between the anus and the anal fin. It has a slight median carina, over which projects an elliptical hook. A rudiment of an escutcheon leans towards the anterior margin of the anal.

The anus opens in a small depression immediately behind the ventrals, at a distance of about two-thirds of an inch from their posterior margin, and one inch and three-sixteenths from the anterior margin of the anal. It is small and surrounded by a cutaneous membrane, bilobed on the posterior side.

The skin over the whole space which the escutcheons do not cover is rough to the touch. Small tubercles with acute points cover uniformly its surface, being every where of equal size and at an equal distance from each other. On the terminal arch of the tail they become lengthened and flattened, and invest the whole space like scales.

The fins are generally small; the dorsal, broader than it is high, is triangular with the upper margin concave. It is composed exclusively of soft rays, with the exception of a fulcrum situated on its anterior margin. The rays are articulated and subdivided only at their extremity.

The upper lobe of the caudal is formed of small bony rays, short and strongly inclined backwards, not reaching the extremity of the fin. The rays of the lower lobe do not differ from those of the dorsal. They bifurcate like these latter, but at the extremity only. The posterior margin of this lobe is notched, in the form of a crescent and elongated in its upper part, along the arch of the tail. The notch is not deep in the lower part.

The anal is opposite to the dorsal, beyond which it extends backwards. It is narrow, elongated, almost twice as high as it is broad. The inner and outer margins are almost straight, the terminal oblique margin slightly curved. The rays are slender, bifurcated at their extremity only.

The ventrals, similar in their form to the anal, are situated at the posterior third of the body. Their structure has nothing that distinguishes them from the anal.

The pectorals are as in the *A. lœvis* the largest of all the fins. Their form is lengthened, the terminal margin is obliquely rounded, and passes to the inner margin by an arch. The anterior and outer margin bears a spinous ray, bent beyond its insertion, and curving inwards a little before the point. It does not reach the extremity of the fin. It is flattened in the horizontal diameter of the fin; its basis is three-sixteenths of an inch broad and terminates in an obtuse point, in the margin of the fin. The surface is striated longitudinally on both surfaces, alternating with small furrows and wrinkles. The soft rays are as in the other fins.

The general color is of a yellowish brown on the upper half of the body, the yellow growing purer on the sides and beneath the belly. A large spot of an intense black, and an elongated quadrangular form occupies, on the middle of the back, the space between the dorsal and lateral series of shields. A second pair of large spots of the same color occupies the same position on the sides of the dorsal fin, on which they even encroach a little. Other small spots are distributed over the sides of the fish from the opercular apparatus (itself included) to the tail and the fins, giving thus to the whole fish a dotted appearance.

P. 1, 43–35; V. 26–28; D. 36; A. 25–28. C. lower lobe more than sixty.

The only specimen of this species which is in my possession was found at Michipicotin on the north-east shore of Lake Superior.

Though this species is very similar in its general characters to the *Acipenser maculosus* Lesueur, from the Ohio, we have not, however, been able to identify it. The description which this author gives of his species is so vague that he does not even tell us the form of the fins. The formula of their rays is far from corresponding with that of our species. Nor is the abdominal series of plates the same; those of the sides and back seem to resemble it more closely. The snout is also more slender; but had not Lesueur mentioned that the species which he saw is of small size, we might have supposed that our specimen was the young, which have generally the snout more pointed than full-grown specimens.

Acipenser rhynchæus, Agass.

This species is very similar to the preceding; it differs from it only in a few characters which we shall here enumerate briefly. The body is more slender and diminishes less abruptly towards the caudal region. The curve of the back is more elliptical; slightly concave at a small distance behind the head, where the third escutcheon is sensibly smaller. The head is contained about four times in the whole length. The face, from the anterior margin of the branchial cavity, forms the fifth part of the length of the trunk, and the snout from the nostrils is in the proportions of one to five. The whole length of the fish is nearly twenty-three inches. The head is slender, elongated, proportionally narrow; its upper surface is very sloping, forming a line feebly broken at the level of the nostrils. A sinus quite deep, widened on both sides, extends along the median line of the skull; narrow at the top, it widens before it disappears upon the snout. The frontal and parietal bones are carinated in their middle. The snout is pointed, but truncated. It is completely covered with small plates which pass before the nostrils and go to join again the bone which terminates the lower and posterior angle of the face. The nostrils open in a bare space which is situated under the eye. Their form and direction are not quite the same as in the preceding species.

The shields of the dorsal series are sixteen in number, cordiform as in the preceding species, but longer than they are broad, approaching however more to a circle. The right and left margins are equally denticulated. An odd plate of medium size is situated behind the dorsal, and behind this latter a pair of much smaller plates fill up the remainder of the space to the anterior margin of the caudal. Both are carinated and provided with a hook.

The lateral series consists of thirty-five pairs of plates, elongated, narrow, irregularly triangular, the most acute point directed upwards, much resembling those of the preceding species.

The abdominal series has from eight to nine plates, generally more irregular, more strongly denticulated, with a strong carina and prominent hook.

The articular bone of the pectoral fin is stronger and more widened. The pectoral fins themselves are longer and more rounded on their posterior margin. The anal is also more narrow. The other fins resemble each other excepting the caudal, which seems to be less furcated. We have not been able to make a fuller comparison of the two species, having had only a dried specimen of the latter in our possession. The following formula of the rays is only an approximation, as the fins are somewhat defective.

P. I, 32 or 33; V. 26; D. 34; A. 25. C. lower lobe one hundred and more.

Very distinct fulcra exist along the anterior margin of all the fins, with the exception of the pectorals.

Habitat, Sault St. Mary.

Acipenser Rupertianus, Richardson.

This species, which we did not find in our excursion, is mentioned here only incidentally, for comparison with those which we have described. Richardson has figured and described it in his Fauna Boreali-Americana. Our comparisons have been made upon a skin from Sault St. Mary, for which I am indebted to Mr. McLeod.

Its head is thicker than it is long, forming one-seventh of the whole length, which is twenty inches; the snout is covered with distinct small plates upon its surface, though it is also granulated. The frontal sinus is broader than deep, and extends over that part of the snout which is contiguous to the skull. The dorsal plates, twelve or thirteen in number, are elliptical; the lateral series number twenty-five or twenty-seven, and resemble somewhat those of the preceding species. The abdominal series have eight or nine plates, longer than broad, whilst the contrary is the case in the *A. rhynchæus*, from the Sault St. Mary. Their circumference is also less. The fins which we have been able to compare show but slight differences in the two species.

Siluridæ.

Whenever we are induced to consider organized beings in their connection rather than by themselves, we perceive at once differences between them, which throw more light upon the laws that regulate their structure, than the most minute investigation of isolated facts. The Siluridæ are fishes which it is difficult to combine with any other group, unless by far-fetched considerations, and afford a striking example of the importance of general considerations in the special study of zoölogy.

Speaking of the sturgeons above, I have already mentioned their affinity to the Goniodonts. It is now a matter of great importance to examine upon what this relation rests, for the systematic position assigned to that family is also decisive for the Siluridæ, which are very closely allied with the Goniodonts. Indeed, Goniodonts and Siluridæ may be united into one family with almost as much propriety as they can be separated, and wherever one of these groups is placed, in a general classification of fishes, the other must follow. That sturgeons belong to the order of Ganoids is now fully ascertained; but whether the affinity of Goniodonts and sturgeons is sufficient to connect the Siluridæ, or whether Siluridæ and Goniodonts are to continue in some connection or other with the many families of Abdominales, with which they have hitherto been combined, remains to be seen. That the position of the ventrals is not sufficient to settle this question is plain, as soon as we consider the position of those fins in the Ganoids, in which they are also placed at the middle of the abdomen. The scales which are wanting in most Siluridæ, would apparently seem, at first sight, to afford little information; let us however remember that there are some genera among Siluridæ, such as Callichthys and Doras, in which scales of a very peculiar character exist, and that several other genera have large bony escutcheons upon their neck. Now these bony plates and scales have the same structure as the enameled scales of the sturgeons, and their position in Doras reminds us strongly of the lateral shields of sturgeons; so much so, that but for the form of the body, we might be led to consider these fishes as closely related. And really, this affinity is not altogether superficial; the development of the jaws and opercular bones is so imperfect, as to show little analogy to the structure of those parts in the common Abdominales, whilst it agrees rather closely to that of the sturgeons. The position of the mouth in Loricaria, below the snout, is another feature which connects the Goniodonts and sturgeons, and the genus Scaphirhynchus may be considered as forming the most natural link between the two families. Again, Goniodonts have pseudobranchiæ and a thick membrane encircling the mouth, which constitute so many more characters connecting them with the sturgeons; although these points are of less value than those already mentioned. I may add, also, that the brain of Siluridæ bears a stronger resemblance to that of the sturgeons, than to that of any of the Abdominales; so that I consider myself justified in referring the families of Goniodonts and Siluridæ to the order of Ganoids, where they may stand as aberrant families, rather than among the other great divisions of the class of fishes.

PIMELODUS.

The genus Pimelodus, as characterized by Prof. Valenciennes, in the Histoire Naturelle des Poissons, seems to me to contain several distinct types, which might with great propriety be considered as distinct genera, characterized by their peculiar teeth, the arrangement of their barbels, and the respective position and extent of their dorsal and anal fins, as well as the form of their caudal. But as the collections now at my command do not contain sufficient materials to limit precisely those genera, I shall only mention that such a revision seems desirable, since, as far as I can now judge, the group of which *P. catus* may be considered as the type, should constitute a first genus, and retain the name of Pimelodus, and that new names should be framed for the other groups of species, of which *P. cyclopum*, *albidus*, *ctenodus*, &c., may be considered as the respective types.

If we now admit the generic sections, which I propose for the numerous species of Pimelodus, their study will be by this very fact much simplified; for when we have once the group to which our species belongs, its comparison with the others will be very easy. Now we have already said that the first group, that which is to retain the name of Pimelodus, will contain the *P. Catus* as its type, and in addition to it the *P. punctulatus* Cuv. and Val., *P. cœnosus* and *borealis* of Richardson, and *P. albidus*, *nebulosus* and *æneus* of Lesueur, besides a new species from Lake Superior, to be described below. All authors have not admitted *P. nebulosus* as a species; the natural history of *P. albidus* and *æneus* leaves also much to be desired, so it is also with *P. punctulatus*. So that we are still in doubt about the real number of species which will compose the genus Pimelodus proper. The *Pimelodus Catus*, which is perhaps the best known, differs considerably from our northern species, so that we need hardly mention the differences; but *P. nebulosus* and *P. albidus* seem to be very closely allied to *P. Catus*, if we judge by the description which we have before us. The *P. æneus* would come near *P. punctulatus*, which in its turn would remind us of *P. Catus*. Hence we may see how important it may be to submit anew these species to a close examination, to study them each in its locality and by minute anatomical as well as zoölogical investigation, to ascertain the value of their characters.

For the present, however, I cannot undertake this comparative study from want of sufficient materials, but I shall attempt to describe the species we brought from Lake Superior, and compare it with *P. cœnosus* and *borealis* of Richardson, from which, though allied to them, it seems however to differ specifically.

PIMELODUS FELIS, Agass.

The general form is that of most species of the genus, neither thick, nor elongated. The abdomen is prominent in the space contained between the branchiostegal apparatus and the ventrals. The curve of the back rises to the height of the dorsal, whence it slopes rapidly upon the head. The body is very compressed from behind the dorsal and ventral fins to the tail. It is completely bare, with a punctulated appearance, caused by the aquiferous holes which open at the surface of the skin, and which are especially numerous on the anterior region and on the head. The lateral line is straight, ascending from the middle of the caudal to the upper angle of the opercular apparatus. The head, from the occiput, forms the fifth part of the whole length, whilst from the posterior margin of the operculum to the end of the snout it constitutes only one fourth. The head is longer than it is broad, and forms a regular oval, truncated behind in the occipital region and elliptical in the anterior circumference. The mouth extends as far back as the eyes; the lips surround it under the form of a fleshy, elastic swelling, in the middle of the jaws only; but at their reunion with the angles of the mouth they grow thinner, widen and flatten, and form a kind of funnel, which enlarges, for a third at least, the opening of the mouth. The teeth are arranged like those of a card, and distributed irregularly upon the circumference of the jaws. They vary in length and size, but are all acute. On the lower jaw they extend much more backwards in the mouth than on the upper jaw, where they do not extend beyond the basis of insertion of the maxillary barbels. These latter, two inches long, reach to the posterior margin of the preoperculum. They follow the upper circumference of the cutaneous funnel at the angles of the mouth for the extent of six-eighths of an inch. Hard, horny and flattened at their basis, they grow gradually softer and more slender towards their termination. The nostrils are situated on the upper surface of the head, at a distance of half an inch from the end of the snout. Their opening, of oblong form, measures one-eighth of an inch in the direction of the greatest diameter. The barbels which arise from their anterior margin, the smallest of the four pairs, have exactly the length of the space contained between them and the anterior extremity of the head. They are soft, flabby and rounded. The eyes, proportionally small and subcircular, are at a distance of one inch from the anterior margin of the head. Their diameter is five-sixteenths of an inch. The four barbels of the lower surface of the head are placed upon an arc of a circle within the branches of the jaw. The two internal ones are more distant from each other than the external ones. These latter are one and one-sixteenth inches long, whilst the former are only seven-eighths of an inch. They are soft upon their whole extent, like those of the nostrils, rounded and elongated.

The opercular apparatus is almost completely hidden under the skin and the muscles; a slight swelling indicates the inferoposterior margin of the operculum. As for the preoperculum, which forms the anterior outline of the apparatus, we can trace its whole margin, which is arched within, and upon which the branchiostegal membrane is fixed. The branchiostegal rays themselves are nine in number; the first two, the most developed, are of about equal size, and follow the outline of the preoperculum, without being attached there otherwise than by the muscles which move them. All are flattened and concave on their outer surface. The humeral apophysis, which we perceive through the skin, is strong and robust. It extends two-thirds the length of the spine of the pectoral fins; its outer margin is wrinkled.

The dorsal fin is composed of a spinous and six soft rays. Its basis measures one and one-eighth inches, the spine is one and one-half inches long; the rays of the centre, one and five-eighths inches. Hence the fin has a quadrilateral form from its insertion to the height of the spinous ray, terminated by an isosceles triangle. The spinous ray itself is slender, slightly arched; its posterior margin has neither furrows nor denticulations. At its upper third is implanted a rudiment of a soft ray which takes an oblique direction upwards. The adipose fin is of medium size, thick at its basis, thin upon its circumference, which extends a little beyond the posterior margin of the insertion of the anal. It is seven-eighths of an inch long. The caudal is subtruncate, almost concave.

It is composed of eighteen articulate, well-developed rays, measuring two inches along the margins and one and eleven-sixteenths inches in the middle of the fin, and of six raylets in the upper margin and ten in the lower margin, hidden in the thickness of the skin. The anal is high and rounded; its insertion is two and a half inches long. It numbers twenty-two rays; those of the centre are one and six-eighths inches high. The ventrals, one and three-sixteenths inches long, are fan-shaped and rounded on their circumference; they have eight soft rays. The pectorals have almost the same form, though less rounded. They are composed of seven soft rays and one spinous, strong and robust, at whose inner side we remark denticulations, varying in their number and form, and extending only along the two upper thirds. The lower third has a carina with a sharp blade. The length of the soft rays is one and three-eighths inches; the length of the spine one and three-sixteenths inches.

The general formula of the rays is as follows: Br. 9; D. I. 6; C. 18; A. 22; V. 8; P. I. 7.

Besides the differences in the number of the rays, as we may estimate by the numbers we have given above, this species differs farther from the *P. cœnosus* and *borealis* in the general form of the fins. Their position upon the body, relatively to each other, affords not less sensible differences when we compare the measures which Dr. Richardson gives for his *P. cœnosus*, setting aside the difference of size of our specimen, which had two inches more for its whole length. Similar differences are remarked between our *P. Felis* and the *P. borealis*, though for this latter we have not been able to make our comparisons upon positive numbers, the celebrated author having neglected to give the numbers of the rays of this species. The proportions and the dimensions of the head are also far from agreeing, being in the *P. cœnosus* two-ninths of the whole length, and in the *P. borealis* as broad as long, whilst we have seen, that in our species its length forms the fourth part of the whole length, and that besides, it is much longer than broad. The spinous ray of the dorsal is more feeble than in *P. cœnosus*, and, besides, unprovided with the deep groove in which the soft ray of this fin is lodged. The spinous ray of the anterior margin of the pectorals, which in *P. borealis* is unprovided with denticulations on its posterior margin, is, on the contrary, in our species, provided with such serratures as is the case in *P. cœnosus*.

Such are the principal features upon which the comparison may rest, while good figures are yet wanting. The differences which we have indicated, however slight they be, do not allow us to identify our species with the one or the other of those mentioned above. The comparison of original specimens would be necessary in order to fix in a sure manner the traits of resemblance, or the differential characters of each of them.

Percopsis, Agass.

In order fully to understand and perfectly to appreciate the characters of this genus, and the interest involved in its discovery, it is necessary to remember various relations of the different types of the whole class, which however do not constitute generic distinctions, although they bear upon the peculiarities of this new type.

In the first place, it is a matter of no little importance that, among the fishes of former ages, we find every where types which differ widely from the forms of our time, and that those forms are the more different, as they belong to older geological deposits. The differences are even so great, that out of the four orders of this class, there are only two which constitute the fauna of fishes in the older formations; two orders, which in our day are comparatively reduced, I mean the Placoids and Ganoids. Moreover, the types are peculiar in all epochs. For instance, the sharks of former days, especially those of older epochs, resemble solely that curious genus of Port Jackson, New Holland, the Cestracion, which is so remarkable among the living fishes as to form a group by itself. The Ganoids, of which there are so remarkably few in the present creation, such as the gar-pike (Lepidosteus) of this continent, are not less peculiar, and in connection with those ancient Placoids, constitute the only representatives of the class of fishes throughout the earlier geological ages down to the deposits of the chalk, when new families of other orders, the Ctenoids and Cycloids, begin to make their appearance, preparatory as it were to the present development of that class, and are successively diversified with the modified adaptations of the whole class. Now the genus Percopsis is as important to the understanding of the modern types of fishes as Lepidosteus and Cestracion are to the understanding of the ancient ones, as it combines characters which in our day are never found together in the same family of fishes, but which in more recent geological ages constituted a striking peculiarity of the whole class. My Percopsis is really such an old-fashioned fish, as it shows peculiarities which occur simultaneously in the fossil fishes of the chalk epoch, which however soon diverge into distinct families in the tertiary period, never to be combined again.

This ancient character of some of the American fishes agrees most remarkably with the peculiarity of the vegetation of this continent, which, as I have shown on former occasions, resembles also the fossil plants of prior ages.

The geographical range of these peculiar, old-fashioned beings is also very remarkable, they living in temperate, or rather cold climates, when their earlier representatives lived in warmer epochs.

The most striking features of the fishes of the tertiary period and those of our time consist in their belonging to two groups of the class only; one, the Ctenoids, with rough, combed scales, in which the respective representatives have also prominent serratures on prominent spines upon the head, in the operculum in particular, and in the fins; the other, the Cycloids, smooth, with simple scales with an entire margin, in which some few types however have also spinous fins.

Now my new genus, Percopsis, is just intermediate between Ctenoids and Cycloids; it is, what an ichthyologist, at present, would scarcely think possible, a true intermediate type between Percoids and Salmonidæ.

The general form of this genus reminds us of the common perches, but it is easily distinguished from them, by the fact that its head and the opercular apparatus are smooth and unprovided with denticulations, as also by the presence of a small adipose fin, as in the salmons. The anterior dorsal is also a small fin, composed of soft branched articulated rays, as in the salmons. The ventral fins are placed at the middle of the abdominal cavity, as in the Abdominales in general. The scales, however, are truly serrated as in the Percoids, a structure which, as far I know, does not occur in any of the Abdominales. The conformation of the mouth is also as in the perches, that is to say, the intermaxillaries form alone the upper margin of the mouth, and the maxillaries stand behind as a second arch, but the vomer and palate are entirely destitute of teeth.

This fish, of which I shall publish a full anatomy, should be considered as the type of a distinct family, under the name of *Percopsides.*

Percopsis guttatus, Agass.

Pl. I., fig. 1 and 2.

This is a fish of small size and slender form, though the back is very much elevated. Its greatest elevation corresponds to the anterior part of the dorsal fin, that is to say, a little nearer the end of the snout than the insertion of the caudal. The tail is proportionally elongated, a little compressed between the adipose fin and the basis of the caudal. The sides are compressed, and diminish gradually in thickness from the front backwards. The ventral line is less prominent than that of the back; it rises more backwards of the anal, to concur in the contraction of the tail. The profile of the head, which is small and compressed like the sides, is regularly conical; the length of the head is contained three times in that of the body, setting aside the lobes of the caudal.

The eyes are large and circular, situated near the upper margin of the face; if a vertical line passed through their centre, it would divide the head into equal parts. The space which separates the anterior margin of the orbits from the end of the snout, is about half an inch. The nostrils open outwards by a double opening, and are very near the eyes. One of these openings has the form of a crescent, whose convexity is turned towards the eye; the other is small, subcircular and situated in the concave space of the preceding. (Fig. 2.) The mouth is small, and appears scarcely larger when opened; the upper jaw extends beyond the lower, and is formed solely by the intermaxillaries, upon which we remark a narrow band of small, excessively fine teeth, arranged like the teeth of a card. The palate is entirely smooth. On the contrary the pharyngeans are covered with similar teeth still more slender, as also the œsophagean shields. On the lower jaw there is a narrow band of teeth, like those of the intermaxillaries. The labials extend a little beyond the intermaxillary to form the angle of the mouth, which corresponds to a vertical line which would pass before the nasal openings. The suborbital bones are very much developed. They are four in number, intimately united, extending from the posterior and lower margin of the eye to the nostrils. The three first, much the smallest, occupy the lower circumference of the orbit; the fourth, almost as large as the three others together, is the strongest and the most robust and protects the lower margin of the nostrils; it sends out a prominent point to the space situated between these latter and the eye.

The opercular apparatus is completely smooth, like the surface of the head itself. The posterior free margins of the bones which compose it, are destitute of any kind of spines or denticulation. The most developed, and at the same time the most robust of the bones of this apparatus, is the preoperculum, which occupies almost the whole width of the face. Its form is triangular; the outer margin of its ascending branch is slightly concave; the lower branch, the most developed, is straight and encircles the lower margin of the face. The operculum is quadrilateral, its four angles are prominent; its upper, hinder and lower borders are notched or concave, its anterior margin is almost straight. The suboperculum, small, narrow, oblong, is lodged in the concavity of the lower margin of the operculum. The interoperculum, which is a third longer than the suboperculum and which it resembles in form, is entirely hidden under the lower branch of the preoperculum. The branchial openings are very large; they continue to the middle of the lower surface of the head, where they are almost contiguous. The branchiostegal membrane is supported by six curved rays; the upper ones, which are the largest, are flattened. There are four branchial arches on whose inner border we remark a double row of shields in relief, covered with small card-like teeth, as we observe on the pharyngeans.

The disposition of the fins is in striking harmony with the form of the fish. The dorsal, which is the largest, is situated at the middle of the back. Its length equals the height of its anterior margin, being more than twice as high as its hinder margin. The upper margin is straight. There are twelve rays. The two first are short and spinous, close together; the third, or first of the soft and articulated rays, is the largest. These latter bifurcate at the middle of their height; every bifurcation subdivides again at its extremity. A small adipose fin is situated at about equal distance between the posterior margin of the dorsal and the basis of the caudal. The caudal is furcated; it has eighteen rays, of which the longest are subdivided three times at their terminal extremity. The anal is situated behind the dorsal. This is a small fin, higher than it is long, with regular and straight margins, composed of eight rays, of which the first, shorter and more slender than the other, is undivided. The second and eighth bifurcate only once, the five middle ones branch so far as to show divisions of the third order. The ventrals are placed perpendicularly to the anterior margin of the dorsal, narrow at their basis; they soon widen to become oval with a regularly rounded circumference. There are eight rays; the four of the centre thrice subdivided, those of the margins twice only, the first being simple. The pectorals arise at a small distance from the branchial opening and occupy almost all the lower part of the body. They are elongated, oval, composed of twelve very slender thread-like rays, subdivided thrice at least at the centre of the fin, the first being simple. Its extremity reaches almost the middle of the dorsal.

Br. 6; D. 2. 10; A. I, 7; C. 8, $18\frac{8}{6}$; V. 8; P. 12.

The scales are large in proportion to the size of the fish. They are little imbricated and of about equal size on the whole surface of the body except under the throat, where they are a little smaller and subcircular. On the sides their height is greater than their breadth. The anterior margin is rounded; their hinder margin forms a very obtuse angle, and under the microscope it exhibits a row of small needles, somewhat distant, and which seem to be implanted in this margin instead of appearing as serratures. This type of scales comes near to that of my *Corniger spinosus*, and to some genera of the cretaceous epoch. The concentric striæ are very distinct, but I could not perceive any radiating striæ.

The lateral line, nearer to the back than to the belly, extends from the upper angle of the operculum, arches slightly upwards towards the dorsal fin, and then descends again insensibly to the middle of the tail, to terminate at the centre of its peduncle.

The ground color is of a yellow, violaceous tint, much darker above the lateral line than below. The back is spread with blackish brown spots, sometimes disposed in two longitudinal rows, sometimes in three, however without great regularity. On the middle of the body extends a silvery ridge tapering slightly from the head to the basis of the caudal. It is not rare to see sometimes blackish spots encroach upon this bright band. The fins are unicolored, and of a transparent whitish tint like that of the abdomen.

We found this fish in great abundance at the Sault St. Mary, at Michipicotin and at Fort William.

Percoids.

Whenever we compare the fishes which occur in a given locality, we are struck with peculiar associations entirely different from those which we may find in other localities. Take the Bay of Massachusetts, for instance, where we have sharks, skates, &c., &c., combined together in numeric proportions, and represented by species altogether different from those which occur on the shores of the Middle States or around Florida and in the Gulf of Mexico. Again, if we compare freshwater fishes, as they occur in any extensive hydrographic basin, for instance, those in the Canadian lakes, or in the Ohio and Mississippi, or those of the lakes and rivers of Europe, with the marine faunæ, we find still more striking differences. Entire families common in the sea under the same latitudes have no representative in fresh water; there are no sharks and no skates, no flounders, soles or turbots, no mackerels, no herrings, as permanent inhabitants of the freshwaters in the latitudes above mentioned; so that a collection of species from the freshwater or from the sea, even if all the species were to be new, could be recognized by an ichthyologist as derived either from the ocean or from some inland water. However different such associations of marine and freshwater species may be, there is nevertheless scarcely any family, whether generally marine or fluviatile, in which there is not some species living in the other element. There are some families again, in which the proportions between marine and fluviatile species are about equal, and there are still others in which the individuals of the same species are alternately at different seasons of the year either marine or fluviatile; this is particularly the case with such as ascend from the sea into the rivers at the spawning season, to deposit their eggs in waters more genial to the growth of their young than those in which they are mainly to live when full-grown.

Percoids belong to those families of which there are certain proportions of strictly freshwater, and certain proportions of strictly marine genera, the number of marine species being however much greater than that of the freshwater ones, and very few of the species having the power of enduring both the freshwater and the sea.

That the family of Percoids, as it is now circumscribed, is in the main a most natural group, cannot be doubted, especially if we remove from it such genera as Trachinus, Uranoscopus, Sphyræna and a few others; there remains however a question, not to be decided here, how far Sparoids and Sciænoids should be considered as distinct. Indeed, at different times, in two editions of the same work, Cuvier in his Animal Kingdom has successively associated them in one great family, and divided them into two distinct groups. The fact is that these fishes are closely related, and it is for future investigations to determine the value of those characters upon which the distinction rests, which consists only in the serrature of the opercular apparatus, the presence or absence of teeth upon the palatine bones, and the degree of development of the so-called mucous canals in the head, characters which have not even been strictly adhered to in the arrangement of individual genera.

Whatever may be their closer or more remote affinities, the Percoids of the Canadian lakes, as well as those of the other fresh waters of North America, are much more diversified than those of the freshwaters under similar latitudes in the Old World. This is not the case with Lake Superior itself, for, on the contrary, that lake furnishes but few true Percoids; but the other great lakes teem with a variety of genera and species of that family, which among themselves, as well as with reference to the common type of the whole family, differ much more from the Percoids than those of Europe; I need only mention the genera Pomotis, Centrarchus, and Huro or Grystes, which all occur in the lower lakes, to show that this is the case; and at the same time to indicate the great difference there is between the fishes of the upper lakes and those of the lower.

A comparative list of the Percoids of the two regions will show better than words, that notwithstanding the free passage there is between all these waters, notwithstanding the great similarity between the waters themselves, there is an organic difference between the ichthyological faunæ of the two regions.

Lower Lakes.	*Lake Superior.*
Centrarchus æneus.	0
Pomotis vulgaris.	0
Huro nigricans.	0
Grystes striatus.	0
Perca flavescens.	Perca flavescens.
Lucioperca americana.	Lucioperca americana.

This list shows not only the great difference there is between the fishes of the upper and lower lakes, but also how closely the ichthyological fauna of Lake Superior resembles that of northern Europe, where the same genera of Percoids have representatives as in the north of this continent, a fact which goes farther to show how much more uniform the fauna of the north is than even the fauna of the temperate zone.

PERCA FLAVESCENS, CUV.

PERCA FLAVESCENS *Cuv.* R. Anim. 1817, II., 133.— *Cuv.* et *Val.* H. N. Poiss. 1828; II., 46. — *Richards.* Fn. Bor. Amer. 1836, III., p. 1., Pl. 74.—*Storer* Rep. 1839, p. 5. — *Ayres* Bost. Journ. Nat. Hist. 1842, IV. 256. — *Dekay* N. York Fauna, 1824, p. 3. Pl. I., f. 1.

BODIANUS FLAVESCENS *Mitch.* Tr. Lit. Phil. Soc. N. Y. 1815, I., 421. — *Kirtl.* Rep. Zoöl. Ohio, p. 169–190.

MORONE FLAVESCENS *Mitch.* Rep. Fish. N. York.

PERCA ACUTA *Cuv.* et *Val.* H. N. Poiss. 1828, II., 49, Pl. 10. *Richards.* Fn. Bor. Amer. 1836, III., p. 4. — *Dekay* N. Y. Fauna 1842, p. 6, Pl. 68, f. 222.

PERCA GRANULATA *Cuv.* et *Val.* Hist. N. Poiss. 1828, II., 48, Pl. 9. *Jard.* Nat. Libr., I., 92, Pl. 1. — *Dekay* N. Y. Fauna. 1842., p. 5, Pl. 48, f. 220. — *Linsl.* Cat. Fish. Conn.

PERCA SERRATOGRANULATA *Cuv.* et *Val.* H. N. Poiss. 1828, II., 47. — *Griff.* in *Cuv.* An. K. X., Pl. 39, f. 1. — *Dekay* N. Y. Fauna. 1842, p. 5, Pl. 22, f. 64.

PERCA GRACILIS *Cuv.* et *Val.* H. N. Poiss. 1828, II., 50. — *Richards.* Fn. Bor. Amer. 1836, III., 4. — *Dekay* N. Y. Fauna, 1842. p. 6.

Closely resembling the European species, the yellow perch of America differs however considerably from it, so that no naturalist after Cuvier, who first distinguished them from each other, has ever thought to identify them. Its several varieties, described first under particular names, seemed then to constitute species quite as distinct from each other as the *Perca flavescens* is from the *Perca fluviatilis*. But at that epoch, when the principle of the constancy or permanence of species had just been placed upon an anatomical foundation, naturalists for a time lost sight of this other fact, that the species common to a fauna are subject to individual variations which run over the whole range of the species. To study these changes, to bring back every variation to its true type, to trace the circle of the species through so many oscillations, was a task whose results could not be anticipated. The principle of the permanence of species has remained in our science as a well-ascertained fact, but naturalists have found that many which had been distinguished as species had to be cancelled as soon as their characters were better understood. Thus, in a series of more than forty individuals of the yellow perch of America, we can no longer trace the limits of separation between the *Perca granulata*, *serratogranulata*, *acuta* and *gracilis*, which all belong as mere varieties to the *P. flavescens*, as Dr. Storer has already determined. A more pointed snout, a more slender form, a more wrinkled head, more marked wrinkles on the operculum, and the denticulation of the opercular bones, are not constant characters, any more than the color, or the number of the transverse bands, which vary with the age of the individual.

We have examined perches from the Sault St. Mary, from Fort

William, from the Pic, and from Lake Huron; we have compared them with specimens from Massachusetts, New York, and Pennsylvania; we have compared again and again all their different characters, and we have seen that the same variations occur in all these supposed species. No difference in the form and relative position of the fins could be noticed; the same arrangement and aspect of the scales characterizes them all. The comparison which we have thus been enabled to make of these different varieties confirms their specific identity. No appreciable difference exists; there are the same crests, the same cavities and sinuosities of the bones of the head, and the same proportions between their different parts.

Pomotis vulgaris, Cuv. et Val.

Pomotis vulgaris *Cuv.* et *Val.* H. N. Poiss. III., 91, Pl. 49;—VII., 464.—*Richards.* Fn. Bor. Amer. III., 24, Pl. 76.—*Storer* Rep. 1839, p. 11.—*Dekay* N. Y. Fauna 1842, p. 31, Pl. 51, f. 166.

I have been able to secure only a few specimens of this species from Lake Huron, about four inches long. By means of comparisons which I have made of specimens from Massachusetts, New York, and Pennsylvania, I have nevertheless been able to ascertain its identity. For more ample details upon this fish I refer to the works quoted above, in which the species is described and figured. I must however remark that I have only mentioned in the synonymy those authors with whose species there remains no doubt in my mind, since I am satisfied that the so-called *Pomotis vulgaris* of the Southern States is not the same species. In order to avoid all confusion, I have left out those synonyms which I was not able to verify directly, quoting only authors who have given minute characters and good figures.

The *Pomotis vulgaris* has been quoted as found in almost the whole extent of the United States. We are sure that it inhabits the Great Canadian Lakes, and the Northern and Central States of the Union. We do not know its western limit, though it is quoted as found in Ohio. Our specimens are from Lake Huron.

Lucioperca americana, Cuv. et Val.

Lucioperca americana *Cuv.* et *Val.* H. N. Poiss. II., 122, Pl. 16.— *Richards.* Fn. Bor. Amer. III., 10. — *Dekay* N. Y. Fauna, p. 17. Pl. 50, f. 153.— *Kirtl.* Rep. Zoöl. Ohio, p. 190.— Bost. Journ. N. H. IV., 237, Pl. 9, f. 2.— *Thomps.* N. H. Verm. 1842, 130 fig.— *Storer* Synops. 1846, p. 24.

This fish has about the same geographical distribution as the *Perca flavescens* northward, but it does not extend so far south. It occurs however in all the great Canadian lakes, and throughout the State of New York and parts of Ohio. It remains still to be ascertained, whether the *Okow* or *Hornfish** belongs or not to the same species.

I do not believe that the *L. canadensis* of Hamilton Smith is even specifically distinct from the *L. americana*, though its author is disposed to view it as a new generic type, because of the presence of five spines on the margin of the operculum, and of the absence of denticulations on the bones of this apparatus. I am satisfied that these opercular spines lose much of their value in this genus. Indeed in two specimens of *L. americana* which I procured about Lake Superior, I have seen that one of them had two small points on the hinder margin of the operculum of the left side only, whilst there was no trace on the right side. The hinder point of the operculum was itself very acute and resembled a third spine a little more robust than the two others. The specimen measured thirteen inches. In a specimen from Lake Michigan, twenty-two inches long, for which I am indebted to Samuel C. Clarke, Esq., of Chicago, the operculum of the left side has equally two spines on its hinder margin, and two very near each other on its upper angle. On the right side there is a single spine observable, but more robust, though very short like the others, and on the upper angle two of equal development.

As for the other bones of the opercular apparatus, the following is what we have observed in other specimens from Lake Superior, as

* *Richards.* Fn. Bor. Amer. III., 14.

also in those of Michigan: the preoperculum is denticulated on its whole circumference; the interoperculum and the suboperculum are equally crenulated or denticulated towards their union, upon the third part at least of their extent. The lower margin of the suboperculum is undulated. The suprascapular bone has fine serratures; the scapular and the humeral are entire. According to Dr. Richardson the crenatures of the margin of the interoperculum are scarcely perceptible, and the suboperculum smooth and straight. The suprascapular should be smooth, like the scapular and humeral, whilst the figure of the Histoire Naturelle des Poissons represents these three latter bones as serrated. This shows great variations in these parts.

The following is the formula for the rays of the fins, as we counted them in our specimens:

Br. 7; D. XIV–II, 19; A. II, 13; C. 5, I., 8, 7, I., 4; V. I., 5; P. 15.

When this fish is young, until it reaches a length of three to four inches, the head resembles still more that of the pike than when full-grown, the snout being then very depressed; but the teeth are all uniform. However, even at this epoch, the whole of its physiognomy reminds us so much of the species described above that we could not hesitate an instant for its determination. The black marblings stand out more distinctly from the ground of the color than in the full-grown; they unite in groups and constitute irregular and vertical zones. Dr. Dekay's *Lucioperca grisea* is also founded upon young specimens of the common pike-perch.

GRYSTES FASCIATUS, Agass.

CICHLA FASCIATA *Lesu.* Journ. Ac. N. Sc. Philad. 1822, II., 216. — *Richards.* Fn. Bor. Amer. 1836, III., 23.

CICHLA MINIMA and OHIOENSIS *Lesu.* l. c. pp. 218 and 220.

CENTRARCHUS FASCIATUS *Kirtl.* Bost. Journ. N. H. 1845, v. 28. Pl. 9, f. 1.

CENTRARCHUS OBSCURUS *Dekay* N. Y. Fauna 1842, 30, Pl. 1, f. 48.

This species is very closely allied to the *Grystes salmoides* of the Southern States,* from which it is however distinguished by the profile of the more raised back, and of course by a broader body. The surface of the skull is uniformly rounded and not depressed as as in *G. salmoides*. The proportions of the head compared with the body are the same as in this latter, but the mouth is less opened and the shorter labials do not reach a vertical line drawn across the hinder margin of the orbits, whilst they exceed such a line in *G. salmoides*. The teeth are arranged like cards, and are similar in both species.

The fins upon the whole seem to be cut on the same pattern as in *G. salmoides*, but when we examine them attentively we see that they are all stabbed like the body itself, the ventrals and pectorals shorter and more widened, the dorsal and anal lower. As for the other details of their structure they are about the same, as we may see from the following formula.

Br. 6 ; D. X. 14 ; A. III, 10 ; C. 7, I, 8, 7, I, 6 ; V. 1, 5 ; P. 16.

The scales are a littlo smaller, but of the same form as in *G. salmoides;* the radiating striæ are perhaps less marked. They cover the opercular apparatus and the cheeks, but at this latter place their smaller size is quite remarkable ; this latter character is very striking when we compare both species.

Our specimens are from Lake Huron ; one of them measures twelve inches, and the other seven. I have also received two specimens from Lake Michigan, through the care of Mr. Samuel C. Clarke, the largest of which measures eighteen inches. Professor Baird forwarded to me specimens from Lake Champlain. Dr. Dekay has found it in Lake Oneida. Finally, this species extends to Pennsylvania, as I was able to convince myself by two specimens collected at Toxburg, and for which I am under obligation to Professor Baird.

* *Grystes salmoneus* does not occur in the Northern nor in the Middle States, although Dr. Dekay mentions it upon the authority of Cuvier, who probably mistook specimens of our *Grystes fasciatus* for the southern species. Having, however, failed to discover this confusion, Dr. Dekay describes the same fish again, under the name of *Centrarchus obscurus*.

Huro nigricans Cuv. is another species of the lower Canadian lakes, which occurs also in Lake Champlain. The generic distinction from Grystes does not, however, rest upon sufficient characters to warrant its preservation in the system of fishes; I shall therefore call it in future *Grystes nigricans*. It is a very common fish in some of the lakes, and highly esteemed as an article of food. Throughout the lake region it is known under the name of black bass, and may be seen in large numbers in the enclosure under the gallery of the Cataract Hotel at Niagara. Dr. Dekay describes it as *Centrarchus fasciatus*, although he copies also Cuvier's description and figure of *Huro nigricans*, but without perceiving their identity.

In the northern lakes there is only one species of true Centrarchus found, the *Centrarchus æneus;* but it does not occur as far north as Lake Superior, though it is common in Lake Huron and the other great lakes.

Cottoids.

As they have been circumscribed by Cuvier, the Cottoids constitute a most natural family, though they contain genera apparently widely distinct. Indeed, between Peristedium and Scorpæna, between Pterois and Aspredophorus, between Gasterosteus and Cottus, there seems to be as great a chasm as can exist in a natural family; however, they all belong to one and the same natural group. But in order to be satisfied that it is so, one should be acquainted with the fact, that animals or plants belonging to one and the same natural division, will in certain cases resemble each other so closely as scarcely to allow distinct subdivisions, as, for instance, the Siluridæ, which, with the same features throughout so numerous a family, run into various extremes of form, in which, however, there is no mistaking the family likeness even in the external appearance; the same is also the case among Cyprinidæ or among Eels. But there are others, whose relations rest upon one particular combination of characters, which will, nevertheless, assume very different features, though preserving throughout that common trait of character. Genera belonging to such families may sometimes at first sight have very little resemblance to each other, they may differ in very different amounts of variation, and nevertheless constitute, at least in the eye of the deeper investigator, a very natural group; such, for instance, is the family of Cottoids, such again is the family of Scomberoids. The difficulty in such cases is not the diversity, but a correct appreciation of the connecting character, which, if misunderstood, might bring together animals widely distinct in structure, but apparently related by external appearance; for instance, the genus Capros among Scomberoids, near Zeus, owing to its form and the dilatability of the mouth, when in truth it belongs to the Chætodonts, in the vicinity of Chelmo.

Taking for granted that the family of Cottoids, as it is now characterized, is in the main a natural one, the question arises at once, what can be done to appreciate correctly the true relations of those remarkable tropical forms, as Pterois, Lynanceia, &c., with the more uniform Cottus, Etheostoma, Gasterosteus, of the freshwaters of temperate regions? To become satisfied that they are truly members of the same family, it is necessary to undertake an extensive comparison of the structure of their head, and especially of the arrangement of their infraorbital bones, when it is seen that frequently the particular development which characterizes, generally, this group, is reduced to a rudimentary state in some of its members, as in Etheostoma and the genera allied to it. This group of small Cottoids having attracted less attention than the larger marine types, we subjoin a synopsis of their genera.

Subfamily of Etheostomata.

Freshwater fishes of medium and small size, somewhat related to the Gobii. Cheeks sometimes covered with scales, sometimes bare. One small suborbital bone only, the anterior. Mouth variable. Head sometimes elongated, sometimes truncated or rounded. Scales proportionally large. No air bladder. No pseudo-branchiæ. Teeth very minute.

Etheostoma, Rafin.

Head elongated, pointed; mouth widely open, not protractile, broad, jaws of equal length. Opercular apparatus and cheeks bare.

Etheostoma blennioides *Raf.*
" notatum.
" third species sent by Prof. *Baird.*

PILEOMA, Dekay.

Head conical, truncated, in form of a hog's snout; opening of the mouth moderate, and in form of an oblique arc of a circle, opening at the end of the snout, very slightly protractile. Lower jaw a little shorter. Operculum and cheeks scaly.

Etheostoma Caprodes *Rafin.*
Pileoma semifasciatum *Dekay.*
" zebra *Agass.* Lake Superior.

PŒCILOSOMA, Agass.

Head short and strong, rounded. Mouth little opened, proportionally broad; it is not protractile, though the maxillary bone be moveable. Opercular apparatus scaly; cheeks bare.

Etheostoma variatum *Kirtl.*
" maculatum *Kirtl.*
" third species sent by Prof. *Baird.*
" fourth species sent by Prof. *Baird.*

BOLEOSOMA, Dekay.

Head very short, rounded in section of a circle; mouth small, horizontal, slightly protractile. Opercular apparatus and cheeks very scaly, neck and sides of the head compressed.

Boleosoma tessellatum *Dekay.*
" tenue *Agass.* Charleston, S. C.
" maculatum *Agass.* Lake Superior.
Etheostoma Olmstedi *Storer.*
" fifth species sent by Prof. *Baird.*

Cottus.

A broad and depressed head, contiguous to a body gradually diminishing towards the tail, is the essential zoölogical character of the genus Cottus, which contains at the same time freshwater and marine species; the former having, as the character of the group, a head generally smoother and less prickly with spines than the marine species, which in their turn are generally larger.

Europe as well as America produces species of both groups. For a long time all freshwater Cotti of central and northern Europe were considered as identical with *Cottus Gobio*, when, twelve years ago, Mr. Heckel * distinguished several species, very similar, it is true, to *Cottus Gobio*, but differing, however, in many respects.

Recently, an American naturalist has attempted to show that all Cotti of Northern America constitute only a single species, and that this species is identical with the *Cottus Gobio* of Europe. However, studying the Cotti which we have collected around Lake Superior, I first recognized two species; then comparing them with the *C. cognatus* Richards. and the *C. viscosus* Hald., I found these two latter not only distinct from each other, contrary to the opinion of Mr. Ayres, but yet distinct from those of Lake Superior. So that the presence of *C. Gobio* in this continent is quite illusive, as also the supposed identity of the Cotti in different regions.

A monograph of the freshwater species of the genus Cottus in Northern America would be a work of very great importance, were its purpose but to rectify the different opinions entertained with regard to them.

Cottus Richardsoni, Agass.

The largest individuals of this species which we have had at our disposal, and on which our description rests, measure four and three-fourths inches with the caudal. The head alone constitutes one and one-fourth inches of this length, of course a little more than the fourth part; its breadth equals three-fourths of its length, and its

* Annalen des Wiener Museums, 1837, II.

height forms a little more than the half. Besides being very depressed and flattened, the head further presents a slight depression on the occiput. The mouth is large, its breadth measures nearly six-eighths of an inch. The jaws are of equal length, bordered with excessively fine teeth, with very hooked points. The upper jaw is slightly protractile. The lips are considerably developed and form a very marked rounded process, on both sides of the lower jaw. The eyes of a circular form, with a diameter which exceeds a quarter of an inch, are placed at a distance of three-eighths of an inch from the end of the snout. The nostrils occupy about the middle of this space. The spine of the preoperculum scarcely forms a projection through the skin; it is strongly bent upwards and backwards. The upper and hinder angles of the operculum terminate in a small process, flat and sharp, which remains hidden in the thickness of the membrane which encircles the free margin of this bone. The branchiostegal rays, six in number, on each side, are slender and cylindrical. The isthmus between the horns of the hyoid bone measures half an inch.

The form of the body is regular, gradually decreasing towards the tail. The line of the back is raised; that of the belly is about straight, forming the continuation to the flattening of the lower surface of the head. The greatest height corresponds to the anterior margin of the first dorsal fin; it measures three-fourths of an inch, whilst the transversal diameter of that same region measures nearly six-eighths of an inch. Above the tail the height is but five-sixteenths of an inch, and the thickness one-eighth. The tail itself is slightly dilated and rounded at the insertion of the caudal.

The fins upon the whole are much developed. The first dorsal has a basis of six-eighths of an inch, and is five-sixteenths of an inch high, and is situated at one and three-eighth inches from the end of the snout. Its upper margin is rounded, the rays of the centre being the longest; they are eight in number and undivided. The second dorsal, twice as long as the first, and one third higher, is composed of eighteen rays, the longest occupying the centre of the fin; a single one of them is dichotomized at its upper end. The caudal, about six-eighths of an inch long, is truncated behind. Its upper and lower margins are slightly rounded. Thirteen rays may be counted there with a few rudiments; the four rays of the centre, bifurcated from the middle of their length, dichotomize anew at their extremity jointly with the two adjacent rays above and below. The anal begins beneath the third ray of the second dorsal and terminates a little before this latter; its form as well as its height is about the same; there are fourteen undivided rays in it. The ventrals contain five simple rays; the first, intimately connected with the second, is a little shorter. Their length is about five-eighths of an inch. The pectorals are large and fan-like; the rays, fifteen in number, are all undivided; the longest occupy the upper third part of the fin. They are only three-fourths of an inchlong, of course much below the length of the head.

Br. 6; D. viii.–18; C. 3-13.1; A. 14; V. I. 4; P. 15.

The anus is situated exactly in the middle of the length, including the caudal, which places it nearer to the insertion of this fin than to the end of the snout; it is bordered behind by a small, triangular, membranous appendage which leans towards the anterior margin of the anal. The body is completely naked and unprovided with scales, as is the case in all species. The lateral line is very distinct, it begins at the upper margin of the operculum, bends slightly downwards, then rises to terminate in a straight line about the middle of the second dorsal after having considerably approached the back. A row of pores is arranged in a straight line, constantly ascending until they are confounded with the back at the hinder margin of the second dorsal, at a distance of three-eighths of an inch from the insertion of the caudal.

The color is a dark olive-colored brown on the whole surface of the head and cheeks and all along the back. The lower half of the sides is of a lighter tint. The abdomen and the lower face of the head have a rather yellowish tint, dotted with very small black spots. The lower jaw is sometimes completely black. The general tint of the fins is the same as that part of the body to which they correspond. The dorsals, caudal,anal and pectorals are barred transversely with blackish spots. The ventrals have the same shade as the abdomen.

The characters which distinguish this species from *C. cognatus* Richardson, are easily made out by comparing the description which that author gives of it. The more distant position of the anus; the proportions in the dimensions of the head and body; the lateral line which terminates before the extremity of the tail; the more anterior position of the anal relatively to the second dorsal, and finally the shorter pectorals in proportion to the length of the head, are the most striking peculiarities.

I have found several specimens of this species in Montreal River. Among the number was one, whose general form has the same aspect, the same tint, the same proportions of the head and body, the same form and structure of the fins, the same mouth, but whose palatine bones bear a small group of teeth like those of the vomer. As yet we know only one freshwater species with palatine teeth, the *C. asper* Rich. From among five other specimens, also from Lake Superior, from Isle Royale, for which I am under obligation to Dr. C. T. Jackson, I have found the same group of palatine teeth in the largest of them, so that I am inclined to consider this peculiarity as an indication of old age, rather than a specific character.

COTTUS FRANKLINI, Agass.

This species is distinguished from the preceding by the following characters: the head retains the same proportions relatively to the body, but the mouth is smaller and less opened, and the teeth are less strong. The body diminishes more abruptly in height beyond the anus, and in its whole length the thickness is proportionally greater. Thence there results a more cylindrical and subconical form. The lateral line is less approximated to the back; it disappears on the sides as in the preceding species, but the row of pores continues as far as above the middle of the insertion of the caudal after a very abrupt depression a little before its termination.

The fins are less developed, but their relative position is the same. The ventrals instead of five rays have only four. The caudal rays alone bifurcate once on the middle of their length. In all other fins they are undivided. They may be reduced to a formula as follows:

Br. 6; D. 8-17; A. 12; C. 1-12.2; V. 1-3; P. 14.

The membranous appendage of the posterior margin of the anus is here only in a rudimentary state, but the position of this orifice is the same as in the preceding species, and this fact excludes, a priori, the idea of an approach to the *C. cognatus* of Richardson. Farther, our species has only four rays in the ventrals and twelve in the anal.

The ground is of a yellow olive color with black spots. The lower side of the head and body and the lower half of the sides are yellowish white. The fins have the color of the region of the body to which they correspond. The ventrals and anal are of one color, the others are barred or simply spotted in transverse rows.

This species is not without some analogy to that of Pennsylvania. The comparison which I have been enabled to make with it by means of specimens, for which I am under obligation to Professor Baird, has shown me differences which I consider as specific.

Found in various localities along the eastern shores of Lake Superior. Prof. James Hall has also sent me specimens collected by him on the southern shores of the same lake.

Boleosoma, Dekay.

This genus has been instituted by Dr. Dekay for a small fresh-water fish of the State of New York. He placed it in the family of Percoids, whence we withdraw it, to associate it to the Etheostomata, which should constitute a distinct group among the Cottoids, and the Gasterostei another near them. The zoölogical characters of this genus may be formulated in the following manner: The form of the body is that of a dart; the head is very short, rounded like an arc of a circle, below which the mouth, generally small and slightly protractile, opens horizontally; the upper jaw sloping over the lower. The neck and the sides of the skull compressed. The opercular apparatus and the cheeks covered with scales.

The species known to me are: the *Boleosoma tessellatum* Dekay, the *B. maculatum* of Lake Superior, the *Etheostoma Olmstedi* of Pennsylvania and the Northern States, which belongs to this genus and not to Etheostoma proper, and a species from South Carolina which I have called *Boleosoma tenue*.

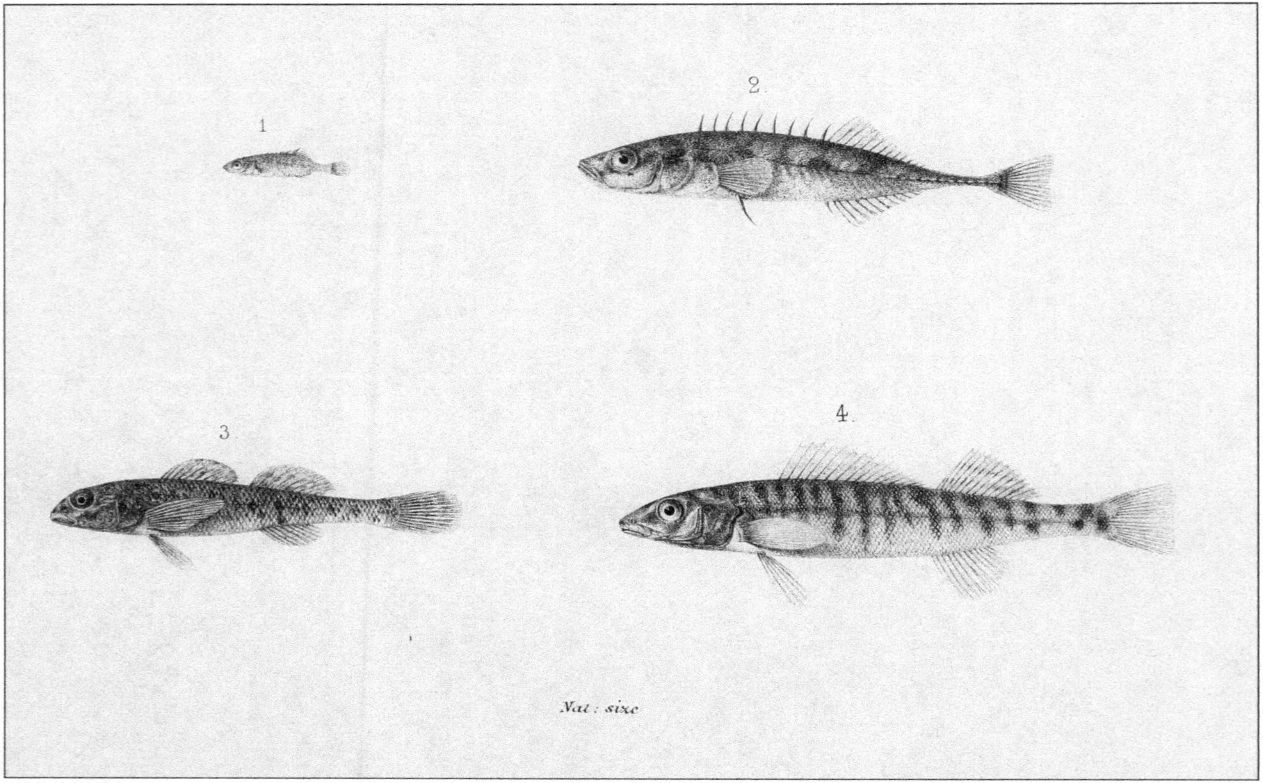

Pl. 4.

1

2.

3.

4.

Nat: size

A. Sonrel on stone from nat.

Lith. of A. Sonrel, Cambridge (Mass)

1. GASTEROSTEUS PYGMÆUS Ag. 2. GASTEROSTEUS NEBULOSUS Ag.

Boleosoma maculatum, Agass.

Plate IV., fig. 3.

The general form of this species is slender. The largest specimens which we have studied measured two and three-eighths inches in their whole length. The occiput and the anterior region of the body, before the first dorsal fin, are sensibly depressed. The space which the dorsal fins occupy forms a slightly convex line, sloping backwards and rising again behind the posterior margin of the soft dorsal and before the origin of the caudal. The ventral line is almost straight; it becomes convex beneath the tail in the same proportion as that of the back is concave. If we add to that a gradual compression of the sides from the front backwards, we shall have for the whole body an oval form, whichsoever be the region upon which we make a transverse section. We shall remark only a gradual decrease of the oval from the head towards the tail.

The head is short and thick; it forms just the fifth part of the whole length, measured from the end of the snout, to the posterior margin of the operculum. The snout grows rounded under the form of an arc of a circle, beneath which the upper jaw is fixed horizontally. It is about semi-elliptical and slopes over the lower jaw on its whole circumference. The latter, by the third part more narrow towards its symphysis than at the origin of its two branches, appears under the form of an acute angle whose summit would be rounded. The mouth is small and surrounded with a lip, continuous, rounded and uniform on its whole circumference. Card-like teeth, excessively small, visible only with the magnifying glass, occupy the margin of the jaws. The vomer also has teeth, but sensibly larger. Upon the pharyngeal bones they become again as slender as upon the jaws. The eyes are large, almost circular, one-eighth of an inch in diameter, situated at the upper margin of the skull, above which they make a regular projection. The distance which separates them from the end of the rostrum is not quite equal to their diameter. The nostrils open in two orifices, both nearer to the orbits than to the end of the rostrum; the upper orifice is twice as large as the lower; this latter is nearest the eye. The cheeks are very prominent and covered with very thin scales, which are hidden in the skin. Those covering the opercular apparatus are larger and more conspicuous. The opercular bones are generally smooth; the preoperculum is rounded; the operculum is triangular, with its summit turned towards the tail, and terminated by two processes, of which one is a cutaneous, thread-like expansion, the other a direct continuation of the bone. The suboperculum is of an irregular elliptical form, extending along the whole lower margin of the operculum. The interoperculum is a quite small triangular plate, lost between the bones above named, which constitute the opercular apparatus. The branchiostegal rays, as usual, six in number, are slender and diminish in length on the side of the isthmus between the horns of the hyoid bone.

The anus is small and a little nearer to the head than to the tail.

The first dorsal, of a roundish form, is generally separated from the second; sometimes, however, a small very low membrane unites the hinder margin of the one to the anterior margin of the other. It is composed of nine or ten spinous rays; the longest occupy the centre of the fin; they measure nearly five-sixteenths of an inch; the first has only the half of this height; the two last, which are still shorter, incline very much on the back. The second dorsal, a little higher than the first, is equilateral, having its upper margin almost straight, and its posterior margin half the height of the anterior margin, where the largest rays are; they are twelve in number, all bifurcated, and a few trifurcated. Its insertion measures about half an inch. The caudal is inserted on a slightly dilated pedicle of the tail; the upper and lower margins, almost straight, diverge a little on their extent; the posterior margin is truncated almost in a straight line; there are seventeen rays, divided from the first third part of their length, which is three-eighths of an inch; on the upper margin we count six, and on the lower five rudiments of rays; the two following on the two margins remain always below the dimensions of the others, nor do they bifurcate, though they be distinctly articulated transversely. The anal is opposite the second dorsal, it is less elevated, equilateral, but its outer margin is rounded; the rays, eleven in number, bifurcate beyond their middle; the ray of the anterior margin remains very short and simple. The ventrals are inserted a little behind the pectorals; they are five-sixteenths of an inch long; their form is lanceolate, narrow at the base and pointed at the extremity; of the six rays which compose it, that of the outer margin is simple, the two central ones are the longest and about equal. The pectorals are the longest of all the fins; their posterior extremity exceeds somewhat the ventrals. Their base, which measures one-tenth of an inch, forms the fourth part of their length. The rays are twelve; the central ones are the most elongated; they diminish regularly to each side, giving thus to the whole of the fin the form of an oval elongated at both ends.

Br. 6; D. IX–12; C. 6–17.5; A. 11; V. I. 5; P. 12.

The posterior margin of the scales is semi-circular and finely pectinated. The lateral line is concave, and median on the tail; it rises perceptibly as it approaches the head. The back and two-thirds of the sides are spotted irregularly with black; excepting a row of larger spots, extending from the posterior margin of the opercular apparatus to the pedicle of the caudal. Below this band, and as far as the under side of the body, it has a uniform yellowish tint. The dorsal and caudal fins, as well as the base of the pectorals, are barred transversely with black; the others have the tint of the belly.

This species was first observed at Fort William; a large number of specimens were also collected at the Pic.

Pileoma, Dekay.

The revision we have made of the species arranged in the genus Etheostoma by authors, has shown the necessity of subdividing this group into several smaller genera, for two of which we have retained names proposed by Dr. Dekay, though he does not seem to have been aware that his species belonged to Rafinesque's old genus *Etheostoma*. Not being able to give at this time a detailed review of this division without further materials which have no reference to the fishes of Lake Superior, I shall limit myself to indicating the general characters of the genus to which I refer the species described below.

The body is slender, fusiform, compressed. The head is conical, truncated, terminated by a kind of hog's snout, which perceptibly exceeds the lower jaw, without, however, sloping over it. The mouth, very slightly protractile, moderately opened, resembles an oblique arc of a circle, and opens at the end of the snout. The opercular apparatus and the cheeks are covered with scales.

Besides the species here described, *Etheostoma Caprodes* Raf., and *Pileoma semifasciatum* Dekay must rank in this genus.

PILEOMA ZEBRA, Agass.

This species is very near the *Etheostoma Caprodes* Raf. (*Pileoma Caprodes* Ag.) from which it differs only in a few peculiarities of the structure of the opercular apparatus, in the direction of the lateral line, and in the proportional size of the eyes. *Pileoma Caprodes* attains larger dimensions than our *P. zebra*, the largest specimens which we have had at our disposal, measuring only about seven inches. Our species is figured Plate 4, figure 4, under the name of *Etheostoma zebra.*

The general form of the species under consideration is elegant and regular. The upper outline of the body describes a slight curve, rising highest at the middle of the first dorsal; it curves more abruptly on the head than on the side of the tail, where it becomes a little concave on the space contained between the hinder margin of the second dorsal and the insertion of the caudal. The abdomen is less convex than the back; from the insertion of the anal, the outline rises and becomes slightly convex beyond this fin. The greatest height perpendicularly above the first dorsal is three-eighths of an inch. The greatest thickness, which corresponds to the same region, amounts to about two-thirds of the height. These proportions of the height and breadth are maintained uniformly along the whole body, from which a regularly compressed form, from the head to the tail, results. The head is conical, more pointed than in the other species of the genus, and forms the fourth part of the length of the body. The surface of the head is smooth. The eyes are large and subcircular, one-seventh of an inch in diameter, and situated at the upper margin; the distance between them exceeds their diameter. The openings of the nostrils are two on each side, placed one before the other, at the extremities of a small furrow, arched outwards. The posterior is the smallest, and occupies the upper and anterior margin of the eye; the second is placed nearer to the snout than to the eye itself.

The scales which cover the opercular apparatus are excessively thin, and allow the form and outlines of the different bones to be distinctly seen, the surface of which presents the same silver-colored reflection as the bare space before the pectorals, which extends also beneath the head. The ascending branch of the preoperculum is almost straight at its hinder margin, which is thinned; the lower angle is rounded. The operculum has the form of a slightly obtuse triangle; the upper angle is armed with a point; the margin forming the hypothenuse is slightly concave or undulated. The suboperculum is proportionally large; a membranous expansion, in which the point of the operculum loses itself, terminates its upper extremity; its lower extremity extends before the operculum in the form of a small hook; the bone itself, like the operculum, is rounded in the form of a stretched and undulated circle, on its circumference. The interoperculum is very small. The cheeks make no projection. The branchiostegal rays, six in number, are bent and flattened. The anus is nearer to the tail than to the head. The lateral line is direct from the centre of the caudal to the head; beyond the anal it approaches nearer the back than the belly. The scales are of middle size; the denticulations of their posterior margin are only visible with the magnifying glass.

Both dorsal fins are distinct and separated from each other. The first begins at three-fourths of an inch from the end of the snout; its insertion is equal to this distance; its greatest height, which is at the anterior third, is about one-fourth of an inch, and diminishes gradually towards its posterior margin. The second dorsal is higher than the first, and has a basis of less than half an inch; it is composed of fifteen bifurcated rays; its anterior and posterior margins are equilateral; its upper margin slopes from before backwards, its greatest height being at the anterior margin. The caudal has seventeen well developed rays—that is to say—articulated and bifurcated; and eight or nine undivided rudiments on each of its sides; its posterior margin forms a slight crescent; its upper and lower margins are straight. The anterior margin of the anal is opposite to that of the second dorsal, but its insertion is an eighth of an inch less, and it is at least as high, if not higher; its terminal margin is more convex; the greatest rays occupy the anterior third part; the first is undivided; the anterior margin is rounded, the posterior short and straight; here are twelve rays. The ventrals have, as usual, six rays, the first undivided; their insertion is a little behind the pectorals; their length exceeds three-eighths of an inch; they are elongated and terminated in a point, which exceeds the posterior extremity of the pectorals. These latter are somewhat longer than the ventrals, and are composed of fourteen rays, the longest of which occupy the centre. The base of these fins measures an eighth of an inch. When expanded, the rays arrange themselves in the form of a fan, with a regularly rounded circumference.

Br. 6; D. XIV–15; C. 9-17.9; A. 12; V. I. 5; P. 14.

The body is barred with black transverse bands, extending from the back towards the sides. They are alternately longer and shorter. None are found on the last third of the sides, which has the color of the abdomen and the lower part of the head. The fins partake of the color of the region of the body to which they belong. Above, the head is finely dotted with black.

The few individuals of this species which we have procured were caught at the Pic.

Gasterosteus nebulosus, Agass.

Plate IV., fig. 4.

The determination of this species has caused us much trouble, from its great resemblance to *Gast. occidentalis* Cuv., *G. concinnus* Richards., and even to *G. pungitius* of Europe, with which the preceding species are compared in the descriptions of authors. Another difficulty occurred to us, and rendered the synonymy of *G. occidentalis* Cuv. very complicated, from Dekay having referred to this fish an analogous species of the State of New York, which differs from it; the same which we find again in Massachusetts, and which Dr. Storer identifies with *G. pungitius* L.. After a minute comparison, we have ascertained that the species of Lake Superior, which we here describe, is a species distinct from all others; that *G. occidentalis* Dekay, and *G. pungitius* Storer, are the same species, differing, however, from the *G. occidentalis* Cuv. This latter will preserve the name which Cuvier gave to it, and the species of New York and Massachusetts will be designated under the name of *G. Dekayi.*

This is not the place to enter into minute details, by means of which to distinguish the species. We shall soon treat of them in a monograph of all the species of North America, limiting ourselves at present to describing the one collected about the Sault of St. Mary.

The body is subcylindrical or compressed, growing thinner from the insertion of the dorsal and anal fins towards the tail, which becomes very thin and slender, widening at the tip for the insertion of the caudal. It is from two inches to two inches and one half long in adult specimens; its greatest height is at the pectorals, and is contained six times in the length. The outlines of the back and belly are slightly convex; the former from behind the occiput to the posterior margin of the dorsal fin, where it descends somewhat; the latter from the lower end of the snout to the posterior margin of the anal, being depressed on the tail. The head, from the end of the snout to the posterior margin of the operculum, is the fourth part of the length, and to the occipital carina one-fifth. The head is subconical, generally pointed forwards; the lower jaw, which somewhat exceeds the upper in the protraction, forms an angle, reëntering in the retraction. The teeth are minute; the fissure of the jaws considerable. The eyes, proportionally large, have a diameter of nearly three-sixteenths of an inch; the distance which separates their anterior margin from the end of the snout is a little longer than their diameter. The nostrils, which open along this space, are very near the orbits.

The suborbital bones, only two in number, are far from covering the cheeks. The first protects the anterior margin of the eyes and the lower margin of the nostrils, leaving a bare triangular space between it and the second suborbital, situated below the vertical line which would pass through the eyeball. It does not exceed the posterior margin of the orbits, and touches the preoperculum only by its lower margin. The rest of the cheek, between the eye and the preoperculum, remains completely bare. They are finely granulated, without spines or denticulations, though their outer circumference presents a few notches. The preoperculum borders the posterior and lower margins of the cheek in the form of an obtuse angle, dilated on the summit, and narrow at its margins. The operculum is triangular with slightly concave sides, the posterior margin rounded, and the surface radiately striated. The suboperculum forms an acute angle; its anterior branch is convex on the side of the operculum, and concave on the side of the interoperculum, which has the form of a small subrectangular triangle.

There are about three equal branchiostegal rays. The branchial fissure itself is well proportioned. The suprascapular and scapular bones are not visible externally; they attach the humeral to the skull. The upper extremity of the humeral forms a small triangle, with granular surface, one side of which extends above the base of the pectorals, thus bounding, at the upper part, the large smooth space which separates these latter from the branchial opening. This smooth space is bordered on its lower circumference by the narrow prolongation of the cubitus on each side, which, at the lower part of the body, forms a triangle, whose summit advances like the point of a gothic arch in the isthmus near to the branchial fissure. The sides extend parallel as far as the ossa innominata, without uniting with them. They thus circumscribe a bare triangular space in the enclosure of the arch, which embraces not quite half of the space, it being a parallelogram for the rest of its extent. The shield under the belly formed by the ossa innominata is triangular, and the basis turned forwards is striated transversely at the outer margin, from which is cut a segment of a circle, which is sometimes obtusely triangular where the bare space disappears, which the branches of the cubitus circumscribe, as we have just mentioned. The hinder point of the triangle is obtuse, and terminates at some distance from the anus. The ventral spine does not quite reach the extremity of the triangle. The ascending branch of the ossa innominata rises at a small distance from the pectorals, inclining backwards. It is somewhat more dilated at its summit than at its origin, forming thus an elongated isosceles triangle, striated at its surface. The anus is situated a little behind the middle of the length.

There are generally nine spines on the back ; a single instance of eight has occurred from among a hundred individuals submitted to our examination ; none contained ten. A small triangular and very low membrane extends from the inferior third and inner part of each of them, to rejoin the back. These spines, of an average height of a tenth of an inch, are thin and bent somewhat backwards ; the last, which is bent a little more than the others, is always independent of the soft dorsal. This latter is generally composed of ten, sometimes eleven, soft rays, upon a base of about two-fifths of an inch ; all are bifurcated, as is the case with the other fins for three-fifths of their length ; at the anterior margin the rays are almost one-fifth of an inch in height, whilst on the posterior margin they are confounded with the line of the back, which gives to this fin the form of a triangle. The anal, which is exactly opposite to it, has somewhat the same form, with a somewhat shorter base, which recedes a little at its anterior margin ; it contains nine rays, and in a few exceptional cases eight ; it is somewhat lower than the dorsal. The caudal is rounded, rather concave on its posterior margin ; there are constantly twelve bifurcated rays, (six in each lobe,) and four rudimentary ones at the upper margin, and as many at the lower ; the inner one has twice the length of the three others ; the largest rays are about one-fifth of an inch in length. The bare space of the upper and lower margins of the tail, which separates the caudal from the termination of the dorsal and anal, varies between one-third and two-fifths of an inch. The pectorals are sometimes as much as three-tenths of an inch long ; they are composed of ten nearly equal rays ; their form is oval, narrowed towards the base. The ventrals are, as in most species, reduced to a spinous ray, inserted on the ossa innominata, with a small membrane from the axilla, at the centre of which a small simple ray is observed. The spinous ray is here very elongated, since it nearly reaches the posterior extremity of the ventral cuirass, against which it leans when at rest. It is about one-sixth of an inch long, slightly curved within, excavated at the inner side of its base, sulcated on its outer surface, thin like those of the back, and with the magnifying glass, traces of fine denticulations may be discerned at its inner margin.

D. IX–10; A. 9; C. 4. 12.4; P. 10; V. I. 1.

The body, besides the bones of the belly, is completely bare and unprovided with scales. On the sides of the tail we remark a small carina, which extends from the hinder third of the dorsal and anal fins to the basis of the caudal. This carina is formed by small bony pieces, upon which rise small depressed hook-like points. The lateral line is continued from the anterior extremity of this carina to the occiput, following the back-bone.

This species has been found in abundance at the Pic. When alive, its color is of an olive brown above, mottled with blackish brown and silvery white below.

Gasterosteus pygmæus, Agass.

Plate IV., fig. 1.

This species is very inferior to the *G. concinnus* in its size, so that we have in it, and not in this latter, the true pigmy of the genus. Its length does not attain eleven-sixteenths of an inch. The head, measured from its anterior extremity to the posterior margin of the operculum, has a little more than one-fourth of it. Its height varies between one-seventh and one-eighth of an inch, and remains nearly the same from the nape of the neck to the anterior fourth of the dorsal. The eyes are proportionally large; the nostrils, situated at the upper margin of the orbits, occupy the middle of the space between this latter and the end of the snout. The head is somewhat sloping. The curve of the back, very elliptical on its middle, descends abruptly towards the tail about the insertion of the soft dorsal; that of the belly is slightly convex, and ascends also very abruptly, to form, in conjunction with that of the back, a narrow contraction on the middle of the peduncle of the tail, which is remarkably short, measuring scarcely one-eighth of an inch from the posterior margin of the dorsal and anal fins to the origin of the caudal. The anus is placed seven-sixteenths of an inch from the head. The body is completely bare; the bones of the head are smooth; the opercular apparatus hidden under the skin; the whole dotted with black. The space between the pectoral fins and the branchial opening is sensibly reduced, and covered by the skin, the aspect of which is the same as on the rest of the body. The thoracic arch is not visible; we have also scarcely found traces of the cuirass formed by the ascending branch of the innominated bones, and about the basis of the ventral spines, which are perceived only with the magnifying glass, under the form of very small hooks.

It was difficult to count the exact number of the rays of the fins, as they are very thin and slender. We have, however, recognized the existence of at least six dorsal spines; the last of which is well developed, and has a small membrane at its posterior margin, arising from the summit of the spine to unite the basis of the soft dorsal. This latter seems to have seven rays, composing a triangular fin, whose posterior angle rests on the tail. The anal has the same form, but is somewhat smaller, opposite to the dorsal, and provided with six rays. The caudal is short, rounded, and has twelve rays, perhaps even fourteen, for the two exterior ones appeared to us almost twice as thick as the others. The pectorals are pointed, and have eight rays of an extreme thinness. As for the ventrals, as we have seen above, they are only visible with the magnifying glass, and all we have been enabled to do was to satisfy ourselves of the presence of the spinous ray common to all species.

Three individuals of this species were found at Michipicotin. Two from among them are only one-quarter of an inch long.

Esocidæ, (*The Pickerels.*)

The family of pickerels is perhaps the least understood of any in the whole class. From the characters assigned to it by Cuvier, it contains a variety of fishes, which can scarcely belong to one and the same natural group, and indeed more recent investigators, as, for instance, Joh. Müller, have divided the Esoces of Cuvier into two families, on the ground of the pseudo-branchiæ; so that we have now the families of Scomberesoces in addition to the true Esoces. Several isolated genera formerly referred also to the family of the Esoces, have either been removed to other natural groups, or become the types of distinct families for themselves, as Lepidosteus and Polypterus.

No species of Scomberesox are found in Lake Superior, nor in any of the lower lakes, although they occur in the Atlantic rivers of these latitudes, where *Belone truncata* is not uncommon, and with it *Scomberesox Storeri*. Without discussing for the present the natural relations of the Esoces and Scomberesoces, I cannot but think that the Scomberesoces are an aberrant type of the great family of Scombridæ, with abdominal ventrals and some other peculiarites.

The true Esoces, as circumscribed by Joh. Müller, are very few; indeed his family contains little else than the true genus Esox, fishes which are all inhabitants of the fresh waters, and occur chiefly in the temperate zone; their structural peculiarities are such that it is difficult to understand their true affinities; their cylindrical, elongated form indicates a low position among abdominales, as does also the composition of their mouth, the maxillary being entirely deprived of teeth, while the palatal bones contain a powerful armature; the connection of the intermaxillaries and maxillaries in one arch places them however in the vicinity of the Salmonidæ. The skeleton, and especially the skull, is remarkably soft in these fishes.

North America seems to be the proper fatherland of the genus Esox, its species being numerous all over this continent, from the great northern lakes, through all the rivers and lakes of the east and west, and as far south even as Florida. In North America, therefore, a deeper study of this family becomes alone possible, in relation both to the knowledge of species and their affinities with the other families of the class.

The species are certainly more numerous than the American authors who have written on the pickerels have recognized; and if we had for examination specimens from all localities of this continent, we might now publish the result of our observations on this family. But, unwilling to introduce in our science unconnected observations, especially on a difficult and controverted subject, we prefer to recur at a future time to this family. We shall limit ourselves here to a description of the species collected from Lake Superior. But its bare description would be without interest, did we not compare it with the species already described from the region of the lakes. Two species are mentioned by Dr. Richardson: an *Esox Lucius* and an *Esox Estor* Lesu. Now the species of Lake Superior is not the *Lucius* of the Fauna Boreali-Americana, as we might infer by comparing the descriptions. In regard to this, we could entertain no doubt. As for the *Esox Estor* of Dr. Richardson, we allow that we have doubts whether or not the author of the Fauna Boreali-Americana had the true *Esox Estor* Lesu., or perhaps my *Esox Boreus*, from Lake Superior. The description which he gives of it* is too incomplete to enable us to recognize it; the more so, as that description is made with reference to *Esox Lucius*, which is found to be quite different. Only two characters occur which may be considered to have some value; but, strange to say, these two characters are found united in none of the species which I know. I mean, first, the form of the scales, which are as high as they are long, a character which we find in the true *Esox Estor* Lesu. But, again, the scales would be much smaller in the species which Dr. Richardson had in view. The *Esox Estor* Lesu. is the species which has the least number of scales on the cheeks and opercula; but Dr. Richardson gives for his *E. Estor* two rows of scales, which descend along the anterior margin of the operculum until they attain the upper angular process of the suboperculum. It is therefore possible that the species referred to *Esox Estor* by Dr. Richardson was neither the *Esox Estor* Lesueur, nor my *Esox Boreus*, but a species distinct from all others, as the small size of its scales seems to indicate.

Esox Boreus, Agass.

When marked external zoölogical characters are wanting in a group, on account of its uniformity, it becomes necessary to resort to another series of facts. When the object is to find the place which a certain family occupies in its order or in its class, comparative embryology and palæontology will often answer the purpose as completely as an anatomical investigation, and even with more precision. If, on the contrary, we have to do with the distinction of species, we may in such cases have recourse to comparative anat-

* Fauna Boreali-Americana, p. 127.

omy. In the present instance, we have had no occasion to hesitate. Having seen by turns the general form, the outlines of the fins, the outer details of the head, and the color, sometimes varying in the same species to a great extent, and at others preserving a monotonous uniformity, we have taken for our guide the structure of the mouth, and particularly that of the palatal bones and of the vomer, and we may say, that whenever we have had series of specimens at our disposal, the general traits of the species have not varied sensibly. We have relied still more confidently on this method, when, after comparing the buccal apparatus, we have seen the extreme variations stop in these limits.

What strikes us, especially in the species here referred to, is the general smallness of the rows of palatal and vomeric teeth. None make a strong projection above the others. The surface of the palatals has a very uniform appearance, and it is only when we examine them closely, that we perceive that the teeth of the inner row alone exceed those of the body of the bone in size by about one-third, though remaining equal among themselves. The palatal bones themselves are slightly bent, with the convexity turned inwards. Their greatest length is one and a half inches, their greatest breadth one-third of an inch, which maintains itself on the anterior two-thirds, diminishing sensibly on the posterior third, the extremity of which terminates in an oblique line, extending from the front backwards. The anterior margin is oblique from behind forwards, as in most species, owing to the curve of the snout. The vomer, including its dilatation and the narrow band, is one and nine-sixteenths inches long. The dilatation is of a triangular form, rounded at the anterior margin, and slightly concave on its sides; its centre is depressed, concave. A certain number of teeth, larger than those of the centre, occupy its circumference. The narrow band of teeth upon the vomer is lanceolate, and terminates in an acute point a little beyond the extremity of the palatals. We barely observe a contraction at the place where it enlarges at its anterior part. In the centre it is one-eighth of an inch broad. The teeth which cover its surface are very small. The intermaxillaries do not measure five-eighths of an inch; they have a single row of teeth as small as those of the vomeric band. The same is the case with the teeth of the lower jaw as with those of the palate. The largest, situated on the posterior two-thirds of the maxillary branches, are uniform among themselves and regularly spaced, slender, flattened, and their acute point is curved either backwards or inwards. At the anterior part, and on the symphysis, the same uniformity exists; and though forming only one single row, they are grouped in pairs. They incline towards the interior of the mouth, and are more conspicuous than on the body of the palatal bones.

The tongue is slightly dilated, laterally rounded, subtruncated at its anterior margin. It has on its middle two contiguous shields, covered with excessively small, card-like teeth. The posterior, of elliptical form, is six-eighths of an inch long, and one-fourth of an inch broad. The anterior, half as long, terminates in a conical point, at a distance of one-third of an inch from the end of the tongue. We remark two small, similar shields on the symphysis of the branchial arches. The pharyngeal bones are furnished with card-like teeth of great uniformity.

The external characters of this species may be indicated in the following manner. In general it is fusiform, the greatest thickness corresponding to the middle of the length, whence the body seems to taper towards both its extremities. The head forms one-fourth of the whole length; its conical form is merely the result of the attenuation of the body forwards, which renders it proportionally small; its upper face is flattened; a medium furrow, with widened margins, occupies the centre of it, between both eyes. The snout is depressed, and terminates in an elliptical curve, which exceeds the extremity of the lower jaw. Numerous and considerably large pores extend on the frontals above the snout; from the occiput they pass beneath the orbits and through the preoperculum on the branch of the lower maxillary. The mouth is moderately opened. The eyes are large and elliptical; their horizontal diameter is eleven-sixteenths of an inch, their vertical diameter nearly five-eighths of an inch. The nasal orifices, two in number on each side, open before and within the eyes; the hinder is separated from the orbit by a space of only one-fourth of an inch; it is crescentic, with the convexity turned towards the eye; a membranous fold shuts its opening; the anterior is ovoid, and has a large opening outwards. The cheeks are completely covered with scales as also the upper half of the operculum. The rest of the opercular apparatus is bare. The preoperculum is narrow, its posterior margin undulated. The operculum is trapezoidal ; its anterior margin concave ; the posterior rounded, and the lower oblique. The suboperculum, somewhat longer than the operculum, is about one-third as broad, being, however, somewhat more narrow behind than in front. The interoperculum is very narrow and elongated, being undulated like the preoperculum on its outer margin. The branchiostegal membrane is narrow ; it contains fifteen rays, of which the first is much the broadest ; all are flattened or compressed ; the longest are two inches ; the shortest five-eighths of an inch long.

The body grows thinner towards the tail from the ventrals, undergoing a considerable contraction behind the dorsal and anal fins. It widens again at the insertion of the caudal.

The dorsal fin has a quadrangular form, its upper margin being only slightly arched ; it is two and three-eighths inches long and two inches high. The rays are twenty-one in number ; the three first are very short, and are applied towards the fourth ; the three last diminish equally in height ; its posterior margin is at a distance of three inches from the rudimentary rays of the caudal. The anal is situated a little farther back than the dorsal, at a distance of two and three-eighths inches only from the basis of the caudal ; its circumference is rounded ; there are ten rays ; the four first near the fifth ; its length is an inch and six-eighths, its height two inches, making it, of course, higher than long. The caudal is composed of eighteen rays ; it is notched ; the breadth at the extremity of the two lobes measures three and a half inches ; the largest rays correspond to the middle of each lobe ; they are two and six-eighths inches long, whilst in the centre they are scarcely one inch and a half ; very small interradial scales extend over a space of three-fourths of an inch for each lobe from their insertion. The ventrals contain eleven rays ; they are somewhat nearer the anal than the pectorals are, and also nearer to the head than to the extremity of the caudal, being situated at ten and six-eighths inches from the snout ; the whole length being nearly one foot eight inches ; their form is broad and rounded on the outer circumference ; their insertion measures about five-eighths of an inch, their greatest breadth one inch and a fifth, and their length two inches. The pectorals, composed of sixteen rays, have the same general form as the ventrals, but still more rounded, longer, and broader by one-fourth of an inch, with a basis of insertion of eleven-sixteenths of an inch.

Br. 15; D. 21; A. 18; C. 28; V. 11; P. 16.

The scales are oblong, longer than broad, and proportionally larger than in the *Esox Estor* Lesu. We may count four of them on the space of three-eighths of an inch. The lateral line is very distinct; it follows the middle of the body from the basis of the caudal to a point in front of the dorsal and anal fins, whence it rises to terminate at the height of the upper third of the operculum.

The upper side of the head, the back, and the upper half of the sides are bluish black, amidst which the scales shine with a metallic azure reflection. The face and the lower half of the sides have a lighter tint, are sprinkled with whitish spots, arranged in horizontal or oblique bands on the face, spherical or ovoid on the sides, and disposed in ill-defined longitudinal rows. The lower side of the head is white; the abdomen is very pale yellow. The fins have an olive-colored tint; the caudal has black spots, elongated in the direction of the rays; these spots affect less regularity on the dorsal and anal, and disappear almost entirely on the ventrals and pectorals.

In the young individual, the spots of the sides do not exist, as such. The general color is more olive, more uniform, and the body is barred vertically with sinuous white bands, which are now and then intercepted. This fish was obtained from various places along the northern shores of Lake Superior.

Gadoids.

The family of codfishes contains numerous species, closely allied, all of which are circumscribed within the colder regions of both hemispheres. The northern seas especially teem with codfishes of various kinds, and the number of individuals of some of the species must be countless, if we judge by the quantity caught annually. Taken as a whole, this family consists of low forms, their body being very much elongated, their vertical fins very large, and the ventrals placed in such a position under the chin, as shows that when they were formed, the vertical fin extended underneath very far forwards. The abdominal cavity extends also far backwards. In some of the genera, the dorsal, caudal, and anal remain continuous; in others, they are slightly divided; in others, they become subdivided into many fins, but in all they extend very far forwards. From their geographical distribution in the colder portions of the northern hemisphere, we need not be surprised at finding a good many of these fishes among the freshwaters, as the northern seas contain less salt than the other portions of the ocean.

The real affinities of the family are still obscure to me. From their peculiar affinities, they stand very much by themselves; however, the large size of the head, the developments of the dorsals, and even the structure of the skeleton, seem to bring them near the Lophioids; and, on the other hand, I cannot but think the Scomberoids somewhat related to them, especially when comparing the Merluccius with Naucrates, etc. In Lake Superior, one single species of that family occurred.

The first account we possess of the Gadoids of North America dates back to the year 1773. At that epoch, J. Reinhold Forster published descriptions of four species of fishes of Hudson's Bay, in a letter addressed to Pennant,* among which a Lota is mentioned, which he identifies with the European species, so well described, he says, by Pennant† himself, that he thought it superfluous to add anything. The sole difference that struck him, was a larger size, and six branchiostegal rays instead of seven. Pennant afterwards inscribes it, in his Zoölogia Arctica, under the same denomination of *Gadus Lota* L.

In 1817 Lesueur published descriptions of two species which he considered as new, under the names of *Gadus maculosus* and *Gadus compressus*,‡ but he cites neither Forster nor Pennant, thinking, no doubt, that they had seen the European species. The same year Dr. Mitchill, though acquainted with the writings of Lesueur, seems not to have been aware that the latter had just named his species, and proposed to call the first *Gadus lacustris*.§ Here

* Philos. Trans., LXIII. 149.
† British Zoölogy.
‡ Journ. Acad. Nat. Sc., Philad., I. 83.
§ Amer. Month. Mag. II. 244.

already begins a discrepancy in the characters assigned to this species. Lesueur says, "*jaws equal,*" and Mitchill, "*upper jaw longest, and receiving the lower.*" He adds: "*The skin is smooth and scaleless.*" The smallness of the scales must have misled him; if not, his *Gadus lacustris* is not the *Gadus maculosus* of Lesueur. Dr. Richardson mentions the *Gadus Lota* in his Journal of the Expedition of Franklin, published in 1823; and in 1836, when publishing the Fauna Boreali-Americana, he describes, under the name of *Lota maculosa*, a species from Pine-Island-Lake, which must be the same he had seen in 1823, since he gives the same synonyms. The description is considerably detailed, but it contains no criterion establishing the perfect identity with the species of Lesueur. He agrees on the point that the jaws are of equal length, but as for the lateral line, Lesueur had said, "*in the middle of the body,*" and Richardson says, "*nearer to the back than to the belly, and is slightly arched till it passes the first third of the anal fin, after which it takes a straight course,*" etc.

In 1839 Dr. Storer* gave a short description of the *Gadus compressus* Lesu., which he places, however, in the genus Lota, without trying to establish a connection between his description and that of Lesueur.

In 1842 Dr. J. P. Kirtland† copies the description of *G. maculosus* of Lesueur, and cites Richardson in the synonyms. He adds a figure. In the same year, 1842, Rev. Z. Thompson‡ describes a species from Lake Champlain, comparing it with the description of *G. maculosus* Lesu., and though retaining for it this name, he remarks certain differences which strike him. Thus, the upper jaw is uniformly longer, and the lateral line, "*anterior to the vent, is much nearer the back than the belly.*" In this sense, the lateral line agrees with the description of Dr. Richardson. Mr. Thompson finds much resemblance between his fish and that described by Dr. Storer under the name of *Lota Brosmiana*, but it differs from it, he says, "*in having the upper jaw longest, in having the snout more pointed and less orbicular.*" He finds that his fish differs as much from the *Lota*

* Rep. etc., p. 134.

† Bost. Journ. Nat. Hist. IV., 24, Pl. 3. f. 1.

‡ History of Vermont, p. 146.

maculosa Lesu. and *Lota Brosmiana* Storer, as these latter differ among themselves; and that they constitute three species or only one. Here, for the first time, we have a critical and comparative examination, but it does not satisfy the writers who follow him, or they seem, indeed, not to have known his account.

As to *Lota compressa* Lesu., Mr. Thompson was not acquainted with it, and, in his turn, he copies the description of Dr. Storer.

The *Natural History of the Fishes of New York* appeared also in 1842. *Lota maculosa* is there inserted with a long list of synonyms, but without comparative criticism. Then characters are noticed, to which nobody had made allusion before. Such are: "*Pectorals long, pointed; their tips reaching nearly to the base of the first dorsal*" — "*first dorsal small, subtriangular;*" and a figure to confirm them. Dr. Dekay says, however, he is acquainted with *Lota compressa* only through the descriptions of Lesueur and Storer, from whom he may have borrowed his. But whence comes his figure, which exists nowhere else, so far as I know? Dr. Dekay describes and figures also another species, which he considers as new, under the name of *Lota inornata* from the Hudson River, and which Dr. Storer considers as synonymous with his *Lota Brosmiana*, of New Hampshire.* Certainly, if this identity is real, it does not exist in the figures which these two authors have published, nor even in their descriptions, since the one, (*Lota inornata* Dekay,) has the upper jaw larger than the lower, while in the other (*Lota Brosmiana* Storer) both jaws are equal. And there are still other differences.

In such a state of things, it was impossible for me to establish the synonymy and to compare critically the species without original specimens for comparison. Possessing myself only such specimens as I procured at Lake Superior, I will describe, provisionally, that species under the name of *Lota maculosa*, without synonymy, and I will limit myself to indicating the analogies and the differences which I have observed, I will not say in the published figures, but in the original descriptions of the authors. The question, thus restored to its true position, may in future lead to further progress.

* Synops. N. Am. Fishes, p. 219.

LOTA MACULOSA.

The description which best coincides with our specimens is that of Mr. Thompson of the Lota of Lake Champlain, and which we have cited above. The wood-cut which he gives of it, though much reduced, sustains this assertion. I will remark one difference only, which is, that the snout is more pointed, and the upper lip slopes more over the lower jaw than in the specimens from Lake Superior. The first dorsal fin seems also to be higher than the second.

Dr. Richardson not having figured the species which he describes, we have compared attentively his description with our specimens, to which it applies in a general way, as also in several peculiarities; nevertheless, we would direct the attention of ichthyologists to the following differences: The head is proportionally more elongated, forming only the fifth part of the whole length; the snout more pointed, the upper jaw somewhat longer than the lower; this latter is besides considerably exceeded by the upper lip. The distance which separates the centre of the orbit from the end of the snout is equivalent to three lengths of the axis of the orbit itself; this axis is contained four times and a half on the space which extends from this same point of departure to the posterior margin of the operculum, being contained seven times and a half in the whole length of the head. The eyes themselves are besides situated at the upper margin of the face, so as to be seen from above. The labials are an inch and a half long, the intermaxillaries one inch. These measures, compared with those which Dr. Richardson gives, show us remarkable differences in the proportions of these bones. The posterior extremity of the labials is besides curved forwards.

Among the fins I find the second dorsal, if not higher than the first, at least as high. The anal is generally lower, though having the same form, and like the second dorsal, rounded and somewhat higher at its termination. The anal terminates a little before the dorsal. The ventrals have seven rays; the second is the longest. Formula:

Br. 7; D. 11–76; A. 64; C. 45; V. 7; P. 19.

The skin which envelopes the fins is thick, a character which we find again in *Lota compressa*, which seems, however, to be a much smaller species.

The head is much depressed. The body is subcylindrical from the occiput to the anus. The tail is also much compressed, and its height diminishes quite insensibly from before backwards.

The color is dark olive brown above, mottled with blackish brown; sowewhat yellow about the lower part of the abdomen, and whitish underneath.

From Michipicotin.

It is very difficult to decide what are the characters which distinguish *Lota compressa* from *Lota maculosa*. It seems that the species is generally smaller. Lesueur gives to it an upper jaw longer than the lower, a character alternately given to it and *L. maculosa* by the authors who have written after him. Whether the body is proportionally shorter is to be verified anew, as also the greater compression of the sides, and the back, which is said to be highest at the basis of the dorsal fins. Lesueur adds, as a character, a more elongated caudal, an equal dorsal and anal.

The description of Dr. Storer, the only one which has been made from nature since Lesueur, as it is not comparative, does not solve the question.

SALMONIDÆ.

So long as the family of Salmonidæ remains circumscribed as it was established by Cuvier, it seems to be a type almost universally diffused over the globe, occurring equally in the sea and in freshwater, so that we are left almost without a clue to its natural relations to the surrounding world. Joh. Müller, working out some suggestions of prince Canino, and introducing among them more precise anatomical characters, had no sooner subdivided the old family of Salmonidæ into his Salmonidæ, Characini and Scopelini, than light immediately spread over this field. Limited now to such fishes as, in addition to the mere general character of former Salmonidæ, have a false gill on the inner surface of the operculum, the Salmonidæ appeared at once as fishes peculiar to the northern temperate region, occurring in immense numbers all around the Arctic Sea, and running regularly up the rivers at certain seasons of the year to deposit their spawn, while some live permanently in freshwater. We have thus in the true Salmonidæ actually a northern family of fishes, which, when found in more temperate regions, occurs there in clear mountain rivers, sometimes very high above the level of the sea, near the limits of perpetual snow, or in deep, cold lakes. That this family is adapted to the cold regions is most remarkably exemplified by the fact that they all spawn late in the season, at the approach of autumn or winter, when frost or snow has reduced the temperature of the water in which they live nearly to its lowest natural point. The embryos grow within the egg very slowly for about two months before they are hatched; while fecundated eggs of some other families which spawn in spring and summer, give birth to young fishes a few days after they are laid. The Salmonidæ, on the contrary, are born at an epoch when the waters are generally frozen up; that is, at a period *when the maximum of temperature is at the bottom of the water*, where the eggs and young salmons remain among gravel, surrounded by a medium which scarcely ever rises above thirty or forty degrees.

It is plain from these statements, and from what we know otherwise of the habits of this family, that there is no one upon the globe living under more uniform circumstances, and nevertheless the species are extremely diversified, and we find peculiar ones in all parts of the world, where the family occurs at all. Thus we find, in Lake Superior, species which do not exist in the course of the Mackenzie or Saskatchawan, and vice versa, others in the Columbia river which differ from those of the Lena, Obi, and Yenisei, while Europe again has its peculiar forms.

Whoever takes a philosophical view of the subject of Natural History, and is familiar with the above stated facts, will now understand why, notwithstanding the specific distinctions there are between them, the trouts and whitefishes are so uniform all over the globe. It must be acknowledged that it is owing to the uniformity of the physical conditions in which they occur, and to which they are so admirably adapted by their anatomical structure, as well as by their instinct. Running up and down the rapid rivers and mountain currents, leaping even over considerable waterfalls, they are provided with most powerful and active muscles, their tail is strong and fleshy, and its broad basis indicates that its power is concentrated; it is like the paddle of the Indian who propels his canoe over the same waters. Their mouth is large, their jaw strong, their teeth powerful, to enable them to secure with ease the scanty prey with which they meet in these deserts of cold water, and nevertheless, though we cannot but be struck by the admirable reciprocal adaptation between the structure of the northern animals and the physical condition in which they live, let us not mistake these adaptations for a consequence of physical causes, let us not say that trouts resemble each other so much because they originated under uniform conditions; let us not say they have uniform habits because there is no scope for diversity; let us not say they spawn during winter, and rear their young under snow and ice, because at that epoch they are safer from the attacks of birds of prey; let us not say they are so intimately connected with the physical world, because physical powers called them into existence; but let us at once look deeper; let us recognize that this uniformity is imparted to a wonderfully complicated structure; they are trouts with all their admirable structure, their peculiar back bones, their ornamented skull, their powerful jaws, their movable eyes, with their thick, fatty skin and elegant scales, their ramified fin-rays, and with all that harmonious complication of structure which characterizes the type of trouts, but over which a uniform robe, as it were, is spread in a manner not unlike an almost endless series of monotonous variations upon one brilliant air, through the uniformity of which we still detect the same melody, however disguised, under the many undulations and changes of which it is capable.

The instincts of trouts are not more controlled by climate than those of other animals under different circumstances. They are only made to perform at a particular season, best suited to their organization, what others do at other times. If it were not so, I do not see why all the different fishes, living all the year round in the same brook, should not spawn at the same season, and finally be transformed into one type; have we not, on the contrary, in this diversity under identical circumstances, a demonstrative evidence that there is another cause which has acted, and is still acting, in the production and preservation of these adaptations; a cause which endowed living beings with the power of resisting the equalizing influence of uniform agents, though at the same time placing these agents and living beings under definite relations to each other?

That trouts are not more influenced by physical conditions than other animals, is apparent from the fact that there are lakes of small extent and of most uniform features, in which two or three species of trout occur together, each with peculiar habits; one more migratory, running up rivers during the spawning season, etc., while the other will never enter running waters, and will spawn in quiet places near the shore; one will hunt after its prey, while the other will wait for it in ambuscade; one will feed upon fish, the other upon insects. Here we have an example of species with different habits, where there would scarcely seem to be room for diversity in the physical condition in which they live; again there are others living together in immense sheets of water, where there would seem to be ample scope for diversity, among which we observe no great differences, as is the case between the Siscowet and the lake trout in the great northern lakes.

If these facts, statements and inductions were not sufficient to satisfy the reader of the correctness of my views, I would at once refer to another material fact, furnished us by the family of Salmonidæ, namely, the existence of two essential modifications of the true type of trouts, occurring everywhere together under the same circumstances, showing the same general characters, backbone, skull, brain, composition of the mouth, intestines, gills, &c., &c., but differing in the size of the mouth, and in the almost absolute want of teeth, these groups being that of the whitefishes, Coregoni, and that of the true trouts, Salmones.

Now I ask, where is there, within the natural geographical limits of distribution of Salmonidæ, a discriminating power between the physical elements under which they live, which could have introduced those differences? A discriminating power which, allotting to all, certain characters, should have modified others to such an extent as to produce apparently different types under the same modification of the general plan of structure. Why should there be, at the same time, under the same circumstances, under the same geographical distribtion, whitefishes with the habits of trout,—spawning like them in the fall, growing their young like them during winter,—if there were not an infinitely wise, supreme Power, if there were not a personal God, who, having first designed, created the universe, and modelled our solar system, called successively, at different epochs, such animals into existence under the different circumstances prevailing over various parts of the globe, as would suit best this general plan, according to which man was at last to be placed at the head of creation? Let us remember all this, and we have a voice uttering louder and louder the cry which the external world equally proclaims, that there is a Creator, an intelligent and wise Creator, an omnipotent Creator of all that exists, has existed, and shall exist.

To come back to the Salmonidæ, I might say, that when properly studied, there is not a species in nature, there is not a system of organs in any given species, there is not a peculiarity in the details of each of these systems, which does not lead to the same general results, and which is not, on that account, equally worth our consideration.

A minute distinction between species is again, above all, the foundation of our most extensive views of the whole, and of our most sublime generalizations. The species of Salmonidæ call particularly our attention from the minuteness of the characters upon which their distinction rests. Their number in the north of this continent is far greater than would be supposed, from the mere investigation of those of the great lakes; but I shall, for the present, limit myself to these.

Salmo fontinalis, Mitch.

Salmo fontinalis *Mitch.* Tr. Lit. and Philos. Soc. N. Y. 1815, I., 435.—*Richards.* Fn. Bor. Amer. 1836, III., 176, Pl. 83, f. 1, and Pl. 87, f. 2.—*Storer* Rep. 1839, p. 106.—*Kirtl.* Rep. Zoöl. Ohio, p. 169; and Bost. Journ. N. H. 1843, IV., p. 305, Pl. 14, f. 2.—*Thomps.* Hist. Verm. 1842, p. 141.—*Dekay* N. Y. Fauna 1842, p. 235, Pl. 38, f. 120.—*Ayres* Bost. Journ. N. H. 1843, IV., 273.—*Storer* Synop. 1846, p. 192.—*Cuv.* and *Val.* H. N. des Poiss. 1848, XXI., 266.

Salmo nigrescens Rafin. Ichth. Ohioens. 1820, p. 45.

Baione fontinalis Dekay N. Y., Fn. 1842. p. 244, Pl. 20, f. 58.

Though this species has been known for a long time and has

often been cited, no satisfactory figure of it has yet been published. Having, to my great disappointment, been unable to supply this deficiency, I will not undertake to give a detailed description of it. Those of my readers who desire to know it, will have to consult the works cited in the synonomy, supplying from one what is not furnished by another. In order to complete the history of this fish with success, it will be necessary to give a figure of it with all the exactness of modern science.

The color varies as much as in the *Salmo Fario* of Europe. To one of the varieties Rafinesque gave the name of *S. nigrescens*. The physiognomy of the young is somewhat different from that of the adult, which has induced Dr. Dekay to make a separate genus of it, which he calls Baione. At that epoch the body is barred vertically with black. There are seven, eight, nine and even ten bands, which grow wider and assume the form of circular spots the more the fish grows. The teeth are all minute and uniform, in these young specimens, and have misled Dr. Dekay to view these fishes as the type of a distinct genus. We have procured several individuals of two and three inches, at Black River, with others of from twelve to fifteen inches.

SALMO NAMAYCUSH, Penn.

SALMO NAMAYCUSH *Penn.* Arct. Zoöl. 1792, II., 139;—Introd. p. cxli.;—*Richards.* Fn. Bor. Amer. 1836, III., 179, Pl. 79 and Pl. 85, f. 1.—*Kirtl.* Rep. Zoöl. Ohio, p. 195; and Bost Journ. N. H., 1842, IV., 25, Pl. 3, f. 2.

SALMO AMETHYSTUS *Mitch.* Journ. Acad. N. Sc. Philad. 1818, I., 410.—*Dekay*, N. Y. Fn. 1842, p. 240, Pl. 76, f. 241.—*Storer* Synops. 1846, p. 193.

SALAR NAMAYCUSH *Cuv.* and *Val.* H. N. Poiss. XXI., 348, 1848.

This species is well known under the trivial name of "Tyrant of the lakes," because of its size and voracity, and is much esteemed for food in the countries which it inhabits. As it has been well known for a very long time, I will not repeat what has been said by my predecessors, but shall limit myself to citing a few observations which I have been able to make on the living animal. The general color varies with the ground on which it is caught. Those found on a muddy bottom are generally grayish, while those from a gravelly bottom are of a reddish color, with much brighter fins. The amethystine color does not show itself distinctly while the fish is swimming, or when first caught, but only after being taken from the water, when the mucus on the surface begins to dry. The sexes differ in shape, the male having a more pointed head than the female, although the jaws are of equal length. The dentition, though somewhat stronger than in the *S. Siscowet*, presents generally the same disposition. The vomer especially has the same structure ; there is a row of teeth on the hinder and rounded margin of the chevron, with a middle row on the body of the bone itself. According to Dr. Richardson, there should be here a double row of teeth. Probably in growing, they are thrown out alternately and obliquely, and thus cause the row to appear double. I should not know how to explain otherwise this divergence, unless the disposition of the teeth upon this bone be subject to great variations, which seems not to be probable. The description of Dr. Dekay is very obscure in relation to the teeth of this species. He speaks of a double row of teeth on the vomer and the palatines, which is an error, especially with regard to the latter. When he says that they are *in two series along the labials, of which the outer is smaller and more numerous*, he evidently speaks of the palatines and upper maxillary together ; therefore, if the upper maxillary and the palatines constitute in his view a single group (labials) of two rows, the palatines cannot at the same time have a double row. This description may have been copied without being understood, like the figure itself, which is taken from the *Fauna Boreali-Americana*.

The small ossicles of the branchial arches are nearly straight and denticulate on their outer margins, as in the *Salmo Siscowet*. The bony shields of the pharyngeals are considerably developed, and the teeth which cover them arranged like cards, and very prominent. There is one behind the tongue, narrow and elongated ; another, but somewhat smaller, which corresponds to it, on the vault of the palate, and behind these two, and surrounding the large throat, two upper and two lower pharyngeal shields.

This species was mentioned by Pennant, towards the close of

Pl. 1.

1

2

3

Sonrel on stone

Cambridge, Lith. of A. Sonrel

the past century, under the name of *S. namaycush*, which must be preserved in spite of the more euphonious name which Dr. Mitchill gave to it twenty-five years later, even if the character to which this latter makes allusion were constant during the whole life of the fish.

Our specimens have been collected all along the northern shores.

SALMO SISCOWET, Agass.

Pl. I., fig. 3.

Along with the two species of salmons above mentioned, Lake Superior furnishes a third, which has not yet been described. The inhabitants of the region designate it under the name of *Siscowet*, a name which I have thought should be preserved in scientific nomenclature. Its general form is stout, broad and thick, more so than any species of salmon except the *S. Trutta* of Central Europe. The height of the body vertically, at the anterior ray of the dorsal, is equal to one-fifth of the whole length. It descends very insensibly towards the head, somewhat more abruptly towards the posterior region; but as far as the anterior margin of the anal it maintains itself in proportions which give to the whole of the body a cylindrical appearance. A considerable inflexion runs along the insertion of the anal, and beneath the tail, whose height exceeds one-third the greatest height of the body. The pedicle of the tail is dilated and subquadrangular.

The head forms one-fourth of the whole length, exclusive of the lobes of the caudal. The frontal line, at first a little inclined, appears broken by a slight depression at the top of the posterior margin of the orbit; thence it descends somewhat rapidly on the snout, which is obtuse and rounded, and forms the principal character of this species.

The lower and upper maxillaries, the intermaxillaries and each of the palatines have a row of conical and acute teeth. The largest are on the lower maxillaries and on the intermaxillaries; they are very slightly curved inwards at their summit. The teeth of the palatines must be enumerated next in the order of their relative size, those of the upper maxillaries being the smallest and the most curved. The teeth of the vomer are of medium size, between those just mentioned, and somewhat more curved at their summit; there is a row of them on the hinder semicircular margin of the chevron, then another row on the middle part of the body of this bone. On the tongue the teeth are disposed in a pair of lateral rows; they are as large as on the palatines, and are the most curved of all.

The small ossicles disposed in rows along the inner margin of the branchial arches are slightly convex within, and finely denticulate on the outer margin of the curve. On the hinder margin of the lingual bone, at the symphysis of the three first branchial arches, there is a small, narrow and elongated shield with card-like teeth. A similar, but triangular shield is contiguous to the lower pharyngeal. Finally, a third shield is applied to the side of the upper pharyngeal.

The eyes are circular and of medium size. Their diameter is contained six and a half times in the length of the head, about one diameter and a half from the end of the snout to the anterior margin of the orbit, and four diameters from the posterior margin. The suborbital is composed of five pieces, which form an uninterrupted chain from the margin of the skull to the front of the nostrils. The first is subtriangular, the summit of the triangle being turned towards the side of the eye. The form of the second is an elongated square of which the greatest diameter is in the direction of the length of the body. The third is more irregular, approaching sometimes to the form of a protracted lozenge in the direction of the length of the fish: it borders the lower and hinder outline of the eye. The fourth is elongated, almost straight, very narrow, and has at its surface a row of pores; it attains the anterior line of the eye. Finally, the fifth is equally perforated, and of a very irregular form; it protects the lower margin of the nostrils and rests upon the intermaxillary. At the anterior and upper margin of the eye is a small superciliary bone.

The openings of the nostrils are apparently equal, and near each other, the hinder being somewhat higher; they are situated at the height of the eye, and nearer to this latter than to the end of the snout; they are protected by two very thin ossicles.

The opercular apparatus differs considerably from that of *S. namaycush;* in the fresh condition it is covered with a thick skin which hides the outlines of its bones. The preoperculum is long, of the form of a very opened crescent, placed almost vertically; its posterior margin is attenuate and entire; its lower branch is more extended than the upper. The operculum of greater height than breadth, is large and notched at the summit, but without prominent processes on the rest of its circumference, which is irregularly circular; the posterior middle part, however, has a tendency to make a projection; the lower margin is denticulate. The suboperculum is one-third smaller than the operculum, irregularly elliptical, pointed at the summit, with an ascending ridge in the form of a fish-hook at its articulation. Finally, the interoperculum has the form of a long square, curved on the posterior side; its height is contained twice in its length.

The branchiostegal rays are thirteen in number, their length diminishing very gradually from the opercular apparatus beneath the throat, where the last is only one-third smaller than the first. This latter can scarcely be distinguished from the interoperculum, so thin and dilated is it; it is only a little more narrow, and we remark that it has a tendency to bend itself. The curve is stronger on the four following, which are still very dilated compared to the eight remaining, which are not larger than ordinary rays, and flattened, with a more marked elbow on their extremity of insertion, which, moreover, is curved inwards.

The fins on the whole are strong and proportioned to the body which they have to support and to move. The dorsal, which is larger than in the *S. namaycush*, is higher than it is long, and occupies exactly the middle of the back; its margins are straight. The adipose, opposite to the posterior margin of the anal, is narrow, lanceolate, with an elliptical summit turned backwards. The caudal is ample and slightly furcate, much less furcated than in *S. namaycush.* The anal is as high as the dorsal, but not as long as this latter, though its rays are more numerous; they are there very dense, and the three first are shorter than the fourth; its terminal margin is straight. The ventrals are inserted beneath the dorsal, vertically, under the seventh ray; they do not reach the anus behind; their outer circumference is oval. In *Salmo namaycush* the ventrals are far more backwards than in *S. Siscowet.* The pectorals are very long, yet still they leave a certain distance between their extremity and the commencement of the dorsal.

Br. 13; D. 12; A. 12–14; C. 6, I., 9, 8, I., 5, V. 9; P. 14.

The scales, generally small, are a little larger on the lower region of the body behind the ventrals. Their general form is elliptical, their greatest diameter in the direction of the length of the fish; their smallest diameter measures one-eighth of an inch on specimens of two feet in length. Those of the lateral line are proportionally more narrow, and perforated with a large canal, which renders this line very conspicuous. It follows the middle of the body upon the caudal region and rises gradually in advancing towards the head, so that in the anterior region it approaches much more to the back than to the belly.

The color varies according to the feeding ground on which it is caught, and is brighter during the breeding season, as is generally the case among all species of this family. The young have transverse bars, which disappear with their growth, like those of other species of salmon.

This also is a fish of high and rich flavor, but so fat as to be almost unfit for food, the greater part of it melting down, as it were, in the process of cooking. This renders its preservation in alcohol very difficult, if not impossible. All the specimens which I brought from our excursion have decomposed. They were caught at Michipicotin, and occur everywhere along the northern shores. They are particularly abundant about Isle Royale.

Coregonus, Artedi.

We shall not treat here of the history and the characters of the genus Coregonus in its whole extent. For this I refer my readers to the twenty-first volume of the *Histoire Naturelle des Poissons.* I shall merely criticise the North American species, which I have been enabled to study in nature, refraining from offering conjectures on those which remain imperfectly known to me. To delay their revision until we possess original specimens, is the only means of preserving their nomenclature intelligible.

The reforms we have proposed to introduce among the species described below, are of a nature to excite the attention of the naturalists of this continent, and to induce those who may find themselves in favorable circumstances to observe minutely, and to collect materials which may some day serve as the basis for a special work on the genus.

The *Coregonus clupeiformis* was described for the first time by two authors simultaneously, who have each given it a particular name. The question of priority might be contested; and what shows that subsequent authors disagreed on this point is, that some adopted the name given by Lesueur, others that of Mitchill. Naturalists have now agreed to adopt the name *clupeiformis*, it having the priority of a few weeks, and being also the more appropriate to this species; and the figure of the Fauna of New York, though leaving still much to be desired, is however sufficient to distinguish it in the present state of science. In the same year, Dr. Kirtland published another figure, which appeared in the Journal of the Natural History Society of Boston, IV., Pl. 9, f. 1. It being much inferior to that of Dr. Dekay, I have omitted it in the synonymy; it seems really to me in contradiction with the other quotations. I have cited the description, because it is literally copied from Lesueur. I should not be surprised however, if the specimens which Dr. Kirtland has had under his eye belonged to another species, though it is impossible to decide this by means of the figure. Richardson also reproduced the original description of Lesueur, not having seen the species himself.

In truth, the history of this species has remained almost what it was in 1818. Dr. Dekay, who has revised the species in nature, does not complete its description, limiting himself to a mention of the most prominent traits. Finally, M. Valenciennes himself is still more brief. I believe, moreover, that he is mistaken when he considers *C. lucidus* Rich. as identical with *C. clupeiformis*. It would rather be with *C. albus* Lesu. that it ought to be compared, and to which it is nearly related; but the position o the eye, a smaller mouth, larger maxillaries, and a different conformation of the opercular apparatus, distinguish it sufficiently.

All authors, after Lesueur, have been mistaken in the *C. albus*; this would not be surprising had they nothing to guide them but the short description of this author; but the figure which accompanies it leaves no doubt about his species, and the most superficial inspection might suffice to give at least an approximate idea of it. In the present state of the science I agree that we may confound our *C. sapidissimus* and *C. latior* in their full-grown condition; but where the question is between so different species as *C. albus* Lesu., and those (for we shall see that there are several) which authors have designated under the same name after Lesueur, we may very naturally ask ourselves, whether the information given by them has been drawn from original sources, or has, perhaps, been published under the belief that the fishes commonly designated under the name of *white-fishes*, must all belong to the same species.

There are two groups of Coregoni; one having the lower jaw longer than the upper, the other having a squarely truncated snout, and the upper jaw overlapping the lower. *C. albus* Lesu. belongs to the first of these groups, whilst the Coregoni described under the same name by subsequent authors, belong to the second group. Let us now review these latter, having no longer to compare them with the species of Lesueur.

Dr. Richardson has described and figured under the name of *C. albus*, a species allied, in certain regards, to our *C. sapidissimus* and *C. latior;* but I think it cannot be identified either with the one or the other, due attention being paid to the differences indicated in our descriptions. Dr. Dekay gives this species as the *C. albus* in his New York Fauna; but not having seen, he says, the species, he borrows his information from Dr. Richardson.

Another species has been mentioned under the name of *C. albus*, by Mr. Thompson. This species is our *C. sapidissimus*.

A third species has hitherto been confounded with the preceding, to which it approaches in several respects. This is our *C. latior*.

Finally, I inquire what may be the *C. albus* of Kirtland? The figure which he gives of it is different at the same time from those published by Dr. Richardson and Mr. Thompson, so that I do not know to which of them to refer it. I should not be surprised to find it the type of a particular species. The details of the head not being minutely given in the figure, do not allow us to make a direct comparison of them.

The presence of small teeth on the surface of the tongue is an almost universal character in Coregonus, though it is more evident in the species in which the lower jaw is longer; this would be another character of this group, which would allow us to associate with it *C. Labradoricus* and *Harengus*, which M. Valenciennes was disposed to discard from it. In the species with a truncated snout, and a longer lower jaw, we remark that the intermaxillaries have a row of teeth. These differences seem to me of sufficient value to justify the formation of two distinct genera for these fishes. I would propose to preserve the name of *Coregonus* for those species in which the snout is prominent, as it was primitively established with reference to such species in Europe. The name of *Argyrosomus* might be applied to the other species, with a truncated snout and a prominent lower jaw.

The species of this continent may be grouped as follows :*

Argyrosomus.	*Coregonus*, proper.
* Coregonus clupeiformis *DeKay.*	* Coregonus sapidissimus *Agass.*
" * albus *Lesu.*	" * latior *Agass.*
" lucidus *Richards.*	" *albus Rich.*
" * Tullibee *Rich.*	" *albus Kirtl.*
" * Harengus *Rich.*	" *otsego Dekay.*
" Labradoricus *Rich.*	" * *quadrilateralis Rich.*

Coregonus clupeiformis, Dekay.

Salmo clupeiformis *Mitch.* Amer. Month. Mag. 1818, II., 321. (*White-fish of the lakes.*)

Coregonus clupeiformis DeKay N. Y. Fna. 1842, p. 248 Pl., 60, f. 198, (*common Shad Salmon.*)—*Cuv.* et *Val.* H. N. Poiss. 1848, XXI., 523, (excl. syn.)

* The names in *italics* indicate species,to be revised. About *C. Labradoricus* we are left in doubt as to its position. We have collected specimens of seven species in Lake Superior, which are marked here with an asterisk (*).

Coregonus Artedi *Lesu.* Journ. Ac. N. Sc. Philad. 1818, I., 231, (*Herring Salmon,*)—*Richards.* Fn. Bor. Am. 1836, III., 203.—*Kirtl.* Bost. Journ., N. H., 1842, IV., 231.—*Storer* Synops. 1846, p. 199.

Possessing only a female individual of this species, our description must not be considered as absolute, and applicable to the males and young, for their form and general outlines. Dr. Dekay has already made the observation that the males are more elongated than the females, and that, besides, the latter *are deeper and more compressed;* which is generally the case in the Salmonidæ.

The general form is regular, spindle-like, neither thick and short, nor slender. The sides are much compressed; the line of the back is nearly straight, somewhat sloping on the nape and the head as likewise on the region of the adipose fin, and raised on the caudal. The curve of the belly is uniform from the lower face of the head to the termination of the anal; the lower side of the tail is straight or slightly concave. The greatest height of the body, taken before the dorsal, is contained five times in the whole length, including most of the caudal fin. The thickness is less than half of the height. It is about the same on the whole abdominal region and the thorax, diminishing gradually towards the tail.

The head is small, compressed like the sides, flattened above, rounded below, pointed before. Its length equals the height of the body, that is to say, it forms one-fifth of the length. The eyes are large and circular, separated from the extremity of the jaw by a diameter of their orbit, and by twice and a half this diameter, from the posterior margin of the opercular apparatus. The nostrils are nearer to the snout than to the orbit. The opening of the mouth is of middle size, of a quadrangular form; the lower jaw considerably exceeds the upper, and rises slightly at its extremity, which is rounded; its margin contains a few fine indentations, which seem to indicate teeth; the intermaxillaries have very fine teeth. The surface of the tongue seems to have two longitudinal rows on its middle shield, if we can call teeth small acute points. The tongue itself is pointed, and does not attain the inner margin of the intermaxillaries. The maxillaries are elongated, of an oblong form, with entire margins; their posterior extremity not attaining a vertical line which would descend through the centre of the eyeball. The mandibles, situated on the inner margin of the maxillaries, are small and narrow, with an undulated outline terminated above by a slender and acute process. The suborbitaries cover two-thirds of the face.

The preoperculum is concave on the middle of its ascending branch; its posterior angle is rounded, and extended to the lower margin of the face, and, conjointly with the lower branch, nearly covers entirely the prolongation of the interoperculum towards the lower maxillary. The part of the interoperculum which remains uncovered, is triangular; the upper angle rises before the operculum. This latter is higher than it is broad above, straight or slightly concave, rounded behind, oblique and straight on the suboperculum, which is the most regular of the bones of this apparatus, being arched on its lower edge, and somewhat more narrow behind than before.

The branchial fissures continue beneath the head, the branchiostegal membrane of the right side unites to that of the left on the region of the isthmus, where they are contiguous, the first jointed beneath the second. The branchiostegal rays, eight in number, are very close, flattened, and almost straight.

The scales are proportionally large, of subcircular form, the inner margin irregular and angular. The largest occupy the middle of the trunk and the abdominal region, where they measure more than a quarter of an inch; they diminish towards the thoracic arch, the back and the tail, where they are smallest. On the middle line of the belly their form is much elongated and elliptical. Their termination is very remarkable on the basis of the caudal, resembling somewhat the fork of this fin by the concave line they form. The lateral line is near the middle, rather near to the back, and is slightly inflected on the abdomen by a very protracted curve.

The dorsal fin, situated on the middle of the back, is much higher than it is long, and its margins are straight; its first ray is short and simple; the second does not reach beyond two-thirds of the height; it is articulated, but not bifurcated. The adipose fin is long and narrow. The anal, longer and less high than the dorsal, is concave on its terminal margin; it somewhat exceeds the adipose fin backwards; its height somewhat exceeds its length. The caudal is deeply furcated ; its lobes are pointed. The ventrals are large, triangular, regular, the outer margin somewhat longer than the inner; their extremity is not an inch from the anus ; they are inserted on the lower face of the body, and very near each other ; their base of insertion is rounded ; the cutaneous prolongation of their upper margin is much elongated. The pectorals are elongated and pointed.

Br. 8 ; D. I, 11 ; A. II, 13 ; C. 7, I, 10, 9, I, 7 ; V. I, 11 ; P. 16.

This species is from the Pic ; but occurs everywhere along the northern shores.

Coregonus albus, *Lesu.*

Coregonus albus *Lesu.* Journ. Acad. N. Sc. Philad. 1818, I., 232 (figured.)

The general form is elegantly elongated, lanceolate, with very regular outlines. The curve of the back is similar to that of the belly, except that the space on the back, which extends from the nape of the neck to the dorsal, is more arched, whilst, on the belly, it is most arched between the ventrals and the anal. However, in young individuals from five to eight inches long, these two lines present the greatest uniformity. The body is regularly compressed ; the greatest height before the dorsal is contained four times and a half in the length, reckoned from the end of the snout to the end of the scales on the caudal. The thickness is equal to half of the height.

The head is conical, pointed at its extremity, and more compressed than the body, attenuated below ; it forms the fifth part of the length, excluding the caudal. The skull is rather flattened than convex ; it is sloping as much as the lower surface is raised. The eyes, very large and circular, are situated at the distance of their diameter from the end of the snout, and of twice and a half this same diameter from the posterior margin of the opercular apparatus. The suborbital bones, very much developed, encroach upon almost the whole face, of which a very small and narrow space is left bare above the anterior branch of the preoperculum as far as the posterior extremity of the maxillaries. The nostrils open on the upper face of the rostrum, at equal distances from its extremity and the anterior margin of the orbit. The mouth is large in comparison to the other species; when open, its form is that of a quadrangular tunnel, measuring seven-eighths of an inch vertically, and one and three-eighths inches transversely: it contains no teeth. But on the other hand, there are two rows of rudimentary teeth on the tongue; in order to see them the membrane of the surface must be removed. The tongue itself is narrow and pointed. The lower jaw is longer than the upper; its extremity is rounded and slightly raised. The intermaxillaries are small; the maxillaries oblong and elongated, attaining, with their posterior extremity, the anterior margin of the eyeball. The labials are one half smaller, and of the same form, having a small point at their anterior extremity.

The outer circumference of the opercular apparatus is rounded and semicircular, and scarcely shows a tendency to undulate in the margin of the suboperculum. The operculum would be triangular were it not for the curve of its upper and hinder margin; the lower margin, contiguous to the suboperculum, is very oblique. The interoperculum attains the lower angle of the operculum; its hinder angle is rounded, subtriangular; its anterior branch is completely covered by the preoperculum, which is very wide at its angle.

The branchiostegal apparatus is little developed, and arranged as in *C. clupeiformis*. There are seven very close, short, and flattened rays.

The scales are proportionally large, easily falling off in individuals fifteen inches long; the largest are those covering the sides near the lateral line, which measure six-eighths of an inch in the longitudinal direction, and somewhat more than four in the transverse. On the abdomen the proportions change; they are somewhat higher than long, and are sensibly oblong with their greatest diameter oblique. Beneath the belly they are, as usual, much elongated. The lateral line is near the middle of the body, somewhat nearer to the back than to the belly: at its origin it rises above the operculum; it is straight along the tail. The termination of the scales on the caudal presents the same peculiarity as in *C. clupeiformis*.

The dorsal is on the middle of the back, its height somewhat exceeding its length, and its upper margin straight. The adipose fin is oblong, and elongated, exactly opposite to the hinder margin of the anal. The anal itself is much longer than high, and the disproportion between the anterior and the posterior margin is greater than on the dorsal; the outer margin is concave. The caudal is furcated; its lobes are pointed. The ventrals are very near each other, and shaped as in *C. clupeiformis*, and the cutaneous prolongation of their upper margin is long and triangular. The pectorals, little longer than the ventrals, are oblong and less pointed than in *C. clupeiformis*.

Br. 7; D. II. 10; A. II. 11; C. 8, I. 9, 9, I. 7; V. 11; P. 17.

Lesueur did not give the dimensions of his fish: those which I have procured do not exceed fifteen and a half inches, though I have seen a numerous series of them. I do not know whether they attain a larger size.

This species is common about the Pic; but I have also secured specimens from various localities along the northern shores of the lake.

Coregonus sapidissimus, Agass.

Coregonus albus *Thomps.* N. H. Verm. 1842, I., 143, (wood-cut)
(*White-fish or Lake shad.*)

We take as the type of this species the description and the figure of Mr. Thompson, which though much reduced, gives a clear idea of it. We have several individuals twenty-two inches in length, the size of those which Mr. Thompson himself has described. A complete series of young individuals enables us to give a full description, and in order to render it more intelligible we shall begin with the adult.

The general form is slender, the sides compressed, the back and belly prominent. The space contained between the anterior margin of the dorsal and the occiput is much arched, convex; and the nape of the neck itself is sometimes very prominent. From the dorsal the line of the back descends abruptly on the tail; it is somewhat depressed immediately behind the adipose fin, and rises somewhat on the insertion of the candal. The ventral line is almost uniformly convex, but the region situated between the ventrals and pectorals is somewhat more prominent. This line becomes very oblique and ascendant beneath the thoracic region and the head. The greatest height of the trunk corresponds to a vertical line along the middle of the space between the pectorals and the ventrals; it is contained about three times in the length, exclusive of the caudal. The thickness at the middle of the trunk corresponds to the height as one to two; it is somewhat less anteriorly, and diminishes gradually towards the caudal region. The head is proportionally small, compressed laterally, pointed. Its upper surface slopes as much as the lower rises, so that in adult individuals it appears disproportioned to the development of the trunk, of which it forms only a very small proportion. Its length, however, is one-fifth of the whole length, the caudal included. The middle surface of the skull on the suture of the frontals, is slightly conical, and causes the two halves of the skull to appear inclined towards the eyes. These latter are large and subcircular; the hinder margin of their orbit is at an equal distance between the end of the snout and the free margin of the operculum. The suborbital bones cover the whole space between the orbit and the upper region of the operculum, but leave bare the lower half of the cheeks; they form a continuous series below the eyes as far as the snout, where this latter elongates itself over the labials, which it receives beneath its lower margin. The nostrils are somewhat nearer to the orbit than to the extremity of the snout. This latter is cut obliquely, and slopes over the lower jaw, which shuts within the intermaxillaries. The mouth is moderate. The intermaxillaries are small, and occupy only the extremity of the rostrum; they have a row of very small teeth, flexible like bristles. The labials are very short, thin, elongated, and attain the anterior margin of the orbit; they have on their termination a small shield, which is bony, pearl-like and included in the skin. The lower jaw seems to be unprovided with teeth, at least we cannot observe any either with the magnifying glass or with the touch. The branches of the lower maxillaries dilate in the form of a very thin blade, which in the state of rest shuts itself up under the suborbital bones. At the anterior margin of this blade we remark a cutaneous expansion, a kind of lip, which is attached to the posterior and terminal margin of the labials, and forms thus the angle of the mouth. The tongue is short and broad, free only on its anterior and lateral outline; its surface, though seeming to be smooth, has some irregular rows of small asperities, which are sometimes perceived only after removing the investing membrane. The operculum is subtriangular and large, when we consider that the upper and hinder margins pass from one to the other by a curve; the lower margin is straight and oblique, and as long as the anterior margin is high. The suboperculum is arched on its whole circumference, and makes a projection beyond the operculum. The interoperculum, almost completely covered by the preoperculum, presents externally only a small triangular surface, and a small narrow band below the lower branch of the preoperculum; though in reality, this bone is as long as the suboperculum, but less broad, having the form of a very acute triangle, of which the summit would be on the anterior side.

The branchial openings are very ample, and join each other at the lower surface of the head. The branchiostegal membrane, whose office it is to shut this fissure conjointly with the opercular apparatus, is proportionally little developed; it contains commonly nine, sometimes ten very crowded, flattened and almost straight rays.

The scales are of middle size in proportion to that of the fish. The largest are situated beneath the belly, the smallest under the throat, the thoracic belt and the caudal region. Those of the lateral line are somewhat smaller than those of the adjacent rows. Their form is generally subcircular or irregularly quadrangular, but their vertical diameter has a slight tendency to surpass the longitudinal diameter. This peculiarity is especially striking on the abdominal region, where really the scales are oblong and of a height sensibly greater than their length; at the same time that their outlines become more regular and nearly oval. Their imbrication has even here something peculiar in being less close; the rows appear independent, and give to the fish a barred aspect. The outlines of those of the lateral line are the most irregular. The outer margin is in all more or less circular and entire. The lateral line itself is nearly straight and nearer to the back than to the belly; it begins from the upper angle of the operculum and extends itself to the middle of the caudal.

The anterior margin of the dorsal fin corresponds to the middle of the space contained between the extremity of the snout and the basis of the caudal; the fourth and fifth rays are the longest; the first two short and rudimentary spines are applied against the third, which is simple but articulated, and almost as long as the following; being higher than it is long, this fin has a triangular form on account of its posterior margin, which is low and inclines on the back. The adipose is broad, covered with small scales on its basis and opposite to the posterior half of the anal, of which it does not attain the extremity. The anal, as long as it is high, occupies the middle of the space between the anus and the basis of the caudal; it has, like the dorsal, two spinous rudimentary rays in its anterior margin, and one soft ray more. The caudal is furcated and ample; small scales encroach upon its basis. The ventrals are large, with their terminal margin straight; they are almost as long as the dorsal is high; the anterior margin opposite to the twentieth ray of the dorsal contains a small spinous rudiment hidden beneath its membrane; the cutaneous appendix of the upper margin is very small. The pectorals are elongated, spindle-like, and proportionally small.

Br. 9; D. II. 11; A. II. 12; C. 7, I. 9, 8, I. 7; V. 12; P. 16.

During the early age, when its size does not exceed eight inches, the slender form is the predominant character of this fish. The line of the back and that of the belly being then very little prominent, and the outline of the head passing in direct continuation to that of the body, there results a harmonious whole in the proportions of these two regions. The compression of the body is already very marked; the head is already pointed and forms one-fifth of the whole length, not including the caudal fin. The rostrum is truncated but rounded, and exceeds the lower jaw. The nostrils are placed at equal distances between its extremity and the eye. The greatest height slightly exceeds the length of the head. The characteristic form of the fins may already be remarked; there being one ray more or less in the one or the other of the fins. The ventrals are placed somewhat more forwards relatively to the dorsal, their anterior margin being perpendicular to the fifth or sixth of its rays. The same complete development is also observed in the opercular apparatus; the operculum alone presents this slight difference, that its height sometimes exceeds a little the length of its lower margin; the breadth of the suboperculum is also subject to some variations. The scales at this period are thin and fall off easily, but we may recognize already the different characters which we have signalized above. The lateral line is straight and nearer to the back than to the belly. But as soon as the individuals attain a length of ten inches, the head becomes declivous, the nape of the neck swells, the back rises, the belly becomes more prominent; but the general form is still slender, the head is in harmonious proportion with the trunk, of which it forms already one-fifth of the length, including half of the caudal. The rostrum becomes somewhat more prominent and more abrupt. The height of the body exceeds however already the length of the head.

When individuals attain fourteen inches the back and the nape of the neck are very convex, and the head very declivous, the belly prominent, and from this moment the head appears disproportioned to the trunk, and is found to form exactly one-fifth of the whole length, the caudal excluded, as we have seen in the adult. The height of the body is contained four times in its length. The scales are still thin and fall off easily, but they already begin to be more adherent than during the preceding stages. The middle surface of the tongue is armed with small asperities as in the adult; and the intermaxillaries have also that row of fine teeth which we have indicated above.

This species is the common white-fish of Lake Superior, of which so large numbers are caught and salted every year. It is one of the most palatable fishes of the freshwaters of the American continent. It is found in large shoals all over the lake.

Coregonus latior, Agass.

Hitherto confounded with the preceding, with which it has a great affinity, this species differs, however, sufficiently to justify its separation, as I hope to show. Possessing young and adult individuals, I shall follow in relation to them the method which I have already adopted, pointing out first the difference existing between adult specimens, and finally adding the peculiar traits of the young. I will here mention that the adults differ in appearance less than the young, —among which, the difference at first sight is most striking.

The adult individual which I have before me measures nineteen inches. The general form reminds us of that of *C. sapidissimus*. As in this latter, the back is arched from the occiput, but the curve is more uniform, the nape of the neck being less prominent, and the belly also less swollen. The body is thicker and stouter than in the *C. sapidissimus*, compressed, fusiform; the greatest height, which is measured vertically at the anterior margin of the dorsal, is contained four times in the length, the caudal included. The lines of the back and belly come near each other on the tail, without abrupt transition; they continue on the head, without rising much on the lower face, and without lowering much on the upper face, though the skull is depressed and slightly sloping. The head, which is thicker and stouter, forms one-fifth of the whole length, including the caudal. It is less pointed than in the preceding species, and the rostrum more obtuse, less exceeding the lower jaw. The mouth is somewhat larger, but constructed in the same manner; that is to say, the ascending branches of the lower maxillary shut themselves up beneath the suborbital bones, and there is a cutaneous appendix at the anterior margin, and a kind of lips, which form the angles of the mouth by uniting with the labials. These latter are broader than long, passing beyond the anterior margin of the orbit. Their terminal extremity has likewise the long and pear-like shield, which we have indicated in *C. sapidissimus*. The lower jaw, again, is surrounded with a folded lip, imitating a border of fringes. We have remarked no trace of teeth on the intermaxillaries, and without deciding upon their absence, they were at least obliterated so as to render them doubtful. The tongue is broad and shows no trace of asperities at its surface. The eyes are large, almost circular, and placed in the same relative position. The nostrils are nearer to the orbits than to the extremity of the rostrum. The suborbitaries present no remarkable difference, unless it be, perhaps, that they encroach less on the cheeks.

In the opercular apparatus, we remark that the operculum is rather quadrangular, and the suboperculum more contracted at its posterior extremity, which renders its lower margin more oblique. The interoperculum is somewhat more uncovered.

The fissure of the gills is the same, but the branchiostegal apparatus is more developed and the rays more bent; their actual number is eight.

The scales are somewhat larger than in the preceding species, and present about the same general form, but their height surpasses their length. Generally more uniform on the different regions, they are, however, larger on the middle of the trunk, those of the middle line being in other respects smaller than the adjacent ones, as is the case for the most of the species. Those of the abdomen affect not a linear disposition, independent from the whole, but all appear as uniformly imbricated. Beneath the belly and the tail they elongate themselves to the form of an ellipsis with tortuous outlines. The lateral line, slightly arched, follows the outlines of the back, to which it is nearer than to the belly. The fins on the whole are much more developed than in the *C. sapidissimus;* their general form and their relative position are sensibly the same. We remark, however, that the height of the dorsal is greater in proportion to its length, and its posterior margin is straighter. The adipose fin, equally covered with small scales on its basis, is opposite the termination of the anal. This latter is triangular, as long as it is high, but less raised than the dorsal. The caudal is deeply furcated. The ventrals, broad and oblong, are rounded on their terminal margin, and contain the strongest rays. The pectorals are elliptical, and longer and broader than in the preceding species, and from the stouter form of the body their terminal extremity is nearer to the ventrals.

Br. 8; D. III., 11; A, II. 11; C. 7, I., 9, 8, I., 7; V., 11; P. 15.

Whoever doubts the validity of this species should only cast a glance on two series of young individuals belonging to both species. We have noted above the peculiar traits of the *C. sapidissimus*, and it will be remembered that we have insisted upon their slender and elongated form. The most striking contrast exists when we compare them with the short, high and stout form of this species.

When this fish has attained the size of seven inches, the height, which exceeds the length of the head, is contained four times in the length of the body, the caudal excluded. The sides are much compressed; the thickness is only one-third of the height. The structure of the head, the form and the development of the fins, are in perfect conformity with the adult. We observe that the rostrum, which is truncated, scarcely exceeds the lower jaw. The form of the buccal opening is quadrangular as in the adult. The intermaxillaries have a row of very fine teeth; there are teeth even on the margin of the lower jaw, but more difficult to perceive even with the magnifying glass. The surface of the tongue is prickled with small, very acute asperities, like the teeth of the intermaxillaries. The eyes are very large; the distance which separates them from the end of the snout does not equal their diameter; the nostrils occupy the middle of this space.

The scales, which are stronger and larger, as we have already seen, easily fall off; we may already signalize in them the same peculiarities which we have seen in the adult. The lateral line is straight and approaching slightly more to the back than to the belly.

When ten inches in length, this fish acquires an increasing height; the height, taken before the dorsal, is contained exactly four times in the length, the caudal included, and the head has almost the proportions of the adult. The body is very compressed and flattened; its thickness is contained three times and a half in the height. The snout is somewhat more prominent, as in the preceding age, though remaining more truncated and shorter, as in the *C. sapidissimus*. The scales grow gradually firmer; those of the upper half of the body somewhat shorter than those of the lower half. The fins themselves grow more prominent. The species is common along the northern shore of Lake Superior, where it is found with *C. sapidissimus*. I have collected a large number of specimens at the Pic.

Coregonus quadrilateralis, Richards.

Among the Coregoni collected at Lake Superior there is one very similar to *C. quadrilateralis* of Dr. Richardson, though I have yet doubts as to its identity. The question can only be decided by comparison of specimens from the localities where the author of the Fauna Boreali-Americana collected his. I have already noticed slight differences in the scales, in the structure of the fins, in the opercular and branchiostegal apparatus, and in the proportions of the body; differences which depend, perhaps, upon the age and size, and which I have not been able to verify in all my specimens, they being below the dimensions which Richardson assigns to his species. I have endeavored to compare them by means of reduction, but I soon perceived that I could not arrive in this way at a precise determination, especially as the proportions of the different regions of the figure of Richardson do not fully agree with the measures which he gives of them in the text. The formula of the fins which I have taken from an individual of fourteen inches, is:

Br. 6 ; D. III, 11 ; A. II, 10 ; C. 7, I, 9, 8, I, 6 ; V. 11 ; P. 16.

The scales of the lateral line, though smaller than the adjacent rows, do not appear to me so absolutely truncated as Dr. Richardson expressly says they are in his species. Their size on the sides equals, if it does not surpass, four-eighths of an inch, and on a surface of an inch square we may count as many as eight. This fact has appeared to me the most prominent.

Richardson reports that when Cuvier sent him the specimens which he had submitted to his examination, the label indicated that he, (Cuvier,) had a related species from Lake Ontario, but we do not find it mentioned by M. Valenciennes in the Histoire Naturelle des Poissons. It is perhaps to this species of Lake Ontario that our specimens ought to be referred. Sir John Richardson, having seen recently the specimen described above, has himself offered doubts respecting its identity with his *C. quadrilateralis*.

CYPRINOIDS.

This is a numerous, but well circumscribed family, whose striking peculiarities are very obvious. I am not aware that any of these fishes have ever been noticed in the waters of the southern hemisphere; nor do they extend anywhere far beyond the limits of the temperate zone, as it is well ascertained that they are most numerous in the rivers and lakes of Central Europe and Central Asia and Northern America. Indeed, it is so much their natural home, that they do not seem to occur in the northernmost freshwater streams, nor anywhere in the tropics, except in very great altitudes, where recently a few have been found in the Andes. The sea is almost entirely destitute of fishes of this family; a few species, however, occur in brackish waters.

The family of Cyprinoids affords another example of the fact, that the species of animals are circumscribed within narrow limits in their geographical distribution. From the great number which have already been described, it is plain that almost every lake and every river has species of its own; but, nevertheless, there is a great uniformity among these fish all over the world; for the carps of China and those of Europe are very similar; so are the little white-fishes of the Nile and those of other basins. But however uniform these fishes may be in the main, we cannot help observing that among them there are peculiar groups, located in particular parts of the world, for instance, the Catostomi, all over the freshwaters of America. The small bearded species are very numerous in Europe, and, in general, in the Old World; species with beards occur there more extensively than on the American continent.* Again, the types with a large dorsal are extensively distributed, but are almost all extra American. The species which occur at great altitudes, as those from the lakes of tropical America, are so peculiar as to differ decidedly from all other Cyprinidæ, being devoid of ventral fins. In Lake Superior and the other Canadian lakes there is a considerable variety of these fishes,—Catostomi mixed with European types, and a genus which has only American representatives.

The little group of Cyprinodonts, which have so universally been connected with Cyprinoids, will be found to differ more from Cyprinoids than has been supposed. We need only compare the structure of their mouths to be satisfied of the difference. There are no representatives of that type in Lake Superior.

How far it might be advisable to subdivide this family into small groups according to their structural differences, remains to be ascertained. The Catostomi, for instance, are very remarkable for the large opening in the centre of their skull, and for the peculiar arrangement of the teeth in the pharyngeal bone.

Rhinichthys, Agass.

I propose to include in the genus Rhinichthys small Catostomi, whose essential character is, as the name indicates, to have a conical

* I would mention, as particularly characteristic of the Old World, the genera Barbus, Cobitis, and the allied types.

prolongation of the rostrum. The mouth is small; the lips which border it are much reduced, smooth, never carunculated, and do not extend themselves on the lower jaw under the form of lobes. This character is well represented on figure 2 of Pl. 2. At the angles of the mouth, the upper lips bend slightly forwards to join the middle of the branch of the lower maxillary; they here form a small tunnel, on whose outer margin is a small barbel, sometimes very difficult to recognize. To this genus we must refer the *Leuciscus atronasus* (Cyprinus atronasus *Mitch.*) and *L. nasutus* Ayres. Though the first of these species has not the character of a very prominent rostrum, the structure of the mouth, and the presence of the barbel, justify this approximation.

There are still other species of this genus found in the United States, yet imperfectly known, which will hereafter also take their place here. Anatomical study will doubtless reveal other characters than those which external conformation already gives, and will also teach us the value of this singular group in the family of Cyprinoids. At present I cannot help considering the Rhinichthys of North America as a diminutive of the group of the Labeos of Africa and the East Indies.

RHINICHTHYS MARMORATUS, Agass.

Pl. II., figs. 1 and 2.

This species is one of the largest of the genus, at least, of those which are as yet known to us. The form is elongated, subcylindrical, compressed. The tail preserves just proportions with the trunk; its two margins are almost straight. The ventral line is a little convex, and rises abruptly at the insertion of the anal. The back is feebly arched from the dorsal fin to the nape of the neck, where the slope continues rapidly from the skull to the snout. The head is entirely smooth; it is small, conical, and well proportioned to the body, in whose whole length it is contained four times. The upper surface is rounded; the eyes are of medium size, and situated near the upper margin of the face, at about an equal distance from the end of the rostrum and the upper angle of the operculum. The nostrils are

Sonrel on stone

Lith of A Sonrel, Cambridge

1 & 2. RHINICHTHYS MARMORATUS Ag. – 3 & 4. CATOSTOMUS AURORA Ag.

very large and near the orbits. The rostrum exceeds the lower jaw by the whole length of the opening of the mouth. This latter is small, semi-elliptical, when the jaws are closed; when opened, it has the form of a crescent whose circumference would be formed by the upper jaw, having below, as a base, the elliptical and rounded outline of the lower jaw. The barbel is about a twelfth of an inch long.

The face and the opercular apparatus are smooth like the head. The preoperculum is hidden beneath the fleshy cheeks. The operculum is large, concave on its anterior margin, rounded on the upper; the lower is straight and oblique, beneath which is the thin and narrow subopercular lamina. The interoperculum is triangular and more robust. The branchial fissures are small, and extend but little to the lower surface of the head, which gives to the isthmus the form of a triangle. The branchiostegal membrane contains three thin rays, of about equal length, bent and flattened.

The dorsal fin occupies exactly the middle of the whole length of the fish; its form is quadrangular, higher than long, and has nearly straight margins. The caudal is obtusely notched, its lobes are rounded. The anal, situated at a small distance backwards from the dorsal, is narrow and elongated; its outer circumference is rounded. The ventrals are inserted somewhat before the dorsal; they are small fins of an oblong form, whose extremity reaches to the anus. The pectorals are placed very low, have an elliptical form, and are more elongated than the ventrals.

Br. 3; D. II, 9; A. II, 8; C. 5. I, 9, 8, I, 4; V. 8; P. 14.

The scales are small and subcircular; the concentric and radiating striæ are easily seen with a lens. Points of black pigment are distributed on their posterior half, and give to the surface of the body a punctulated appearance. The lateral line is in the middle; it is only feebly inflected on the abdomen.

The ground color is a reddish brown mottled with black, orange and dark green. The black marbling is predominent. A large spot of this color occupies the basis of the caudal, where it radiates on the rays of this fin. The lips, the margin of the branchiostegal membrane, the basis of the pectorals, ventrals and anal are of an intense orange-red, which prolongs itself on the rays. The ground of the fins is light orange.

Fig. 1 represents this species of its natural size.

Fig. 2 is the lower surface of the head magnified, to show the configuration of the mouth.

From the Sault St. Mary, where it seems not to be infrequent.

Catostomus, Lesueur.

The study of the species of the genus Catostomus has become quite as difficult as that of the genus Leuciscus, and for the same reason; the multiplicity of species. There are about thirty described or mentioned, very few of which are accessible for comparison. Hence, we are left, either to identify species which have only distant analogies, or to separate, on the other hand, some which have the closest affinities. Which of these two obstacles is the most injurious to science? Doubtless the first; since it leaves science in a state of equivocal stability, during which no advance is attempted, satisfied, as we are then, with our present attainments.

In endeavoring to determine the different Catostomi from Lake Superior, I began by comparing them with species already known from the same geographical zone to which they would have the nearest relations. One had been known for three quarters of a century as an inhabitant of the gulfs of Hudson's Bay, and was described by Forster under the name of *Cyprinus Catostomus*, which, forty-four years later, became the type of the genus Catostomus, with the specific appellation of C. *Hudsonius*, the author of this reform not having known the fish otherwise than through the description and the figure of Forster.

In 1823, that is to say, about fifty years after Forster, Dr. Richardson gave a detailed description of the *C. Hudsonius*. He described also another under the specific name of *Forsterianus*, and referred to it as a synonymous variety of the preceding, indicated by Forster himself. His specimens were from Lake Huron and from Slave Lake.

Among the species of Catostomi which I have brought from Lake Superior, there are two which have a very great analogy, in their general traits, with *C. Hudsonius* and *Forsterianus*. However, in comparing them attentively and singly with the descriptions of Dr.

Richardson, I was convinced of some differences, respecting the first, which I consider as specific. Respecting the second, the question becomes more difficult to solve, as Dr. Richardson had specimens from two very different localities, from which his description was made.

This complication caused me to hesitate for a long while respecting these species; and even now, though describing the second species under a new name, I am still in doubt upon the following points: Are there really two species of Catostomi with red bands on the sides? This would not be extraordinary, if we do not allow specific diagnoses to rest upon color. As soon, however, as the existence of two species is demonstrated by ultimate researches, it is evident that that of Lake Huron will be the same as our *C. aurora*, whilst that of Slave Lake will be the *C. Forsterianus*, the same which Forster had in view.

However, upon consulting the original Memoir of Forster, I am almost tempted to consider his second variety as the very species I describe hereafter, under the name of *C. Forsterianus*, and which, as we shall see, is nearly related to *C. Hudsonius*. It has that red tint of the lateral line, with the same general ground color. If that be the case, the name of *Forsterianus* would be ill applied, for the name would remind us of one species, whilst the description would apply to another.

Catostomus aureolus, Lesu.

I cannot do more than mention this species, as I possess only a few specimens, and all very young, between three and four inches long. The general characters of the species are, however, already well indicated upon them. A thick and stout head, almost as high as long, truncated in front; the considerable development of the operculum at the expense of the suboperculum; the sides, the scales, their uniformity upon all the regions of the body, and their rhomboidal form, such are the traits which characterize it.

The species would thus extend farther northwards than has been known heretofore. It is, however, still important to verify the fact, either by comparing young *C. aureolus* of Lake Erie with these, or by procuring large specimens from Lake Superior, to compare them with specimens of the other lakes.

CATOSTOMUS FORSTERIANUS, Agass.

I possess a complete series of individuals of this species, from the size of eleven inches up to seventeen. My description was made principally from the largest, to bring it nearest to that of *C. Hudsonius;* but I must, at the outset, remark that the characters noticed are the same in all. Not possessing a specimen of *C. Hudsonius*, I have referred to the description Dr. Richardson has given in establishing the points of comparison.

The general form of the body is very regular; the dorsal and ventral lines circumscribe an elongated oval, approaching to a cylinder towards the head, and to a parallelogram along the tail. The greatest circumference taken on the line of the greatest height, that is to say, before the dorsal, is nine inches and a half. The sides are compressed; the body passes to the head, or, we might rather say, the head passes to the body, without any enlargement on the nape of the neck. The greatest height of the body does not become double the greatest thickness, this latter being taken at the very origin of the trunk; thence it diminishes gradually and insensibly towards the caudal region, and the proportion begins to become progressively stronger in favor of the height from the posterior margin of the dorsal.

The head itself is very smooth, and covered with a thick skin; it is rather conical than quadrangular, on account of the declivity of the upper surface, which continues from the nape of the neck to the obtuse and rounded snout. It forms about the fifth part of the whole length, or rather less; its height forms three-quarters of its length, in which the breadth between both eyes is contained twice. The eyes are subcircular, and situated near the upper surface of the head; the anterior margin of their orbit is at equal distances from the end of the snout and the posterior extremity of the operculum; in other terms, the diameter of the orbit is contained twice in the space which separates it from the margin of the operculum, and thrice in that which extends between it and the rostrum. The nostrils are large, and at a distance of one-fourth of an inch from the anterior margin of the orbits; their structure varies little in different species.

The mouth is placed immediately beneath the extremity of the rostrum; it is of medium size, very protractile; its opening is subcircular, and easily receives the largest finger beyond the first phalanx. Its lips are caruncuIate; the upper is thin, and of equal breadth on the whole circumference of the jaw; it dilates itself from the angle of the mouth, to pass to the thickened and rounded lobes, with fringed circumference of the lower jaw; these fringes are equally visible on the margin of the upper lip; the two lobes are united on the symphysis of the jaw, by a narrow cutaneous slip; the caruncles which cover their surface are scarcely more marked than those of the upper lip. On the head we remark several rows of pores similar to those of *C. Hudsonius* and other species. These rows are perfectly distinct in individuals preserved in alcohol. One of them is the continuation of the lateral line of the body; it passes along the upper margin of the operculum, descends beneath the orbit, and terminates on the end of the snout, describing some undulations on its passage. The second row begins at the nostrils, and terminates on the occiput, a little before the union of the head with the body, on which point of union we observe a third single row, united transversely by its two extremities to the first double row. Finally, a fourth row is situated upon the face, and follows the outer margin of the preoperculum.

The opercular apparatus differs from that of *C. Hudsonius*, as described by Dr. Richardson, in two of its bones, the preoperculum and the interoperculum. This latter, in the species which is here referred to, has exactly the length of the suboperculum, though it is more robust and of more irregular form. It has a median carina on its anterior angle, whose extremity reaches that of the preoperculum in contact with the lower maxillary; the posterior part, contiguous to the operculum and suboperculum, is triangular, and rises to one-third of the height of the anterior margin of the operculum. The preoperculum is more slender, more elongated, and narrower than the interoperculum; its form is that of a very opened crescent.

The branchial fissures are very large, and somewhat approximated on the isthmus, where the membrane passes to the integuments of the abdomen, appearing somewhat like a transverse furrow.

The intestinal canal measures twice the length of the body. The lower pharyngeals form a complete ring around the œsophagus. Each bone, taken by itself, resembles in its form a sickle; that is to say, a crescent with a stalk. With this short, robust and flattened stalk the two bones unite, by means of a muscular bridge, which modify constantly the separation of which they are capable. The crescent presents two distinct sides; one, the inner, is compact, rounded and smooth, and is only the continuation of the stalk; the other, or outer, is widened, embracing only the circumference of the crescent; it is composed of vertical laminæ, of which the teeth are the continuation, with the exception of two lower ones, which are implanted on the very body of the bone. There are about thirty teeth; the lower are much developed, strong, and compressed laterally, surmounted by a crown which slopes over their inner side. From the middle of the crescent the teeth diminish abruptly towards its summit, and are reduced to feeble laminæ, which are lost in the body of the bone, which is also subject to a gradual diminution from the stalk to its upper angle.

The air bladder is composed of two compartments; the anterior is pear-shaped, and not quite half the length of the posterior, whose form is cylindrical.

The color of this fish is bluish gray on the back, the head and the sides; upon the sides an orange-colored red tint, with a very fine reflection, combines itself with the main color; the belly and the lower side of the head are whitish. The pectoral and ventral fins are gray, on an orange-colored ground; the caudal has the tint of the back, as also the dorsal; the anal is sometimes whitish, like the belly, sometimes gray like the ventrals.

This species is very common along the northern shores of Lake Superior.

Catostomus Aurora, Agass.

Pl. II., fig. 3 and 4.

Catostomus Forsterianus *Richards.* Frankl. Journ. 1823, p. 720; Fn. Bor. Amer. III., 1836, 116.—*Cuv.* et *Val.*, Hist. Nat. Poiss. 1844, 463.—*Storer* Synops. 1846, p. 167.

Mithomapeth *Pen.* Arct. Zoöl. Introd. ccxcix.

We have stated above, when speaking of the generic characters, the reasons which have induced us to change the name of this species, and to work out again its synonymy. Therefore, nothing more remains to be said on this point, and we proceed to give a full description of it, also comparing it with the above species, and regretting that we have been unable to compare it in nature with the *C. Hudsonius*. As described by Dr. Richardson, his *C. Forsterianus*, which is our *Aurora*, is rather compared with that species than described in detail, and as these two species are very different from each other, the comparison has not been made in its most minute peculiarities.

The body is subcylindrical, compressed. Its general form, less thick and stout than in the preceding species, presents the same regularity of outlines, and the same harmony of the regions among themselves. The greatest height corresponds also to the anterior margin of the dorsal, and forms the fifth of the whole length, the caudal excluded; this height forms five-sevenths of the greatest thickness of the body, which corresponds to the immediate back of the head. The diminution is gradual towards the tail. The head forms exactly the fifth of the whole length, and it is of course contained four times in that of the body, the caudal included. It is almost as compressed as in the preceding species, but less rounded on the upper surface, more elongated, more conical, and the rostrum more prominent. The skull is, however, declivous. The nostrils are very large. The position of the eyes, opposite the rostrum and the margin of the operculum, has the same relations as in the preceding species. The mouth is larger, and seems to be placed more backwards, on account of the developement of the nose, but the upper lip, when we extend it, easily reaches to its extremity. The lips are more developed, and covered with more prominent caruncles. The two lobes especially are more extended, and are not at all attached to each other on the maxillary symphysis, as they are in the preceding species, being in this respect more independent of each other. (Pl. 2, f. 4.)

The surface of the head is covered with a smooth skin, through which the rows of pores open, upon the whole, similar to those which we have described in the preceding species.

The opercular apparatus is smaller and more convex than in the preceding species, and all the bones are so, proportionally, I having, however, been careful to take two individuals of the same size for the purpose of comparison. The operculum is as broad as high, though narrower at the upper margin than at the lower, which is oblique; the posterior margin is almost straight. The suboperculum is more regular, on account of its lower margin being less convex. The interoperculum is less extended on its posterior extremity, which emits no processus along the anterior margin of the operculum. The outer surface is very convex, and almost smooth. The preoperculum is longer and more slender than the interoperculum, and proportionally broader than in the preceding species.

The branchial fissures are large also; the branchiostegal membrane is strong and thick; it contains three rays. The dorsal fin is quadrangular, its posterior margin equals in height two-thirds of its anterior margin, where we observe two or three small rudimentary rays, without articulations. Its upper margin is almost straight or subconcave. The anal is long, and attains the base of the caudal in the male, whilst it is shorter in the female; its anterior and posterior margins are parallel on the first two-thirds; beyond which they approach each other to form a triangle, and to terminate the fin in a more or less obtuse point. The caudal is notched; the scales advance more on the base of the lower lobe, which predominates slightly over the upper; but this character is not constant; I have even observed it only on the single individual which I have had figured; there is one, sometimes two, rudimentary rays at the anterior margin. The ventrals are broad and expanded, like an equilateral fan in the male; while in the female the inner margin is shorter, which changes the aspect of the outer circumference, which is straight and more uniform in the male. Generally, we observe the rudiments of a ray at the anterior margin, which corresponds to the fifth ray of the dorsal, the rudiments excluded. The pectorals are long and of an irregularly elliptical form, or oblong, sometimes pointed at their terminal extremity. The anterior ray is strong and robust; the fifth is the largest.

Br. 3; D. III, 11; A. II, 8; C. 5, I, 8, 8, I, 5; V. I, 10; P. 17–18.

The scales, very small at the anterior part of the trunk, increase in size towards the tail, without, however, attaining to the dimensions of the species above mentioned, nor even to those of *C. Hudsonius*. This increase of the scales from the head to the tail is real, and agrees with the imbrication. Their form is irregular and very variable, though we may say that they are generally oblong, of greater length than height, with convex margins, which are undulated, and never parallel and straight, like the upper and lower margins of the scales in the preceding species. Now and then we may find a few circular ones, but they are exceptions. Those which cover the shoulders are still much larger than those situated between the pectoral fins on the lower surface of the abdomen. The lateral line is median, slightly inflected on the abdomen before the dorsal. It rises a little on the pedicle of the caudal. The abdominal walls are covered with a blackish pigment. The length of the intestinal canal is contained twice and a half in that of the body. The pharyngeal bones, though having the same structure as in the preceding species, are, however, much more slender, and their teeth are much more feeble, thinner, and sharper on their extremity.

The air bladder, equally divided into two compartments, presents this difference, that, instead of being cylindrical, the posterior compartment terminates in a pointed cone. The size and the relative proportions remain almost the same in the two species.

The color is an olive yellow, very dark on the back and head, where it passes to the green on the sides. Following the course of the lateral line there is a band of a very brilliant carmine red, without precise outlines circumscribing it. In the females the red is less lively, and the belly remains white. The dorsal, caudal, and pectoral fins are colored like the back; the ventrals and the anal like the abdomen, but of a more intense yellow. The rays are of an olive-colored green.

This species occurs frequently along the northern shores of Lake Superior. I secured, however, most of my specimens at the Pic.

Genus Alburnus, Heck.

This genus has been known only in the Old World, until I discovered the species described below, which was caught at the Sault of St. Mary. The species described before are about equally divided between Europe and Syria. The principal character of the genus is to have the mouth opening upwards, the lower jaw exceeding a little the upper (Pl. 3. figs. 2 and 3.) The dorsal is narrow; the anal slightly broader. The body is compressed.

Alburnus rubellus, Agass.

Pl. III., figs. 1–3.

This is as yet the only species of the genus found in North America. The body is compressed; its form is elegant, slender, the back somewhat more convex than the belly; the tail is contracted. The greatest height of the body corresponds to the anterior third, or the region situated between the pectorals and the ventrals, and is contained six times in the length, exclusive of the caudal fin. The head, small, conical and compressed, like the sides, is somewhat less than the fifth of the whole length. The upper surface continues the declivous line of the back towards the end of the snout. The eyes are large and circular, approaching the upper region of the head, and at an equal distance from the end of the snout and the posterior extremity of the opercular apparatus. The suborbital ossicles are three in number; two are contiguous to the posterior and lower margin of the orbit, the other at the anterior margin, covering the whole space between the nostrils and the lower maxillary. The nostrils, proportionally large also, are nearer to the eyes than to the extremity of the snout, and opening into two apparently equal orifices. Fig. 2, which represents the upper surface of the head, shows only the anterior orifice, the posterior being covered by the intermediate membrane which separates them from each other. The mouth is moderately opened; its angles reach behind a vertical line which would pass before the eyes. The lower jaw slightly exceeds the upper (figs. 2 and 3.)

The preoperculum is rounded at its posterior margin. The lower margin of the operculum is straight and oblique. The suboperculum is narrow, and terminates behind in a point; its upper margin, contiguous to the operculum, is straight; its lower margin forms a slight elliptical curve. Scarcely can we distinguish the lower mar-

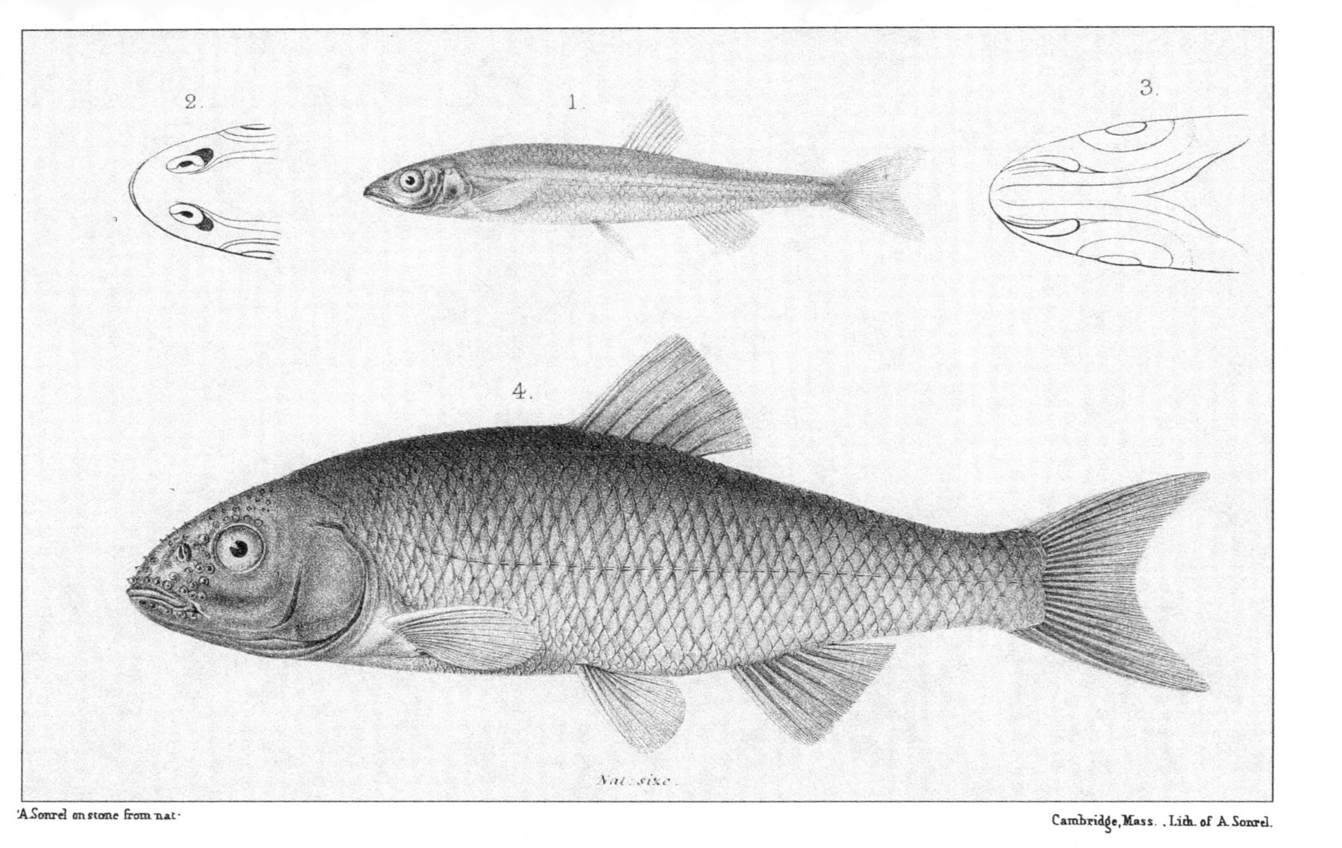

A. Sonrel on stone from nat.

Cambridge, Mass., Lith. of A. Sonrel.

1-3. ALBURNUS RUBELLUS Ag. — 4. LEUCISCUS FRONTALIS Ag.

gin of the interoperculum, this bone being hidden behind the preoperculum. The branchiostegal rays, three in number, are flattened and excessively thin, almost equal in form and in size, and slightly arched.

The dorsal fin is higher than long, and situated about on the middle of the back. Its anterior margin is twice as high as its posterior. The upper margin is straight. There are ten rays, of which the anterior is short and undivided; the bifurcation is repeated to the third degree on the central rays. The caudal is long and furcated; the rays are twice bifurcated; the largest only have slight indications of a three-fold division. The anal, placed behind the dorsal, is broad, but less high than this latter; its margins are straight; it contains eleven rays, of which two are rudimentary and undivided at the anterior margin. Those of the centre show the traces of a triple bifurcation. The ventrals, narrow at their base, extend considerably at their circumference, which is rounded; they are situated before the dorsal, and contain eight rays, the first being simple, the five following subdivided to the third degree. The pectorals, narrower and more elongated than the ventrals, are inserted behind the suboperculum at a small distance from this bone. There are eleven rays; the first does not bifurcate at all, though it is articulated; the six following are articulated on their last third only; the five remaining are very short.

Br. 3; D. I. 9. A. II., 10; C. 4. I. 9. 8. I. 4; V. 8; P. 11.

The scales are of medium size, and about equal on all regions of the body. Their form is subcylindrical; the concentric and radiating striæ are visible only under the microscope. The lateral line is slightly inflected from the upper angle of the opercular apparatus upon the abdomen, to rise again opposite the dorsal, and thence continues in a straight line towards the tail, following the middle of the sides.

The back is of a yellowish green, with the outlines of the scales black. The upper surface of the head and the snout are of a darker tint. The face, the opercular apparatus and the sides have a brilliant silvery reflection, with a more marked median band. There are some reddish spots on the face and the opercular apparatus, fading sometimes into a uniform reddish tint all over the head and shoulders. The iris is gold-colored; the fins are of a uniform color, a transparent, pale yellow.

Fig. 1 represents the fish of natural size. Figs. 2 and 3 are enlarged, to show the characters of the mouth and the jaws.

This species is very common at the Sault of St. Mary; specimens were also obtained from the Pic.

GOBIO PLUMBEUS, Agass.

This species is widely distinct from *Gobio cataractæ*; the only species of that genus found in North America which has hitherto been described. The body is elongated, subcylindrical, compressed; its greatest length is about seven inches. The head is contained somewhat more than four times in this length, and the height of the body forms exactly the fifth of it. The back is very slightly convex; the belly describes a very marked curve; the tail beyond the anal fin straightens almost abruptly. The head itself is conical, irregularly quadrangular, the upper surface being very flattened, sometimes even concave on the middle line, and the lower surface plain. The eye is situated at the upper region of the face; its diameter is one fourth of an inch. The nostrils are large also, and situated in circular cavities at the upper part of the face. The anterior opening is oblong; its canal is oblique from behind forwards; its posterior margin, when extended, forms a cover to the second opening, which is the largest, perforated like the first, and placed a little more outwards. The snout is flattened. The upper jaw exceeds the lower, and thus removes the mouth to the lower side of the head. At the angles of the mouth there is a very small barbel, still more slender than in the *G. cataractæ*. It needs a very attentive examination to notice it.

The posterior margin of the operculum is notched in the form of a small crescent at whose margin is a process of this bone. The lower margin is oblique and slightly concave, bordered on its whole length by the suboperculum, a small, thin, narrow and elongated lamina. The interoperculum and the preoperculum are hidden beneath the fleshy skin of the cheeks. The branchiostegal membrane contains three rays; it is continued upon the opercular valve.

The dorsal is situated exactly on the middle of the whole length, somewhat farther back than in *G. cataractæ ;* it is higher than long. The caudal is notched ; its lobes are pointed. The anal is somewhat smaller than the dorsal, but it has the same form. The ventrals, situated somewhat in front of the dorsal, are rounded on their circumference. The pectorals are narrower than these latter ; they are also more elongated and more rounded on their circumference ; their form is oblong.

Br. 3 ; D. I., 9 ; A. I., 9 ; C. 5, I., 9, 8, I., 4 ; V. II., 8 ; P. 16.

The scales are large ; we can scarcely count sixty rows from the gills to the caudal ; somewhat oblong on the sides, they are subcircular on the back and belly. We readily perceive with the magnifying glass the concentrical and radiating striæ. The lateral line is deflected on the abdomen into an open curve, and recovers its direct line beyond the dorsal, towards the tail. It is almost central in its whole course.

The head, the back, and the upper half of the sides are ash-gray. A narrow lead-colored band extends along the upper side of the lateral line. The abdomen is yellowish white, interspersed with small gray points on the scales. The lower side of the head and belly is of a uniform color. The dorsal, caudal, and pectorals are gray, the ventrals and the anal yellow. The largest specimens of this species are from Lake Superior. We have also a few from Lake Huron.

I am well aware that the position of this species in the genus Gobio is not natural, as it has neither the particular cut of the outline of the head which characterizes the European species of Gobio, nor their narrow dorsal, nor their projecting barbel, nor their pharyngeal teeth, but I am unwilling to establish a new genus for it before I have organized the American Cyprinidæ more extensively. I will only add that were it not for the barbel this species might be very properly placed in the genus Leuciscus. But the European Leucisci have not rudiments of such appendages on the sides even of the mouth ; while all the species of Cyprindo of North America, which have been referred to the genus Leuciscus, have, as far as I know, such short barbels. I am therefore inclined to believe that these species will have to be removed from that genus, Leuciscus, and constitute by themselves a distinct genus, to which my *Gobio plumbeus* will also belong, as it is not to be separated generically from *Leuciscus pulchellus* and other American species.

Leuciscus frontalis, Agass.

Pl. III., fig. 4.

At first sight this species reminds us of *L. cornutus* of New England, to which it bears a close resemblance. Its general form is short and stout. Its sides are much compressed. The back is very convex. The height of the body is proportionally great, and is contained only four times in the whole length, from the anterior extremity of the head to the termination of the caudal. It has thus a corpulent form, and is even higher than *L. cornutus*. The tail also loses its dimensions less abruptly. The head itself participates of the abbreviated form of the body, being somewhat less than a quarter of its length. Its upper surface is rounded, very declivous, and descends abruptly on the snout, which renders it very obtuse, rounded, and, as it were, prominent. The eyes are large and circular, proportionally larger than in *L. cornutus*, and approach less to the top of the head. They are situated but little nearer to the end of the snout than to the posterior margin of the opercular apparatus. The lower margin of their orbit corresponds to a horizontal line traced along the middle of the face. The nostrils open by a double opening in a circular depression situated before the eyes, and nearer to these latter than to the terminal margin of the head. The anterior opening, which is the smallest and of subcircular form, is bordered behind by a small membrane which applies itself like a cover on the posterior opening, rendering its form crescentic. The mouth is of medium size, but shorter cleft; its angles attain a vertical line which would descend from the nostrils; it is terminal and oblique; the lower jaw is somewhat shorter than the upper.

The opercular apparatus has nothing remarkable. The bones which compose it are all hidden beneath a thick skin through which we scarcely distinguish their outlines. All are rounded on their outer margin, and give to the extended outline of the whole operculum the form of a crescent on whose convexity the branchiostegal membrane is continued to the upper margin of the operculum.

The branchial fissures are large. There are three strongly developed branchiostegal rays, flattened and arched. The two outer on each side may approach very near to each other on the middle line of the lower surface of the head, where they are parallel for a short distance. The branchiostegal membrane is endowed with great elasticity.

The rays of the centre of all the fins are bifurcated to the third degree. In front of the dorsal, of the anal and of the ventrals we remark the rudiment of a spinous ray, often very difficult to recognize. The following ray is never bifurcated, though distinctly articulated as the remaining ones; this is also the case with the ray of the anterior margin of the pectorals, and with the great outer ray of the lobes of the caudal, which for this reason is stouter.

The anterior margin of the dorsal fin corresponds exactly to the middle of the length of the body, excluding the caudal; so that it extends behind the most prominent part of the back, along the curve of the posterior half of the body; its length nearly equals the height of its anterior margin; its upper margin is very slightly rounded. The anal is both lower and shorter than the dorsal, but its length equals its height. Its outer margin is almost straight. The caudal is admirably regular; its posterior margin is notched by a subcircular crescent; the ventrals are oblong, rounded, when extended; their outer circumference equals three widths of their base; their posterior extremity passes somewhat beyond the anus. The pectorals have precisely the general form of the ventrals, but they are larger; their terminal extremity is almost contiguous to the base of insertion of the ventrals.

Br. 3; D. I., 9; A. 10; C. 3, I., 9, 8, I., 3; V. I., 8; P. 14.

The scales cover more than half of each other by imbrication; they are oblong in the vertical direction, and seen in their natural position, they represent lozenges which vary a little according to the regions; the largest occupy the middle region of the body as far as the pedicle of the tail; but on this latter region they are broader in proportion to their height. On the back they have almost the size and the form of those of the tail. On the belly they are much smaller and subcircular. The lateral line curves slightly on the abdomen as far as the height of the anterior margin of the dorsal, whence it continues almost directly towards the tail, approaching nearer, however, to the lower line of the body.

Small circular shields with depressed surface, surmounted with very small conical and acute points, cover the surface of the head, the snout and the back, as far as the dorsal fin. A row of five or six of the largest border the lower jaw; those of middle size cover the extremity of the snout and the space situated before the eyes. On the back they are excessively small.

The head and the back are of a bluish black, the sides and the abdomen of a gold-colored yellow, everywhere with a metallic reflection. The fins are of uniform color and participate of the tint of the regions to which they belong.

From Montreal River on the eastern shore of Lake Superior.

Leuciscus gracilis, Agass.

There is still another Leuciscus which, at first sight, one might be disposed to confound with *L. cornutus* or with the *frontalis* above described. And it must be confessed that it has much analogy with those two species, between which it must be placed in a natural series.

In a family so numerous in species as that of the Cyprinidæ, it is only by minute study that we can succeed in making out the history of each of them. Here, as in Europe, the species, though belonging often to different genera, gradually pass from one genus to another, in their general appearance; the type of the family, that of the genus itself, seems to predominate in all; and by reason of the multiplicity, and also the diversity of forms under which these characters manifest themselves, the species appear to be mere varieties. These difficulties occur also in all genera which have numerous species in other families of this and other classes, but, far from impressing naturalists merely with the monotony to be overcome, they should render them attentive to the most minute details which characterize, in a permanent manner, natural groups in the animal kingdom. In the case of this species and the two others mentioned in connection with it, I am satisfied that they should constitute a distinct genus, characterized chiefly by their scales, which are so much higher than long, besides the particular form of their head and body and their pharyngeal teeth. There are some more species of this genus yet undescribed, which have been discovered in Pennsylvania by Prof. Baird ; but I do not know one from Europe.

Though the length of this species is the same as that of *L. frontalis*, its general form shows a marked difference. It is fusiform, rather slender but very compressed, the curve of the back being very elliptical, and the abdomen making a stronger projection. The height is somewhat less than a quarter of the whole length. The head is small and conical; its upper surface rather flattened than convex, with a less marked declivity. The anterior part, less developed than in the *L. frontalis*, renders the head more pointed, though the snout be obtuse. The eyes are somewhat larger, and nearer the upper margin of the skull. The face is less developed, both jaws are of equal length. The opercular and branchiostegal apparatus are less robust. The head forms about the fifth of the entire length, and this slight difference in the proportions, when compared with *L. frontalis*, accounts for the differences of the general form, which we have noticed above. Again, as the consequence of a more slender body, smaller fins are required to sustain it, and there being space for separation between them they become more distant from each other. Thus is the distance enlarged between the extremity of the pectorals and the base of the ventrals, and between the extremity of the ventrals and the anus. All the fins, taken together, are smaller than in *L. frontalis*. Thus the pectorals and the ventrals are less widened, while the length is the same. The dorsal is higher than it is long; the anal lower than the dorsal, but also higher than long. The caudal is narrower, a natural consequence of a smaller tail.

Br. 3 ; D. I, 9 ; A. I, 10 ; C. 4, I, 9, 8, I, 4 ; V. 8 ; P. 15.

The rays of the dorsal, caudal, and pectoral fins, present bifurcations of the second degree only; slight indications of three-fold bifurcation are observed on the central rays of the ventrals and anal, but with less regularity than in the preceding species.

The scales are larger than those of *L. frontalis*, and are less extensively imbricated, showing, however, the same proportions on the different regions, which we have given for the preceding species. The lateral line is apparently the same; only the curve inflected on the abdomen seems wider.

The back and the head are greenish-brown; the lower face of the head and the abdomen are of a very pale golden yellow, with a very brilliant silvery reflection of the scales. The operculum is gold colored. The rays of the dorsal, caudal, and pectoral fins, have a gray tint on a yellowish ground. The ventrals and the anal are of a golden yellow, like the abdomen.

The head is smooth; we notice only on the space between the eye and the occiput some rudiments of tubercles hidden beneath the skin, perceptible only to the touch.

This species is distinguished from *L. cornutus*, not only by the color of its fins and the absence of armature on the head, but also by differences in the general form and structure of the fins, analogous to those which we have pointed out in *L. frontalis*.

From Lake Huron.

Leuciscus Hudsonius, Dekay.

Leuciscus Hudsonius *Dekay*. N. Y. Fn. 1842, p. 206, Pl. 34, fig. 109.

Clupea Hudsonia *De Witt Clinton*, An. Lyc. N. H. N. Y., I., 1824, 49, Pl. 2, fig. 2.

The resemblance of this species to the Clupea is only superficial, and does not require a long examination to be refuted. With the exception of the general outline, it has not one of the essential characters of organization of that family. The external conformation of the mouth could not leave us for a moment in hesitation as to which natural group it belongs. It is of the family of Cyprinidæ, where it has been placed by the author of the Zoölogy of New-York. Already DeWitt Clinton, though arranging it in the genus Clupea, entertained some doubts in this respect, on account of the absence of a ventral serrature.

The species is tolerably well described by the authors whom we have just cited, so that we have only to refer our readers to them.

We must, however, remark that the figures which they give of it are rather incomplete. The oldest is still the best for the general outlines, and the species is there more easily recognized than by that of the Fauna of New York, where the fins are too stiff and too rectilinear, and the scales drawn in an inverse direction from what they are in nature, the posterior margin being turned towards the head.

The formula for the fin rays is as follows:

Br. 3; D. II. 9; A. II., 9; C. 4, I. 9, 8, I., 4; V. 8; P. 15.

A very slight difference in the dorsal and anal may be noticed, but we consider it of little importance here. Their rays bifurcate to the third degree, with a few unsymmetrical indications of a three-fold bifurcation on one of the rays of the anal, and on some of the central ones of the lobes of the caudal. The rays of the pectorals subdivide only once. As for the branchiostegal rays, we find only three of them, though DeWitt Clinton has counted four; perhaps he counted the suboperculum. Dr. Dekay does not mention them. There is also something to be corrected respecting the lateral line; the former says it is obsolete; the latter describes it as straight. On the individuals which we have had under notice, it is almost median; arising from the upper angle of the operculum, it is deflected upon the abdomen to rise again gradually beyond the dorsal fin, and finally to extend straight towards the extremity of the tail.

From Lake. Superior and Lake Huron. Very common about Fort William and the Pic.

This is another form of the group of Leucisci, of which there is no representative in Europe. It is likely to become the type of a distinct genus; for it has many striking peculiarities. I have, however, refrained from establishing it until I shall have ascertained whether the specimens found in different localities are specifically identical or not.

Such a critical revision of the fishes of Lake Superior, and the other great Canadian lakes, was the first necessary step in the investigation I am tracing, in order to ascertain the natural primitive relations between them and the region which they inhabit. Before drawing the conclusions which follow directly from these facts, I should introduce a similar list of the fishes living in similar latitudes, or under similar circumstances, in other parts of the world ; and more particularly of the species of Northern Europe. But such a list, to be of any use, should be throughout based upon a critical comparative investigation of all the species of that continent, which would lead to too great a digression. The comparison of the freshwater fishes of Europe, which correspond to those of North America, has been carried so far, that I feel justified in assuming, what is really the fact, that all the species of North America, without a single exception, differ from those of Europe, if we limit ourselves strictly to fishes which are exclusively inhabitants of freshwater.

I am well aware that the salmon which runs up the rivers of Northern and Central Europe, also occurs on the eastern shores of the northern part of North America, and runs up the rivers emptying into the Atlantic. But this fish is one of the marine arctic fishes, which migrates with many others annually further south, and which migratory species is common to both continents. Those species, however, which never leave the freshwaters, are, without exception, different on the two continents. Again, on each of the continents, they differ in various latitudes ; some, however, taking a wider range than others in their natural geographical distribution.

The freshwater fishes of North America, which form a part of its temperate fauna, extend over very considerable ground, for there is no reason to subdivide into distinct faunæ the extensive tracts of land between the arctics and the Middle States of the Union. We notice over these, considerable uniformity in the character of the freshwater fishes. Nevertheless, a minute investigation of all their species has shown that Lake Superior proper, and the freshwaters north of it, constitute in many respects a special zoölogical district, sufficiently different from that of the lower lakes and the northern United States, to form a natural division in the great fauna of the freshwater fishes of the temperate zone of this continent.

We have shown that there are types, occurring in all the lower lakes, which never appear in Lake Superior and northwards, and that most of the species found in Lake Superior are peculiar to it ; the Salmonidæ only taking a wider range, and some of them covering almost the whole extent of that fauna, while others appear circumscribed within very narrow limits.

Now, such differences in the range which the isolated species take in the faunæ is a universal character of the distribution of animals; some species of certain families covering, without distinction, extensive grounds, which are occupied by several species of other families, limited to particular districts of the same zone.

But, after making due allowance for such variations, and taking a general view of the subject, we arrive, nevertheless, at this conclusion; that all the freshwater fishes of the district under examination are peculiar to that district, and occur nowhere else in any other part of the world.

They have their analogues in other continents, but nowhere beyond the limits of the American continent do we find any fishes identical with those of the district, the fauna of which we have been recently surveying. The Lamprey eels of the lake district have very close representatives in Europe, but they cannot be identified. The sturgeons of this continent are neither identical with those of Europe nor with those of Asia. The cat-fishes are equally different. We find a similar analogy and similar differences between the perches, pickerels, eelpouts, salmons, and carps. In all the families which occur throughout the temperate zone, there are near relatives on the two continents, but they do not belong to the same stock. And in addition to these, there are also types which are either entirely peculiar to the American continent, such as Lepidosteus and Percopsis, or belong to genera which have not simultaneously representatives in the two worlds, and are therefore more or less remote from those which have such close analogues. The family of Percoids, for instance, has several genera in Europe, which have no representatives in America; and several genera in America which have no representatives in Europe, besides genera which are represented on both continents, though by representatives specifically distinct.

Such facts have an important bearing upon the history of creation, and it would be very unphilosophical to adhere to any view respecting its plan, which would not embrace these facts, and grant them their full meaning. If we face the fundamental question which is at the bottom of this particular distribution of animals, and ask ourselves, where have all these fishes been created, there can be but one answer given which will not be in conflict and direct contradiction with the facts themselves, and the laws that regulate animal life. The fishes and all other freshwater animals of the region of the great lakes, must have been created where they live. They are circumscribed within boundaries, over which they cannot pass, and to which there is no natural access from other quarters. There is no trace of their having extended further in their geographical distribution at any former period, nor of their having been limited within narrower boundaries.

It cannot be rational to suppose that they were created in some other part of the world, and were transferred to this continent, to die away in the region where they are supposed to have originated, and to multiply in the region where they are found. There is no reason why we should not take the present evidence in their distribution as the natural fact respecting their origin, and that they are, and were from the beginning, best suited for the country where they are now found.

Moreover, they bear to the species which inhabit similar regions, and live under similar circumstances in Europe and Asia, and the Pacific side of this continent, such relations, that they appear to the philosophical observer as belonging to a plan which has been carried out in its details with reference to the general arrangement. The species of Europe, Asia, and the Pacific side of this continent, correspond in their general combination to the species of the eastern and northern parts of the American continent, all over which the same general types are extended. They correspond to each other on the whole, but differ as to species.

And again, this temperate fauna has such reference to the fauna of the Arctic, and to that of the warmer zones, that any transposition of isolated members of the whole plan, would disturb the harmony which is evidently maintained throughout the natural distribution of organized beings all over the world. This internal evidence of an intentional arrangement, having direct reference to the present geographical distribution of the animals, dispersed over the whole surface of our globe, shows most conclusively, that they have been created where they are now found. Denying this position were equivalent to denying that the creation has been made according to a wise plan. It were denying to the Creator the intention of establishing well regulated natural relations between the beings he has called into existence. It were denying him the wisdom which is exemplified in nature, to ascribe it to the creatures themselves, to ascribe it even to those creatures in which we hardly see evidence of consciousness, or worse than all, to ascribe this wonderful order to physical influences or mere chance.

As soon as this general conclusion is granted, there are, however, some further adaptations which follow as a matter of course. Each type, being created within the limits of the natural area which it is to inhabit, must have been placed there under circumstances favorable to its preservation and reproduction, and adapted to the fulfilment of the purposes for which it was created. There are, in animals, peculiar adaptations which are characteristic of their species, and which cannot be supposed to have arisen from subordinate influences. Those which live in shoals cannot be supposed to have been created in single pairs. Those which are made to be the food of others cannot have been created in the same proportions as those which feed upon them. Those which are everywhere found in innumerable specimens, must have been introduced in numbers capable of maintaining their normal proportions to those which live isolated, and are comparatively and constantly fewer. For we know that this harmony in the numerical proportions between animals is one of the great laws of nature. The circumstance that species occur within definite limits where no obstacles prevent their wider distribution, leads to the further inference that these limits were assigned to them from the beginning, and so we should come to the final conclusion, that the order which prevails throughout the creation is intentional, that it is regulated by the limits marked out on the first day of creation, and that it has been maintained unchanged through ages, with no other modifications than those which the higher intellectual powers of man enable him to impose upon some few of the animals more closely connected with him, and in reference to those very limited changes which he is able to produce artificially upon the surface of our globe.

VII.

DESCRIPTION OF SOME NEW SPECIES OF REPTILES FROM THE REGION OF LAKE SUPERIOR.

Hylodes maculatus, Agass.

Pl. VI., figs. 1, 2, 3.

This species is so characteristic as to leave no difficulty in distinguishing it from those already known belonging to the same genus. Its form is narrow, elongate ; and its head smaller, in proportion to the body, than in any other species. The length of the head is contained twice in the length of the body, thus forming one-third of the whole length. The body is oblong, rounded, somewhat broader than high, tapering towards its posterior extremity. The head is elliptical, tapering towards the snout, somewhat distinct from the trunk by a slight contraction of the neck ; its greatest width is behind the eyes ; its upper surface is depressed so that the head appears rather flat. The eyes, of a medium size, are turned upwards near the margin of the head, but are hardly prominent. The nostrils are lateral, and very near the extremity of the snout. The tympanic circle is small, and near the angle of the mouth. The mouth is widely split ; the lower jaw is overlapped by the upper, and the snout slightly prominent. The palatal teeth are arranged in pairs, upon two small, very narrow bones ; they are extremely minute. Those of the upper jaw, still less developed, occur only on the middle third of its arch. The tongue is broad, and fills the whole floor of the mouth ; it is free upon two-thirds of its posterior extremity, the margin of which is obtusely bilobed ; the anterior margin and the sides are hardly free.

1 2 3 4 5 6 7 8

O. Wallis on stone from nat. Lith. of A. Sonrel.

1-3 HYLODES MACULATUS Ag. — 4 & 5 RANA NIGRICANS Ag. — 6-8 CROTALOPHORUS

The limbs are very slender; the fingers very slim, and free for their whole length. The carpus and tarsus are hardly broader than the forearm and leg. The posterior extremities exceed the length of the body by the length of the longest finger. All the fingers are turned in one direction, bent outwards. The anterior limbs, half as long as the posterior, have the two outer toes turned outwards, while the two others are arched inwards.

The upper surface of the head is smooth, as are also the back and the legs; but the sides are covered with minute cutaneous tubercles, which extend over the whole lower surface of the body, where they increase in size; they extend, also, over the thigh and forearm; the lower jaw and extremities of the limbs, alone, being perfectly smooth underneath.

The color is of a bluish gray, irregularly speckled with small black dots, which are partly oblong, partly circular, and very well circumscribed in their outlines, so that they show distinctly, notwithstanding the slight difference in color. The lower surface is of a yellowish white, dark upon the sides, lighter and purer under the head and along the margin of the lower jaw. A very narrow white band extends along the margin of the upper jaw, as far back as the insertion of the arm, upon which it encroaches somewhat.

Figs. 1, 2, represent the species of the natural size; the first, in the natural attitude of the animal; the second, as seen from below. Fig. 3 represents a tadpole, remarkable for the great length the tail still preserves, the legs being already very far advanced in their development. Whether they undergo their metamorphoses in one season, or spend the first winter in an intermediate state between their larval and adult form, has not been ascertained.

Rana nigricans, Agass.

Pl. VI., figs. 4, 5.

This species is intermediate, with reference to its size and the development of its limbs, between R. *clamitans* and R. *halecina*. It differs from both by its color, and by the form of its legs; the hind foot being more extensively palmate, and their membrane extending to the base of the last fingers. The fingers, however, are comparatively more slender, and those of the anterior foot more unequal when compared to each other.

The head is rather prominent, the snout, however, being rounded. The nostrils, which are very small, open at its extremity. The eyes are circular, and of medium size, slightly prominent. The upper eyelid rises to the greatest height of the head. The tympanic circle is very large, and very near the orbit. The mouth, widely split, is provided with acute teeth upon the whole margin of the upper jaw. There is also a small group of teeth, in pairs, upon the palatal bones. The tongue is broad, oblong, pear-shaped, lining the whole floor of the mouth from the symphysis of the lower jaw; it terminates backwards in two obtuse lobes.

The body is proportionally long, ovate, the head forming one-third of the whole length. A cutaneous keel, of the same color as the main hue of the back, extends on both sides from the posterior angle of the orbit to the anus. The posterior limbs are longer than the whole body by the whole length of the feet. The thighs are comparatively thick and short. The anterior limbs bear the same proportion to the size of the whole body that are usually observed in the various species of frogs. Figs. 4 and 5 give, not only an accurate idea of the general appearance of the animal, but the proportional thickness and length of the toes are drawn with the greatest minuteness.

The largest specimens I have collected are about one-fourth larger than the figures. The color is of a blackish brown upon the whole upper surface of the body, head and limbs. Irregular, deep black spots, of an angular form, are dispersed over this whole surface; they are very small upon the head, but larger upon the back, and largest upon the hind legs. In large specimens, the general color is more uniform, somewhat darker, and the spots less distinct. The whole lower surface is either uniformly whitish, or with a slight yellowish tint towards the hind extremity, and frequently with small blackish or brownish spots along the sides. The outline of the lower margin is bordered with white. Specimens of this species were caught in various localities along the northern shores of Lake Superior.

Crotophorus.

Pl. VI., figs. 6 to 8.

I abstain from giving a specific name to this species, from fear of adding a useless synonym to its nomenclature. It is, indeed, very closely allied to, and probably identical with *C. tergeminus*. Its head, however, is rather elliptical than triangular, and the spots which cover it differ, as may be seen on comparing our figure with that of Dr. Holbrook.* The snout is truncate. Having no authentic specimen of *C. tergeminus* to compare with mine, I shall only point out the differences I have noticed between my specimen and the description and figure of Dr. Holbrook, leaving it to future comparisons to settle the question of the specific identity or difference.

The general color is the same as that of *C. tergeminus*, but the two brown bands which exist along the neck on each side, and converge upon the back, are shorter. The bands of the same color, which arise from the eyes, extend beyond the angle of the mouth, and nearly meet the other bands, where they unite with the first spot on the back. The width of these bands covers three rows of scales. The white band below this is much narrower, and covers but one single row of scales, and is bent at the angle of the mouth. Along the back there are thirty oblong transverse spots, deeply emarginate on the anterior side, and slightly concave on the posterior side backwards. They appear like a pair of spots united. Upon the tail there are five quadrangular, oblong, transverse spots, in advance of the caudal plates. Upon the sides there is a double row of smaller spots, of an oblong or subcircular form, varying in size, and alternating with each other, while in *C. tergeminus* there is only one small lateral row. The lower surface of the body is mottled with black and white, with very minute gray dots. There are one hundred and thirty abdominal plates, apparently broader than those of *C. tergeminus;* and, in addition, in advance of the anus, they are of a semicircular form. The caudal plates are twenty-eight in number, twenty-five of which are entire, and three, in advance of the rattle, bilobed.

* **North American Herpetology, vol. III., Pl. 5.**

The lobes of the rattle have the same dimensions as those of *C. tergeminus*. The whole length of the body is two feet two inches; the head measures one inch and a quarter; the tail, three inches and five-eighths of an inch. There are other slight differences in the proportional length of the body and of the tail, corresponding to the differences noticed in the greater number of caudal plates and the greater width of the abdominal plates.

The specimen was caught on the southern extremity of Lake Huron.

Besides those species, the following reptiles occur about Lake Superior:

Tropidonotus sirtalis,
" erythrogaster,
" a species allied to rigidus, from Lake Huron,
Bufo Americanus,
Rana halecina,
" sylvatica.

These three species occur as far north as Neepigon Bay, and a circumstance, which has struck me very forcibly, is the remarkable size of the specimens observed in these high latitudes.

Plethodon erythronotus Bd.
Menobranchus maculatus. This species does not properly occur in Lake Superior, but is found in Muddy Lake, below Sault St. Marie.

No turtles are found any where on the northern shores of Lake Superior, as far as I know.

VIII.

REPORT OF THE BIRDS COLLECTED AND OBSERVED AT LAKE SUPERIOR,

BY J. E. CABOT.

The striking scarcity of birds and quadrupeds about the lake has already been noticed in the Narrative. In the case of the granivorous and frugivorous species, this might be accounted for from the scarcity of their proper food. To the insectivorous birds, however, this reasoning certainly could not apply. One would have expected to find the warblers, especially, breeding in abundance in this region. But the only birds that could be called tolerably abundant (except in special localities) were Zonotrichia pennsylvanica, and in a less degree, Parus atricapillus and Ampelis cedrorum. Something, no doubt, must be attributed to the season, many birds having passed further northward, and others being engaged in incubation. Then all birds are more silent at this season, and less inclined to locomotion. On the other hand, we found a great abundance and variety of birds at the Sault, much greater than would be found in Massachusetts at that season. And whenever we came to a trading post, we found a great difference in this respect, although the Indians, whether from scarcity of food or from wantonness, destroy great numbers even of the smaller species. It would seem, that apart from a more abundant supply of nourishment, the neighborhood of man is in some way attractive to birds,—partly perhaps from the greater freedom of such situations from beasts and birds of prey. As to the water-birds, the nature of the country would at once indicate that none but piscivorous species were to be expected. In the annual migrations, it is said large numbers of ducks, and particularly of geese, alight, for a day or two, in the streams and pools of the shore. But the deep, cold waters of the lake, permitting no growth of water-plants, except occasionally in a sheltered cove, possess no attractions further. Accordingly, the only water-birds we saw were Larus argentatus, Colymbus glacialis and Mergus cucullatus, all which we usually saw in small numbers every day, and one specimen of Colymbus septentrionalis. In the neighborhood of Detroit we saw black terns in abundance, and heard that some of the light-colored species bred about St. Joseph's Island, but we saw none of them beyond the St. Clair.

Seeing the importance that is beginning to be given to even minute details of geographical distribution, I have subdivided the following list of species observed, so as to present first the species of most extensive range, and afterwards those of more confined localities.

From the Sault to Fort William.

Corvus cedrorum.
Ampelis cacalotl.
Parus atricapillus.
Regulus satrapa.
Vireo olivaceus.
Mniotilta coronata.
Hirundo bicolor.
" rufa.
Zonotrichia pennsylvanica.
Ectopistes migratorius.
Tringoides macularia.
Larus argentatus.
Colymbus glacialis.
Mergus cucullatus.

From the Sault to the Pic, and at Fort William.

Bonasa umbellus.
Zonotrichia melodia.

From the Sault to St. Ignace.

Turdus migratorius.
Mniotilta virens.
Fringilla hiemalis.
Carpodacus purpureus.
Tinnunculus sparverius.
Halietus leucocephalus.

From the Sault to the Pic.

Sialia Wilsoni.
Mniotilta æstiva.
Setophaga ruticilla.
Sitta canadensis.
Fringilla pinus.
Zonotrichia socialis.
Pandion Carolinensis.

From the Sault to Michipicotin.

Corvus Americanus.
Cyanocorax cristatus.
Mniotilta maculosa.

From Michipicotin to Fort William.

Tetrao canadensis.
Myiobius Cooperi.

From the Pic to Fort William.

Perisoreus canadensis.
Parus Hudsonicus.
Loxia americana.
" leucoptera.
Picus villosus.
" pubescens.
Picoides arcticus.
" hirsutus.
Totanus melanoleucus.

At the Sault.

Agelaius phœniceus.
Vireo noveboracensis.
Mniotilta maritima.
" Pennsylvanica.
Trichas Philadelphia.
Setophaga Wilsonii.
Guiraca ludoviciana.
Zonotrichia Savanna.
Syrnium nebulosum.
Colymbus septentrionalis.

Neighborhood of Mamoinse.

Chordeiles Virginianus.
Mniotilta striata.

At the Pic.

Colaptes auratus.
Turdus brunneus.

At the Pic.

Mniotilta peregrina (and young).
Myiobius nunciola.
" virens.
Zonotrichia pusilla.
" Lincolnii.

Neighborhood of St. Ignace.

Falco peregrinus (unfledged).
Surnia ulula.

At Fort William.

Cotyle riparia.
Ceryle alcyon.
Tringa Schinzii.
Totanus flavipes.

At the Sault and Fort William.

Setophaga canadensis.

IX.

DESCRIPTIONS OF SOME SPECIES OF LEPIDOPTERA, FROM THE NORTHERN SHORES OF LAKE SUPERIOR.

BY DR. THADDEUS WILLIAM HARRIS.

Pontia oleracea H.

Pl. VII., fig. 1.

Pontia oleracea Harris, New England Farmer, vol. VIII., p. 402 (1829).—Discourse before the Massachusetts Horticultural Society, p. 7, 21 (1832).—Catalogue of Insects of Massachusetts, in Hitchcock's Report, 1st ed. p. 589 (1833).—The same, 2d ed. p. 590 (1835).—Report on Insects of Massachusetts injurious to Vegetation, p. 213 (1841).—Kirby, Fauna Boreali-Americana, Part IV., p. 288 (1837).

Pieris oleracea Boisduval, Spécies Gén. des Lépidoptères, tome I., p. 518 (1836).

Alis subrotundatis integerrimis albis; anticis basi costaque nigricantibus, subtus apicem et posticis, infra, luteis fusco-venosis.

Alar. exp. 2 unc.

Body black above. Antennæ black, annulated with white, and rufous at the tip. Wings yellowish white; the anterior pair dusky on the front edge and base; tip, beneath, pale yellow, with dusky veins. Under side of the hindwings pale yellow, with broad, dusky veins, and a saffron-yellow spot on the humeral angle.

The tip of the forewings is often marked with two or three little dusky stripes, in the males. The dusky veining of the under side of the hindwings is less distinct in the females than in the other sex,

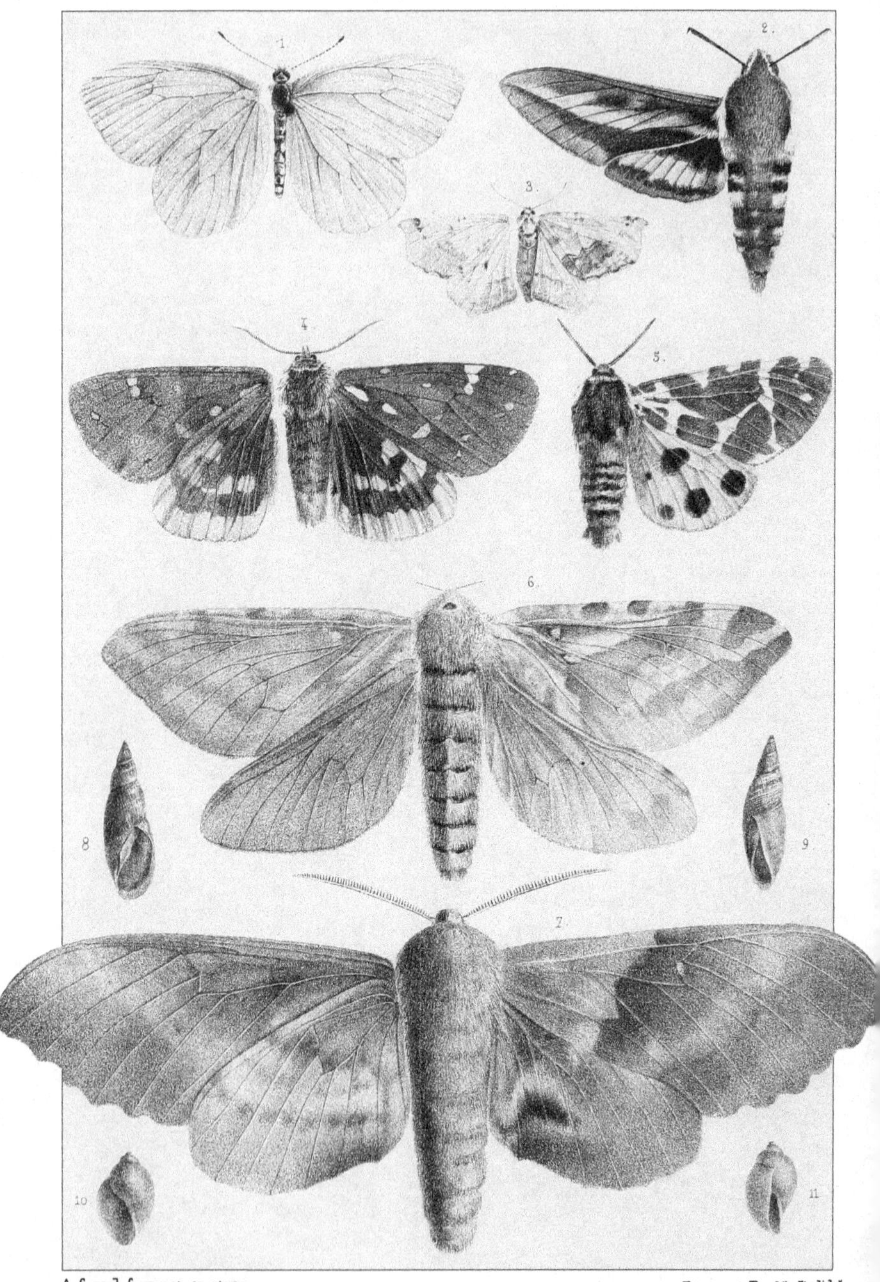

A. Sonrel from nat. on stone.   Tappan & Bradford's lith.

1. Pontia oleracea Harr
2. Deilephila Chamænerii Harr.
3. Ennomos macularia Harr
4. Arctia Parthenos Harr
5. Arctia americana Harr
6. Hepiolus argenteomaculatus Harr.
7. Smerinthus modestus Harr.
8 & 9. Limnea lanceata Gould.
10 & 11. Physa vinosa Gould.

and is sometimes entirely wanting. Specimens of the females have been seen, though rarely, with one or two dusky spots on the upper side of the forewings, towards the outer margin.

The eggs of this insect are pyriform, longitudinally ribbed, and of a yellowish color. The *larva* is pale green, very minutely sprinkled with darker dots, and with a darker dorsal line. It grows to the length of one inch and a quarter. Its natural food is unknown, but it is found abundantly on the leaves of the mustard, turnip, radish, cabbage, and other cultivated oleraceous plants, to which it is often very injurious. The *pupa* is pale green or white, regularly and finely spotted with black. There is a conical projection on the front, and a securiform one on the thorax; and the sides of the body are angular and produced in the middle. Length of the pupa eight-tenths of an inch. The pupa state lasts about eleven days in the summer, and continues through the winter; there being two broods of the larva in the course of one season.

This species rarely extends further south than the latitude of New Hampshire. It has not been figured before. Mr. Kirby's *Pontia casta* may, perhaps, be only a variety of it.

DEILEPHILA CHAMÆNERII H.

Pl. VII., fig. 2.

Sphinx Epilobii Harris, Cat. Ins. Mass. in Hitchcock's Report, 1st ed., p. 590 (1833).—The same, 2d ed., p. 591 (1835).
Deilephila Chamænerii Harris, Catalogue of North Amer. Sphinges. Amer. Journ. Science, vol. 36., p. 305 (1839).

Olivaceo-brunnea; capite thoraceque linea laterali alba; alis primoribus vitta duplici intermedia, apice attenuata, parte exteriori dentata pallidé ochracea, parte interiori flexuosa fusca; secundariis nigrofuscis, fasciâ lata macula rubra includente rosea, intus, ciliisque albis; abdomine punctis sex dorsalibus albis, lateribus fasciis duabus nigris et albis prope basin, duabusque albis posterioribus abbreviatis.

Alar. exp. $2\frac{3}{4}$—3 unc.

Olive-brown, with a white lateral line, extending from the front above the eyes on the sides of the thorax, where it is margined above with black. Palpi white below. Forewings with a black spot at base and another adjacent to a white dash within the middle of the outer edge; a flexuous buff-colored stripe, beginning near the base of the inner margin, indented externally, extends to the tip, and is bounded within by a dark brown tapering stripe. Hindwings blackish, or dusky brown, with a broad sinuous rosy band including a deep red spot, and uniting with a white one near the inner angle. Fringes of the hindwings, and inner edge of the forewings white. Abdomen with a dorsal series of six white dots; two black and two alternating white bands on each side of the base, and two narrow transverse white lines near the tip; ventral segments edged with white. Legs brown; the tibiæ edged externally with white.

This species, which occurs abundantly in New Hampshire, was taken on the northern shore of Lake Superior, and is now figured for the first time. It is the American representative of *Deilephila Galii.* Mr. Kirby's *D. intermedia*, which has the stripe on the forewings of a pale rose-color, and wants the dorsal series of white dots, may possibly be a local variety of *D. Chamænerii.* The larva of our species lives on the *Epilobium angustifolium.* It is bronzed green above, and red beneath, with nine round cream-colored spots, encircled with black on each side, and a red caudal horn.

Smerinthus modesta H.

Pl. VII., fig. 7.

Smerinthus modesta Harris, Catalogue of North American Sphinges. Amer. Journ. Science, vol. 36., p. 292 (1839).

Olivaceo-ochracea; capite parvo non cristato, masculorum antennis subtus transversé biciliatis; alis primoribus crenatis, strigâ flexuosa transversa basali virguloque stigmaticali pallidis, fasciâ lata undulata media, strigisque duabus crenatis posterioribus, saturaté olivaceis; secundariis medio basique purpureis, maculâ transversa nigra fasciâque abbreviata fusca prope angulum analem sitis.

Alar. exp. 5 unc.

Olive-drab; head very small, and without a prominent crest; antennæ of the males transversely biciliated beneath. Forewings scalloped, with a transverse sinuous pale line near the base; a whitish comma-shaped stigma on a broad undulated dark olive-colored central band, and two transverse undulated lines towards the tip; under side purple in the middle of the disk. Hindwings purple in the middle and at base, with a transverse black spot, and an abbreviated dusky blue band near the anal angle. Body very robust, and with the legs immaculate.

One of the largest species of the genus. A single male was taken on the northern shore of Lake Superior in the summer of 1848, and a fine female was captured in Cambridge, Mass., on the 20th of July, 1849, which have afforded the means for a more full and correct description than has heretofore been given. This species appears to be rare, and has not before been figured. It is the representative of the European *S. Tiliæ* and *Quercûs*.

Hepiolus argenteomaculatus H.

Pl. VII., fig. 6.

Hepialus argenteomaculatus Harris, Catalogue in Hitchcock's Report, 1st ed. p. 591 (1833).—The same, 2d ed. p. 592 (1835).—Report on Insects injurious to Vegetation, p. 295 (1841).—Gosse, Canadian Naturalist, p. 248 (1840).

Fusco-ochraceus vel cinereo-brunneus; alis primoribus pallidis, ochraceo vel brunneo fasciatis, guttisque duabus prope basin argenteis; secundariis rubro-vel cinereo-ochraceis, immaculatis.

Alar. exp. 2¾, 3¾ unc.

Only two specimens of this fine insect have fallen under my observation. They differ much in size and color. The smallest, apparently a male, was taken in Cambridge, Mass., many years ago. When at rest, the wings are very much deflexed, and form a steep roof over the back. The body is light brown; the forewings are of a very pale ashen brown color, variegated with darker clouds and oblique wavy bands, and are ornamented with two silvery white spots near the base, at the inner angles of the discoidal cells; the anterior spot being round and the posterior and larger one triangular. The hindwings are light ashen brown at base, passing into dusky ochre-yellow. The large specimen is a female, and was taken by Professor Agassiz on the northern shore of Lake Superior. The body is of a dusky ochre-yellow color, tinged on the sides and on the legs with red. The forewings are light rosy buff, with brownish ochre clouds and bands, two silvery spots near the base, and a whitish dot near the tip. The hindwings, above, and all the wings beneath, are of a deep ochre-yellow color, tinged with red.

The empty pupa-skins of this or of an allied species are sometimes found on our sea-beaches.

Arctia Parthenos H.

Pl. VII., fig. 4.

Alis primoribus fusco-brunneis, maculis sparsis lactifloreis; secundariis fulvo-flavis, basi, maculâ media triangulari, fasciâque postica undata nigris; abdomine supra fusco apice fulvo.

Alar. exp. unc. 2½.

Head brown, with a crimson fringe above and between the black antennæ. Thorax brown above, margined before with an arcuated cream-colored band, which is continued on each side of the outer edge of the shoulder-covers; upper edge of the collar crimson-red. Forewings dusky brown, with three small cream-colored spots on the outer edge; four spots of the same color in a line near the inner margin, and several more scattered on the disk. Hindwings deep ochre-yellow, with the base, the basal edge of the inner margin, a triangular spot in the middle, adjoining the basal spot, and a broad indented band behind, of a black color. Abdomen dusky above, tawny at tip and beneath. Legs dusky, thighs and tibiæ fringed with crimson-red hairs.

This fine species was taken on the northern shore of Lake Superior. It belongs to the same group as the European *Caja*, from all the known varieties of which it differs in having the arcuated white line on the thorax, and the black band on the hindwings. The situation of this band is not so far back as the black spots found on the hindwings of the allied species. The banded hindwings, with the entirely black or dusky antennæ, will sufficiently distinguish this species from the *Arctia Americana*, a description of which is here added for the purpose of comparison.

ARCTIA AMERICANA H.

Pl. VII., fig. 5.

Arctia Americana Harris, Report on Insects injurious to Vegetation, p. 246 (1841).

Alis primoribus brunneis, maculis, rivulisque albidis; secundariis fulvo-flavis, maculis unica media reniformi, tribusque posticis rotundis nigris; abdomine fulvo, dorso nigro-quadrimaculato.

Alar. exp. unc. 2½.

Head brown, antennæ white above, with brown pectinations. Thorax brown above, margined before with an arcuated yellowish white band, which is continued on the outer edge of the shoulder-covers; upper edge of the collar crimson-red. Forewings coffee-brown, with three yellowish white spots on the outer edge, and crossed by irregular anastomozing yellowish white lines. Hindwings bright ochre-yellow, with a large reniform central black spot, two round black spots behind, a third smaller spot near the anal angle, and a black dot between the middle and the inner margin. Abdomen tawny, with four blackish dorsal spots. Legs dusky, the thighs and anterior tibiæ fringed with red hairs; the hindmost tarsi whitish, annulated with black.

This species, which is now for the first time figured, was taken by Mr. Edward Doubleday, near Trenton Falls. From the *Caja* it is distinguished, like the *Parthenos*, by the arcuated white margin of the thorax, &c. The arrangement of the white spots and rivulets on the forewings is the same as in the European species.

Ennomos macularia H.

Pl. VII., fig. 3.

Flava; alis angulatis subdentatis, anticis apice sinuato-truncatis, prope basin apicemque brunneo maculato-fasciatis; omnibus postice macula magna rhomboidea brunnea marginem posticum angulumquè analem attingente.

Alar. exp. $1\frac{1}{4}$ unc.

This pretty Geometer has the form of *Ennomos* (*Eurymene*) *dolabraria*, and perhaps belongs to the same subgenus. It is found in Massachusetts as well as on the northern shore of Lake Superior.

The antennæ are brown, and are pectinated only in the males. The tongue is half as long as the body, which, with the upper side of the forewings, is citron-yellow; the hindwings and under sides are somewhat paler. The forewings have a rust-brown costal spot near the shoulders, a transverse row of spots near the base, a stigmatical dot, three little spots near the tip, and a very large lozenge-shaped spot at the anal angle, of the same brown color, the large spot being bordered before and behind with darker brown. The hindwings have a central brownish dot, and a large pale brown spot, bordered before and behind with a darker line at the anal angle, which also is deeply tinged with brown.

List of Lepidopterous Insects, taken by Professor L. Agassiz on the northern shore of Lake Superior.

I. Papiliones.

Pontia Oleracea *Harris*.
Colias Pelidne ? *Boisduval*.
" Chrysotheme ? *Esper*. var.? *Boisd*.
Polyommatus.
Limenitis Arthemis *Drury*.
Danaus Archippus *F*.
Argynnis Aphrodite *F*. (nec Daphnis, *Cr*., nec Cybele, *F*.)
Melitæa Myrina *Cramer*.
" Cocyta *Cr*.

Vanessa J. album *Boisd.*
" Cardui *L.*

II. Sphinges.

Ægeria exitiosa *Say.*
Deilephila Chamænerii *H.*
Sphinx (Lethia *Hübn.*) Kalmiæ *Smith—Abbott.*
Smerinthus modesta *H.*
Alypia octomaculata *F.*

III. Phalænæ.

1. *Bombyces.*

Lithosia (Eubaphe *Hübn.*) aurantiaca *Hübn.*
Arctia Parthenos *H.*
Clisiocampa silvatica *H.* var.
Hepiolus argenteomaculatus *H.* var.

2. *Noctuæ.*

Apatela.
Agrotis devastator *Brace.*
"
"
Noctua clandestina *H.*
Hadena amica *Stevens.*
"
"
Mamestra.
"
Heliothis.

3. *Geometræ.*

Crociphora transversata *Drury.*
Ennomos macularia *H.*
Zerene ?
"
Melanippe.
Cidaria ?
"
Also three more *Geometræ*, of undetermined genera.

4. *Pyralides.*

Macrochila pulveralis, *H.* Cat. ms.
Anania octomaculata ? *L.*

5. *Tortrices.*

Two species, undetermined.

6. *Tineæ.*

(Crambidæ.)

Crambus.

7. *Alucitæ.*

Pterophorus.

The collections of insects of other orders made during our excursion have not yet been sufficiently worked out to allow us to give an account of their contents. A considerable number of Neuroptera and Orthoptera have, however, been collected; Hymenoptera, Diptera, and Hemiptera, have also not been neglected, though of the latter chiefly Hydrocorisæ have been found.

The Crustacea, crawfishes, and other small freshwater shrimps, as well as the leeches and other worms, have also attracted our attention, and some interesting species have been collected; but the difficulty of establishing their synonymy induces me to postpone the publication of their description. L. A.

X.

THE ERRATIC PHENOMENA ABOUT LAKE SUPERIOR.

So much has been said and written within the last fifteen years, upon the dispersion of erratic boulders and drift, both in Europe and America, that I should not venture to introduce this subject again, if I were not conscious of having essential additions to present to those interested in the investigation of these subjects.

It will be remarked by all who have followed the discussions respecting the transportation of loose materials over great distances from the spot where they occurred primitively, that the most minute and the most careful investigations have been made by those geologists who have attempted to establish a new theory of their transportation by the agency of ice.

The part of those who claim currents as the cause of this transportation has been more generally negative, inasmuch as, satisfied with their views, they have generally been contented simply to deny the new theory and its consequences, rather than investigate anew the field upon which they had founded their opinions. Without being taxed with partiality, I may, at the outset, insist upon this difference in the part taken by the two contending parties. For since the publication of Sefstroem's paper upon the drift of Sweden, in which very valuable information is given respecting the phenomena observed in that peninsula, and the additional data furnished by de Verneuil and Murchison upon the same country and the plains of Russia, the classical ground for erratic phenomena has been left almost untouched by all except the advocates of the glacial theory. I need only refer to the investigations of M. de Charpentier, Escher, Von Derlinth and Studer, and more particularly to those extensive and most minute researches of Prof. Guyot in Switzerland, without speaking of my own and some contributions from visitors, as the Martins, James Forbes and others, to justify my assertion that no important fact respecting the loose materials spread all over Switzerland has been added by the advocates of currents since the days of Saussure, DeLüc, Escher and Von Buch ; whilst Prof. Guyot has most conclusively shown that the different erratic basins in Switzerland are not only distinct from each other, as was already known before, but that in each the loose materials are arranged in well-determined regular order, showing precise relations to the centres of distribution, from which these materials originated; an arrangement which agrees in every particular with the arrangement of loose fragments upon the surface of any glacier, but which no cause acting convulsively could have produced.*

The results of these investigations are plainly that the boulders found at a distance from the central Alps, originated from their higher summits and valleys, and were carried down at different successive periods in a regular manner, forming uninterrupted walls and ridges, which can be traced from their starting point to their extreme peripheric distribution.

I have myself shown that there are such centres of distribution in Scotland and England and Ireland. And these facts have been since traced in detail in various parts of the British Islands by Dr. Buckland, Sir Ch. Lyell, Mr. Darwin, Mr. McLachlan and Professor James D. Forbes, pointing clearly to the main mountain groups as to so many distinct centres of dispersion of these loose materials.

Similar phenomena have been shown in the Pyrenees, in the Black Forest, and in the Vosges, showing beyond question, that whatever might have been the cause of the dispersion of erratic boulders, there are several separate centres of their distribution to be distinguished in Europe. But there is another question connected with this local distribution of boulders which requires particular investigation, the confusion of which with the former has no doubt

*A comparison of the maps showing the arrangement of the moraines upon the glacier of the Aar in my *Système Glaciaire*, with the map which Prof. Guyot is about to publish of the distribution of the erratic boulders in Switzerland, will show more fully the identity of the two phenomena.

greatly contributed to retard our real progress in understanding the general question of the distribution of erratics.

It is well known that Northern Europe is strewed with boulders, extending over European Russia, Poland, Northern Germany, Holland and Belgium. The origin of these boulders is far north in Norway, Sweden, Lapland and Liefland, but they are now diffused over the extensive plains west of the Ural Mountains. Their arrangement, however, is such that they cannot be referred to one single point of origin, but only in a general way to the northern tracts of land which rise above the level of the sea in the Arctic regions. Whether these boulders were transported by the same agency as those arising from distinct centres, on the main continent of Europe, has been the chief point of discussion. For my own part, I have indeed no doubt that the extreme consequences to which we are naturally carried by admitting that ice was also the agent in transporting the northern erratics to their present positions, has been the chief objection to the view that the Alpine boulders have been distributed by glaciers.

It seemed easier to account for the distribution of the northern erratics by currents, and this view appearing satisfactory to those who supported it, they at once went further, and opposed the glacial theory even in those districts where the glaciers seemed to give a more natural and more satisfactory explanation of the phenomena. To embrace the whole question it should be ascertained.

First, Whether the northern erratics were transported at the same time as the local Alpine boulders, and if not, which of the phenomena preceded the other; and again, if the same cause acted in both cases, or if one of the causes can be applied to one series of these phenomena, and the other cause to the other series. An investigation of the erratic phenomena in North America seems to me likely to settle this question, as the northern erratics occur here in an undisturbed continuation over tracts of land far more extensive than those in which they have been observed in Europe. For my own part, I have already traced them from the eastern shores of Nova Scotia, through New England and the North Western States of North America and the Canadas as far as the western extremity of Lake Superior, a region embracing about thirty degrees of longitude. Here, as in Northern Europe, the boulders evidently originated farther north than their present location, and have been moved universally in a main direction from north to south.

From data which are, however, rather incomplete, it can be further admitted that similar phenomena occur further west across the whole continent, everywhere presenting the same relations. That is to say, everywhere pointing to the north as to the region of the boulders, which generally disappear about latitude 38°.

Without entering at present into a full discussion of any theoretical views of the subject, it is plain that any theory, to be satisfactory, should embrace both the extensive northern phenomena in Europe and North America, and settle the relation of these phenomena to the well-authenticated local phenomena of Central Europe.

Whether America itself has its special local circumscribed centres of distribution or not, remains to be seen. It seems, however, from a few facts observed in the White Mountains, that this chain, as well as the mountains of north-eastern New York, have not been exclusively—and for the whole duration of the transportation of these materials—under the influence of the cause which has distributed the erratics through such wide space over the continent of North America. But whether this be the case or not, (and I trust local investigations will soon settle the question,) I maintain that the cause which has transported these boulders in the American continent must have acted simultaneously over the whole ground which these boulders cover, as they present throughout the continent an uninterrupted sheet of loose materials, of the same general nature, connected in the same general manner, and evidently dispersed at the same time.

Moreover, there is no ground, at present, to doubt the simultaneous dispersion of the erratics over Northern Europe and Northern America. So that the cause which transported them, whatever it may be, must have acted simultaneously over the whole tract of land west of the Ural Mountains, and east of the Rocky Mountains, without assuming anything respecting Northern Asia, which has not yet been studied in this respect; that is to say, at the same time, over a space embracing two hundred degrees of longitude.

Again, the action of this cause must have been such, and I insist strongly upon this point, as a fundamental one, the momentum with which it acted must have been such, that after being set in motion in the north, with a power sufficient to carry the large boulders which are found everywhere over this vast extent of land, it vanished or was stopped after reaching the thirty-fifth degree of northern latitude.

Now it is my deliberate opinion that natural philosophy and mathematics may settle the question, whether a body of water of sufficient extent to produce such phenomena can be set in motion with sufficient velocity to move all these boulders, and nevertheless stop before having swept over the whole surface of the globe. Hydrographers are familiar with the action of currents, with their speed, and with the power with which they can act. They know also how they are distributed over our globe. And, if we institute a comparison, it will be seen that there is nowhere a current running from the poles towards the lower latitudes, either in the northern or southern hemisphere, covering a space equal to one-tenth of the currents which should have existed to carry the erratics into their present position. The widest current is west of the Pacific, which runs parallel to the equator, across the whole extent of that sea from east to west, and the greatest width of which is scarcely fifty degrees. This current, as a matter of course, establishes a regular rotation between the waters flowing from the polar regions towards lower latitudes.

The Gulf Stream on the contrary runs from west to east, and dies out towards Europe and Africa, and is compensated by the currents from Baffin's Bay and Spitzbergen emptying into the Atlantic, while the current of the Pacific, moving towards Asia and carrying floods of water in that direction, is maintained chiefly by antarctic currents, and those which follow the western shore of America from Behring's Straits. Wherever they are limited by continents, we see that the waters of these currents, even when they extend over hundreds of degrees of latitude, as the Gulf Stream does in its whole course, are deflected where they cannot follow a straight course.

Now without appealing with more detail to the mechanical conditions involved in this inquiry, I ask every unprejudiced mind acquainted with the distribution of the northern boulders, whether there was any geographical limitation to the supposed northern current to cause it to leave the northern erratics of Europe in such regular order, with a constant bearing from north to south, and to form, on its southern termination, a wide, regular zone from Asia to the western shores of Europe, north of the fiftieth degree of latitude, before it had reached the great barrier of the Alps? I ask whether there was such a barrier in the unlimited plains which stretch from the Arctic seas uninterrupted over the whole northern continent of America as far down as the Gulf of Mexico?

I ask, again, why the erratics are circumscribed within the northern limits of the temperate zone, if their transportation is owing to the action of water currents? Does not, on the contrary, this most surprising limit within the artic and northern temperate zones, and in the same manner within the antarctic and southern temperate zones, distinctly show that the cause of transportation is connected with the temperature or climate of the countries over which the phenomena were produced. If it were otherwise, why are there no systems of erratics with an east and west bearing, or in the main direction of the most extensive currents flowing at present over the surface of our globe?

It is a matter of fact, of undeniable fact, for which the theory has to account, that in the two hemispheres the erratics have direct reference to the polar regions, and are circumscribed within the arctics and the colder part of the temperate zone. This fact is as plain as the other fact, that the local distribution of boulders has reference to high mountain ranges, to groups of land raised above the level of the sea into heights, the temperature of which is lower than the surrounding plains. And what is still more astonishing, the extent of the local boulders, from their centre of distribution, reaches levels, the mean annual temperature of which corresponds in a surprising manner with the mean annual temperature of the southern limit of the northern erratics.

We have, therefore, in this agreement a strong evidence in favor of the view that both the phenomena of local mountain erratics in Europe and of northern erratics in Europe and America have probably been produced by the same cause.

The chief difficulty is in conceiving the possibility of the formation of a sheet of ice sufficiently large to carry the northern erratics into their present limits of distribution; but this difficulty is greatly removed when we can trace, as in the Alps, the progress of the boulders under the same aspect from the glaciers now existing, down into regions where they no longer exist, but where the boulders and other phenomena attending their transportation show distinctly that they once existed.

Without extending further this argumentation, I would call the attention of the unprejudiced observer to the fact, that those who advocate currents as the cause of the transportation of erratics, have, up to this day, failed to show, in a single instance, that currents can produce all the different phenomena connected with the transportation of the boulders which are observed everywhere in the Alps, and which are still daily produced there by the small glaciers yet in existence. Never do we find that water leaves the boulders which it carries along in regular walls of mixed materials; nor do currents anywhere produce upon the hard rocks *in situ* the peculiar grooves and scratches which we see everywhere under the glacier and within the limits of their ordinary oscillations.

Water may polish the rocks, but it nowhere leaves straight scratches upon their surface; it may furrow them, but these furrows are sinuous, acting more powerfully upon the soft parts of the rocks or fissures already existing; whilst glaciers smooth and level uniformly, the hardest parts equally with the softest, and, like a hard file, rub to uniform continuous surfaces the rocks upon which they move.

But now let us return to our special subject, the erratics of North America.

The phenomena of drift are more complicated about Lake Superior than I have seen them anywhere else; for, besides the general phenomena which occur everywhere, there are some peculiarities noticed which are to be ascribed to the lake as such, and which we do not find in places where no large sheet of water has been brought into contact with the erratic phenomena. In the first place, we notice about Lake Superior an extensive tract of polished, grooved and scratched rocks, which present here the same uniform character which they have everywhere. As there is so little disposition, among so many otherwise intelligent geologists, to perceive the facts as they are, whenever they bear upon the question of drift, I cannot but repeat, what I have already mentioned more than once, but what I have observed again here over a tract of some fifteen hundred miles, that the rocks are everywhere smoothed, rounded, grooved and furrowed in a uniform direction. The heterogeneous materials of which the rocks consist are cut to one continuous uniform level, showing plainly that no difference in the polish and abrasion can be attributed to the greater or less resistance on the part of the rocks, but that a continuous rasp cut down everything, adapting itself, however, to the general undulations of the country, but nevertheless showing, in this close adaptation, a most remarkable continuity in its action.

That the power which produced these phenomena moved in the main from north to south, is distinctly shown by the form of the hills, which present abrupt slopes, rough and sharp corners towards the south, while they are all smoothed off towards the north.

Indeed, here, as in Norway and Sweden, there is on all the hills a lee-side and a strike-side. As has been observed in Norway and Sweden, the polishing is very perfect in many places, sometimes strictly as brilliant as a polished metallic surface, and everywhere these surfaces are more or less scratched and furrowed, and both scratches and furrows are rectilinear, crossing each other under various angles: however, never varying many points of the compass on the same spot, but in general showing that where there are deviations from the most prominent direction, they are influenced by the undulations of the soil. It has been said, that the main direction of these striæ was from north-west to south-east, but I have found it as often strictly from north to south, or even from north-east to south-west; and if we are to express a general result, we should say that the direction, assigned by all our observations to the various scratches, tends to show that they have been formed under the influence of a movement from north to south, varying more or less to the east and west, according to local influences in the undulations of the soil. It is, indeed, a very important fact, that scratches which seem to have been produced at no great intervals from each other, are not absolutely parallel, but may diverge for ten, fifteen, or more degrees.

There is one feature in these phenomena, however, in which we never observe any variation. The continuity of these lines is absolutely the same everywhere. They are rectilinear and continuous, and cannot be better compared than with the effects of stones or other hard materials dragged in the same direction upon flat or rolling surfaces; they form simple scratches extending for yards in straight lines, or breaking off for a short space to continue again in a straight line in the same direction, just as if interrupted by a jerk. There are also deeper scratches of the same kind, presenting the same phenomena, only, perhaps, traceable for a greater distance than the finer ones. These scratches, instead of appearing like the tracing of diamonds upon glass, as the former do, would rather assume the appearance of a deeper groove, made by the point of a graver, or perhaps still more closely resemble the scratches which a cart-wheel would produce upon polished marble, if the wheel were chained, and coarse sand spread over the floor. The appearance of the rock, crushed by the moving mass, is especially distinct in limestone rocks, where grooves are seldom nicely cut, but present the appearance of a violent pressure combined with the grooving power, thus giving to the groove a character which is quite peculiar, and which at once strikes an observer who has been familiar with its characteristic aspect. Now, I do not know upon what the assertions of some geologists rest, that gravel moved by water under strong heavy currents will produce similar effects. Wherever I have gone since studying these phenomena, I have looked for such cases, and have never yet found modern gravel currents produce anything more than a smooth surface with undulating furrows following the cracks in the rocks, or hollowing their softer parts; but continuous straight lines, especially such crushed lines and straight furrows, I have never seen.

When we know how extensive the action of water carrying mud and gravel is on every shore and in every water current,—when we can trace this action almost everywhere, and nowhere find it similar to the phenomena just described, I cannot imagine upon what ground these phenomena are still attributed to the agency of currents. This is the less rational as we have at present, in all high mountain chains of the temperate zone, other agents, the glaciers, producing these very same phenomena, with precisely the same characters, to which, therefore, a sound philosophy should ascribe, at least conditionally, the northern and Alpine polished surfaces, and scratched and grooved rocks, or at least acknowledge that the effect produced by the action of glaciers more nearly resembles these erratic phenomena than does that which results from the action of currents. But such is the prejudice of many geologists, that those keen faculties of distinction and generalization, that power of superior perception and discrimination which have led them to make such brilliant discoveries in geology in general, seem to abandon them at once as soon as they look at the erratics. The objection made by a venerable geologist, that the cold required to form and preserve such glaciers, for any length of time, would freeze him to death, is as childish as the apprehension that the heavy ocean currents, the action of which he sees everywhere, might have swept him away.*

Now that these phenomena have been observed extensively, we may derive also some instruction from the limits of their geographical extent. Let us see, therefore, where these polished, scratched and furrowed rocks have been observed.

In the first place they occur everywhere in the north within certain limits of the arctics, and through the colder parts of the temperate zone. They occur also in the southern hemisphere, within parallel limits, but in the plains of the tropics, and even in the warmer parts of the temperate zone we find no trace of these phenomena, and nevertheless the action of currents could not be less there, and could not at any time have been less there than in the colder climates. It is true, similar phenomena occur in Central Europe and have been noticed in Central Asia, and even in the Andes of South America, but these always in higher regions, at definite levels above the surface of the sea, everywhere indicating a connection between their extent and the colder temperature of the places over which they are traced.

More recently, a step towards the views I entertain of this subject, has been made by those geologists who would ascribe them to the agency of icebergs. Here, as in my glacial theory, ice is made

* Berlin Academy, 1846.

the agent; floating ice is supposed to have ground and polished the surfaces of rocks, while I consider them to have been acted upon by terrestrial glaciers. To settle this difference we have a test which is as irresistible as the other arguments already introduced.

Let us investigate the mode of action, the mode of transportation of icebergs, and let us examine whether this cause is adequate to produce phenomena for which it is made to account. As mentioned above, the polished surfaces are continuous over hills, and in depressions of the soil, and the scratches which run over such undulating surfaces are nevertheless continuous in straight lines. If we imagine icebergs moving upon shoals, no doubt they would scratch and polish the rocks in a way similar to moving glaciers. But upon such grounds they would sooner or later be stranded, and if they remained loose enough to move, they would, in their gyratory movements, produce curved lines, and mark the spots where they had been stranded with particular indications of their prolonged action. But nowhere upon arctic ground do we find such indications. Everywhere the polished and scratched surfaces are continuous in straight juxtaposition.

Phenomena analogous to those produced by icebergs would only be seen along the sea-shores; and if the theory of drifted icebergs were correct, we should have, all over those continents where erratic phenomena occur, indications of retreating shores as far as the erratic phenomena are found. But there is no such thing to be observed over the whole extent of the North American continent, nor over Northern Europe and Asia, as far as the northern erratics extend. From the arctics to the southernmost limit of the erratic distribution, we find nowhere the indications of the action of the sea as directly connected with the production of the erratic phenomena. And wherever the marine deposits rest upon the polished surfaces of ground and scratched rocks, they can be shown to be deposits formed since the grooving and polishing of the rocks, in consequence of the subsidence of those tracts of land upon which such deposits occur.

Again, if we take for a moment into consideration the immense extent of land covered by erratic phenomena, and view them as produced by drifted icebergs, we must acknowledge that the icebergs of the *present period* at least, are insufficient to account for them, as they are limited to a narrower zone. And to bring icebergs in any way within the extent which would answer for the extent of the distribution of erratics, we must assume that the northern ice fields, from which these icebergs could be detached and float southwards, were much larger at the time they produced such extensive phenomena than they are now. That is to say, we must assume an ice period; and if we look into the circumstances we shall find that this ice period, to answer to the phenomena, should be nothing less than an extensive cap of ice upon both poles. This is the very theory which I advocate; and unless the advocates of an iceberg theory go to that length in their premises, I venture to say, without fear of contradiction, that they will find the source of their icebergs fall short of the requisite conditions which they must assume, upon due consideration, to account for the whole phenomena as they have really been observed.

But without discussing any farther the theoretical views of the question, let me describe more minutely the facts as observed on the northern shores of Lake Superior. The polished surfaces, as such, are even, undulating, and terminate always above the rough lee-side turned to the south, unless upon gentle declivities, where the polished surfaces extend in unbroken continuity upon the southern surfaces of the hills, as well as upon their northern slopes. On their eastern and western flanks, shallow valleys running east and west are as uniformly polished as those which run north and south; and this fact is more and more evident, wherever scratches and furrows are also well preserved and distinctly seen, and by their bearings we can ascertain most minutely, the direction of the onward movement which produced the whole phenomena. Nothing is more striking in this respect than the valleys or depressions of the soil running east and west, where we see the scratches crossing such undulations at right angles, descending along the southern gentle slope of a hill, traversing the flat bottom below, and rising again up the next hill south, in unbroken continuity. Examples of the kind can be seen everywhere in those narrow inlets, with shallow waters intersecting the innumerable highlands along the northern shores of Lake Superior, where the scratches and furrows can be traced under water from one shore to the other, and where they at times ascend steep hills, which they cross at right angles along their northern slope, even when the southern slope, not steeper in itself, faces the south with rough escarpments.

The scratches and furrows, though generally running north and south, and deviating slightly to the east and west, present in various places remarkable anomalies, even in their general course along the eastern shore of the lake. Between Michipicotin and Sault St. Marie we more frequently see a deflection to the west than a due north and south course, which is rather normal along the northern shore proper, between Michipicotin and other islands, and from the Pic to Fort William; the deep depression of the lake being no doubt the cause of such a deviation, as large masses of ice could accumulate in this extensive hollow cavity before spreading again more uniformly beyond its limits. To the oscillations of the whole mass in its southerly movement, according to the inequalities of the surfaces, we must ascribe the crossing of the straight lines at acute angles, as we observe also at the present day under the glaciers, as they swell and subside, and hence meet with higher and lower obstacles in their irregular course between the Alpine valleys.

In deep, narrow chasms, however, we find now and then greater deviations from the normal direction of the striæ, where considerable masses of ice could accumulate, and move between steep walls under a lateral pressure of the masses moving onwards from the north. Such a chasm is seen between Spar Island and the main land opposite Prince's Location, south of Fort William, where the furrows and scratches run nearly east and west. But here also, there is no tumultuous disturbance in the continuation of the phenomena, such as would occur if icebergs were floated and stranded against the southern barrier. The same continuity of even, polished surfaces, with their scratches and furrows, prevails here as elsewhere. The angles which these scratches form with each other are very acute, generally not exceeding 10°; but at times they diverge more, forming angles of 15°, 20° and 25°. In a few instances, I have even found localities where they crossed each other at angles of no less than 30°; but these are rare exceptions. It may sometimes be noticed that the lines running in one direction form a system by themselves, varying very little from strict parallelism with each other, but crossing another system, more or less strongly marked, of other lines equally parallel with each other. At other times, a system of lines, strongly marked and diverging very slightly, seem to pass over another system, in which the lines form various angles with each other. Again, there are places,—and this is the most common case,—where the lines diverge slightly, following, however, generally one main direction, which is crossed by fewer lines, forming more open angles. These differences, no doubt, indicate various oscillations in the movement of the mass which produced the lines, and show probably its successive action, with more or less intensity, upon the same point at successive periods, in accordance with the direction of the moving force at each interval. The same variations within precisely the same limits may be noticed in our day on the margin of the glaciers produced by the increase or diminution of the bulk of their mass, and the changes in the rate of their movement.

The loose materials which produced, in their onward movement under the pressure of ice, such polishing and grooving, consisted of various sized boulders, pebbles and gravels, down to the most minute sand and loamy powder. Accumulations of such materials are found everywhere upon these smooth surfaces, and in their arrangement they present everywhere the most striking contrast when compared with deposits accumulated under the agency of water. Indeed, we nowhere find this glacial drift regularly stratified, being everywhere irregular accumulations of loose materials, scattered at random without selection, the coarsest and most minute particles being piled irregularly in larger or smaller heaps, the greatest boulders standing sometimes uppermost, or in the centre, or in any position among smaller pebbles and impalpable powder.

And these materials themselves are scratched, polished and furrowed, and the scratches and furrows are rectilinear as upon the rocks *in situ* underneath, not bruised simply, as the loose materials carried onward by currents or driven against the shores by the tides, but regularly scratched, as fragments of hard materials would be if they had been fastened during their friction against each other, just as we observe them upon the lower surface of glaciers where all the loose materials set in ice, as stones in their setting, are pressed and rubbed against underlying rocks. But the setting here being simply ice, these loose materials, fast at one time and movable another, and fixed and loosened again, have rubbed agamst the rock below in all possible positions; and hence not only their rounded form, but also their rectilinear grooving. How such grooves could be produced under the action of currents, I leave to the advocates of such a theory to show, as soon as they shall be prepared for it.

I should not omit here to mention a fact which, in my opinion, has a great theoretical importance, namely, that in the northern erratics, even the largest boulders, as far as I know, are rounded, and scratched and polished, at least, all those which are found beyond the immediate vicinity of the higher mountain ranges; showing that the accumulations of ice which moved the northern erratics covered the whole country; and this view is sustained by another set of facts equally important, namely, that the highest ridges, the highest rugged mountains, at least, in this continent and north of the Alps in Europe, are as completely polished and smoothed as the lower lands, and only a very few peaks seem to have risen above the sheet of ice; whilst, in the Alps, the summits of the mountains stand generally above these accumulations of ice, and have supplied the surface of the glaciers with large numbers of angular boulders, which have been carried upon the back of glaciers to the lower valleys and adjacent plains without losing their angular forms.

With respect to the irregular accumulation of drift-materials in the north, I may add that there is not only no indication of stratification among them, such unquestionably as water would have left, but that the very nature of these materials shows plainly that they are of terrestrial origin; for the mud which sticks between them adheres to all the little roughnesses of the pebbles, fills them out, and has the peculiar adhesive character of the mud ground under the glaciers, and differing entirely in that respect from the gravels and pebbles and sands washed by water currents, which leave each pebble clean, and never form adhering masses, unless penetrated by an infiltration of limestone.

Another important fact respecting this glacial drift consists in the universal absence of marine as well as freshwater fossils in its interior, a fact which strengthens the view that they have been accumulated by the agency of strictly terrestrial glaciers; such is, at least, the case everywhere far from the sea-shore. But we may conclude that these ancient glaciers reached, upon various points, the sea-shore at the time of their greatest extension, just as they do at present in Spitzbergen and other arctic shores; and that therefore, in such proximity, phenomena of contact should be observed, indicating the onward movement of glacial material into the ocean, such as the accumulation within these materials of marine fossil remains, and also the influence of the tidal movements upon them. And now such is really the case. Nearer the sea-shores we observe distinctly, in some accumulations of the drift, faint indications of the action of the tide reaching the lower surface of glaciers, and the remodeling, to some extent, of the materials which there poured into the sea. A beautiful example of the kind may be observed near Cambridge, along Charles River, not far from Mount Auburn, where the unstrati-

fied glacial drift (*a*) presents in its upper masses strictly the characters of true terrestrial glacial accumulation, but shows underneath faint indications (*b*) of the action of tides. Above, regular tidal strata (*c*) are observed, formed probably after the masses below had subsided. The surface of this accumulation is covered with soil (*d*).

The period at which these phenomena took place cannot be fully determined, nor is it easy to ascertain whether all glacial drift is contemporaneous. It would seem, however, as if the extensive accumulation of drift all around the northern pole in Europe, Asia and America was of the same age as the erratics of the Alps. The climatic circumstances capable of accumulating such large masses of ice around the north pole, having, no doubt, extended their influence over the temperate zone, and probably produced, in high mountain chains, as the Alps, the Pyrenees, the Black Forest, and the Vosges, such accumulations of snow and ice, as may have produced the erratic phenomena of those districts. But extensive changes must have taken place in the appearance of the continents over which we trace erratic phenomena, since we observe in the Old World, as well as in North America, extensive stratified deposits containing fossils which rest upon the erratics; and as we have all possible good reasons and satisfactory evidence for admitting that the erratics were transported by the agency of terrestrial glaciers, and that therefore the tracts of land over which they occur, stood at that time above the level of the sea, we are led to the conclusion that these continents have subsided since that period below the level of the sea, and that over their inundated portions animal life has spread, remains of organized beings have been accumulated, which are now found in a fossil state in the deposits formed under those sheets of water.

Such deposits occur at various levels in different parts of North America. They have been noticed about Montreal, on the shores of Lake Champlain, in Maine and also in Sweden and Russia; and, what is most important, they are not everywhere at the same absolute level above the surface of the ocean, showing that both the subsidence, and the subsequent upheaval which has again brought them above the level of the sea, have been unequal; and that we should therefore be very cautious in our inferences respecting both the continental circumstances under which the ancient glaciers were formed, and also the extent of the sea afterward, as compared with its present limits.

The contrast between the unstratified drift and the subsequently stratified deposits is so great, that they rest everywhere unconformably upon each other, showing distinctly the difference of the agency under which they were accumulated. This unconformable superposition of marine drift upon glacial drift is also beautifully shown at the above mentioned locality near Cambridge. (See Diagram.) In this case the action of tides in the accumulation of the stratified materials is plainly seen.

The various heights at which these stratified deposits occur, above the level of the sea, show plainly, that since their accumulation, the main land has been lifted above the ocean at different rates in different parts of the country; and it would be a most important investigation to have their absolute level, in order more fully to ascertain the last changes which our continents have undergone.

From the above mentioned facts, it must be at once obvious that the various kinds of loose materials, all over the northern hemisphere, have been accumulated, not only under different circumstances, but during long-continued subsequent distinct periods, and that great changes have taken place since their deposition, before the present state of things was fully established.

To the first period,—the ice period, as I have called it,—belong all the phenomena connected with the transportation of erratic boulders, the polishing, scratching and furrowing of the rocks and the accumulation of unstratified, scratched, and loamy drift. During that period, the main land seems to have been, to some extent at least, higher above the level of the sea than now; as we observe, on the shores of Great Britain, Norway and Sweden, as well as on the eastern shores of North America, the polished surfaces dipping under the level of the ocean, which encroaches everywhere upon the erratics proper, effaces the polished surfaces and remodels the glacial drift. During these periods, large terrestrial animals lived upon both continents, the fossil remains of which are found in the drift of Siberia, as well as of this continent. A fossil elephant recently discovered in Vermont adds to the resemblance, already pointed out, between the northern drift of Europe and that of North America; for fossils of that genus are now known to occur upon the northernmost point of the western extremity of North America, in New England, in Northern Europe, as well as all over Siberia.

To the second period we would refer the stratified deposits resting upon drift, which indicate that during their deposition the northern continent had again extensively subsided under the surface of the ocean.

During this period, animals, identical with those which occur in the northern seas, spread widely over parts of the globe which are now again above the level of the ocean. But, as this last elevation seems to have been gradual, and is even still going on in our day, there is no possibility of tracing more precisely, at least for the present, the limit between that epoch and the present state of things. Their continuity seems almost demonstrated by the identity of fossil shells found in these stratified deposits, with those now living along the present shores of the same continent, and by the fact that changes in the relative level between sea and main land are still going on in our day.

Indications of such relative changes between the level of the waters and the land are also observed about Lake Superior. And here they assume a very peculiar character, as the level of the lake itself, in its relation to its shores, is extensively changed.

All around Lake Superior we observe terraces at different levels; and these terraces vary in height, from a few feet above the present level of the lake, to several hundred feet above its surface, presenting everywhere undoubted evidence, that they were formed by the waters of the lake itself.

As everywhere the lake shores are strewed with sand and pebbles stranded within certain limits by the waves, the lowest accumulations of loose materials remain within the action of heavy storms, and within such limit they are entirely deprived of vegetation.

Next, another set of beaches is observed, consisting generally of coarser materials, forming shelves above the reach of even the severest storms, as shown by the scanty cryptogamous vegetation, and a few small herbaceous plants which have grown upon them.

Next, other beaches, retreating more and more from the shores, are observed, upon which an older vegetation is traced, consisting of shrubs, small trees, and a larger number of different plants, among which extensive carpets of wonderful lichens sometimes spread over large surfaces of greater extent. And the gentle slope of some of the terraces shows that the lake must have stood at this level for a longer time, as higher banks rise precipitously above them, consisting also of loose materials, which must have been worn out and washed away, for a considerable time, by the action of the waves from the lake. In such a manner, terrace above terrace may be observed, in retreating sheltered bays or along protected shores, over extensive tracts; sometimes two or three in close proximity, perhaps within twenty to fifty feet of each other; and again, extensive flat shores, spreading above to another abrupt bank, making the former shore, above which other and other terraces are seen ; six, ten, even fifteen such terraces may be distinguished on one spot, forming, as it were, the steps of a gigantic amphitheatre. The most remarkable of all the amphitheatres has been sketched by Mr. Cabot, and forms the frontispiece to this volume. Its height has been determined by Mr. Logan, in his Geographical Report of Canada, page 10, where it is minutely described. I therefore refer to this account for further details. I would only mention here, that the first shelf, within the reach of the lake, consists of minute sand, and forms a narrow strip of sterile ground along the water-edge ; next, we have a slope of about 10°, followed by a flat terrace, extending for nearly fifty paces to a seconl very steep slope, about 26° and 30° inclination ; then, a sloping terrace with an inclination of near 16°, stretching for eighty to a hundred paces, above which rises another steep slope of 20°, beyond which an extensive flat, slightly sloping, extends for several hundred paces, crowned by some irregular ridges at its summit, and along the rocky ledges which form the bay at the bottom of which this high gravel bank rises.

In connection with these lake terraces, we must consider also the river terraces which present similar phenomena along their banks all around the lake, with the difference that they slope gradually along the water courses, otherwise resembling in their composition the lake terraces, which are altogether composed of remodeled glacial drift, which, from the influence of the water and their having been rolled on the shores, have lost, more or less, their scratches and polished appearance, and have assumed the dead smoothness of water pebbles. Such terraces occur frequently between the islands, or cover low necks connecting promontories with the main land, thus showing, on a small scale, how by the accumulation of loose materials, isolated islands may be combined to form larger ones, and how, in the course of time, by the same process, islands may be connected with the main land.

The lake shores present another series of interesting phenomena, especially near the mouth of larger rivers emptying into the lake over flats, where parallel walls of loose materials, driven by the action of the lake against the mouth of the river, have successively stopped its course and caused it to wind its way between the repeated accumulations of such obstacles.

The lower course of Michipicotin River is for several miles dammed up in that way by concentric walls, across which the river has cut its bed, and winding between them, has repeatedly changed its direction, breaking through the successive walls in different places. The largest and lowest of these walls, a kind of river terrace near the margin of the lake, shuts at present the factory from the immediate lake shore and the river, which has cut its way between the rocks to the right and the walls, has left a bold bank in this dam on its left shore.

An important question now arises, after considering these facts, how these successive changes in the relative level of the lake and its shores have been introduced. Has the water been gradually subsiding, or has the shore been repeatedly lifted up? Merely from the general inferences of the more extensive phenomena described above, respecting the relative changes between land and sea, I should be inclined to admit that the land has risen, rather than to suppose that the waters have gradually flowed out. But there are about the lake itself sufficient proofs, which leave in my mind not the slightest doubt that it is the land which has changed its level, and not the lake which has subsided.

In the first place, to suppose that the lake had once stood as high as the highest terraces, it would be necessary to admit that its banks were, all round its shores, sufficiently high to keep the water at that highest level, or, at least, that there were, at the lower outlets, bars to that height, which have been gradually removed since. But neither is the main land sufficiently high, at the western extremity and along the southern shores, to admit of such a supposition, nor is there about the outlet of the lake, between Gros Cap and Cap Iroquois, an indication of a barrier which has been gradually removed. There, as everywhere along the lake shores, the loose movable materials consist of the same drift, the accumulation of which, at various levels, we are aiming to account for. If, therefore, we consider this same drift as the barrier under whose protection the lake modeled other parts of its mass, we should be compelled to admit another cause to remove the barrier, a supposition for which there is not the slightest indication in the geological structure of the country. But if, on the contrary, we suppose the lake to have removed the barrier, there is no cause left for its accumulation, and the changes in the comparative level of the main land and the terraces remain equally unaccounted for.

Indeed, the terraces are so unequal in their absolute level when compared to each other, that a gradual subsidence of the lake removing a barrier of loose materials at its outlet could never explain their irregularity. But if we suppose that the innumerable dykes which cross, in all directions, the rocks which form the shores of the lake, have at various intervals lifted up these shores, we have at the same time a cause for the change of the relative level between the terraces and the lake, and also for the change of its absolute level, as it removed larger and larger portions of materials accumulated at its eastern extremity.

That these dykes have produced such changes will not be doubted by any one who may study the phenomena described in the following chapter respecting the origin of the present outlines of the lakes, as produced by the intersection of all the dykes traversing the metamorphic and plutonic rocks of the northern shores.

We should therefore conclude that, as there has been a general gradual change between the relative level of the main land and sea, so there has also been a gradual local change in the relative level of the lake and its shores; and hence the local phenomena would only corroborate the induction derived from more general geological facts.

Systems of Dykes.

No. 1 System of Michipicoten E-W.
No. 2 System of the Pic N 30° W.
No. 3 System of Neepigon N-S.
No. 4 System of Black Bay N. 30° E.
No. 5 System of Thunder Cape E. 30° N
No. 6 System of Isle Royale E 45° N

Glacial Scratches
Lines & arrows indicating their direction

No. 3 No. 3
No. 4
No. 4
No. 2
No. 1
No. 5
No. 6
No. 6
No. 1
No. 1
No. 6
Neepigon B.
Black B.
Thunder B.
Fort William
Thunder C.
Pt. Porphyry
Pigeon R.
Isle Royale
Pic R.
Slate Is.
Otter I.
Michipicoten Harb.
Michipicoten I.
C. Choyer
C. Gargantua
Montreal Is.
Montreal R.
Mica B.
Mamainse
Batchewanaung B.
Fond du Lac
La Pointe
Pt. Keewawanona
Gulais B.
White Fish Pt.
Falls of St. Mary
Mud Lake
I. of St. Joseph
Michilimackinac

A. Sonrel on stone

Tappan & Bradford's lith.

XI.

THE OUTLINES OF LAKE SUPERIOR.

SINCE it has been ascertained that the present form of the surface of our globe, and the distribution of land and water and their relative level, and the general outline of their contact, is the result of the successive geological changes which our globe has undergone, the efforts of geologists have more or less had in view to ascertain the order of succession of these phenomena, and their mutual dependence. One result is already established beyond question, namely, that the changes which have brought about the present physical state of our globe have been successive and gradual, and have followed each other at more or less remote epochs. So that its present configuration, far from being the result of one creative act, must be considered as the combination of a series of successive changes; fa from being moulded like a bell at one furnace, it has been built up by successive superstructures. This is not merely a view adopted in accordance with our theories and preferences, but it is actually shown by geological evidence, that the solid parts which constitute the crust of our globe have been consolidated at different epochs, and have been lifted to the surface above the level of the sea at long distant intervals; so that continents are known to have been built up by the successive rise of groups of islands, combining, by their gradual elevation above the level of the sea, into larger tracts of main land, until they have assumed their present definite outline and general relations.

The modes in which these changes have taken place have been quite diversified. We have indications of large tracts of land extending in horizontal continuity over great extents at considerable heights above the level of the sea.

We have in other instances, ridges of mountain chains intersecting the plains and forming prominent walls in various directions across the more level country. We have again isolated peaks rising like pyramids above the surrounding country,—shallow waters covering large flats,—deep excavations extending over considerable parts of the ocean,—or narrow chasms, precipitous holes increasing the diversity of the bottom of the sea, as mountain chains, volcanic cones, high plateaus, deep valleys, rolling hills, and flat plains modify the aspect of the main land. And all these differences, all these peculiar features have been introduced gradually and successively by the combined action of the elevation of the land, and recession of the sea; by the uplifting of the solid crust by volcanic and plutonic action, and by the abrading influence of water currents, and the regular undulations of the ocean tides.

Taking the whole globe in its general appearance, we can thus trace to the agency of a few influences, repeated at long intervals in different ways, all the phenomena we observe upon its surface. And the order of succession of the isolated events which have thus modified the surface of our globe has been ascertained with such unexpected precision, that at present, the relative age of the different geological events is established with as much certainty as the great periods in the history of mankind.

There is, however, one direction in which these investigations need to be followed out still farther. The secondary events of minor extent and less prominent importance have to be studied with the same precision, and perhaps with even more detail, than the general phenomena have been, up to the present time. After working out the general history of our globe, we have, as it were, to write its memoirs, the anecdotic part of the relation, and try to contribute in this minute investigation to a fuller illustration of its history. After ascertaining, in a general way, that the elevation of mountain chains, the rise of extensive tracts of land, have marked out the general outlines of continents and their limits with reference to the ocean; knowing, for instance, that the Scandinavian Alps determine the general form of Norway and Sweden; that Spain is separated from France by a high mountain range; that it owes its square form to the direction of its mountain chains precisely as Italy derives its form from the direction of its mountains; after having satisfied ourselves that the existence of an almost unbroken chain of the highest mountains, over the centre of Europe and Asia, constitutes the main difference in the physical features of the Old World, when contrasted with those of America, where the principal mountains run north and south; after having thus ascertained the intimate relation there is in general, between geological phenomena and the geography of continents, the physical features of the different parts of the world, it is a subject worthy of our attention to investigate how far the particular features we may distinguish in a given circumscribed locality may be ascribed to similar agencies, and to subordinate influences depending upon the same general principles, which have been active in the production of the general frame.

Are the Swiss lakes, for instance, with their peculiar form, as naturally the consequence of geological phenomena as the general features of the country? Are the numerous fiords of Norway and Maine owing to the same cause? Is there any connection which can be appreciated with any degree of precision between the general course of rivers on one continent, or in various parts of the same continent? And can a single lake, for instance Lake Superior, be analyzed, so as to refer the bearings of its outlines to precise geological phenomena?

The knowledge I had before visiting Lake Superior, of the direct connection of many of these apparently subordinate features in the physical aspect of a country, with the main geological phenomena upon which it rests, led me, during my excursions on this continent, to keep this subject constantly in view. I had seen how the Lakes of Neuchatel and Bienne were excavated at the junction of the Jura, and the tertiary deposit at its base; I had noticed that the Alpine lakes followed fissures at right angles with the axis of elevation of the Alps. I was aware that some of these lakes consist of two distinct parts, probably formed at different periods, but now united by the sheet of water filling them.

With such intimations, the great Canadian lakes, which form so naturally a boundary between the Northern United States and the British possessions upon this continent, could not but strongly call for an investigation of their natural features; some running east and west, others straight north and south, and others forming a regular crescent, with its convexity turned northwards. Their absolute position is at once characteristic. They are excavated chiefly between the plutonic masses rising north, and the stratified deposits south of the primitive range.

Lake Superior, especially, fills a chasm between the northern granitic and metamorphic range, and the oldest beds deposited along their southern slopes in the primitive age of this continent. Lake Ontario and Lake Erie, on the contrary, run between the successive layers of different sets of beds of the same great geographical period; while Lakes Huron and Michigan fill up the cracks which run at right angles with the main northern primitive range, and which, no doubt, owe their origin to the elevation of the chains north of Lake Huron and Lake Superior; repeating, on a large scale, what has been said above of the dependence of the Swiss lakes upon their geological positions and relation to the mountain chains which encircle them.

Besides this general relation of the lakes in connection with their shores, I have been able to trace a more intimate connection of the outlines of their shores and their geological structure, especially in Lake Superior.

As a whole, that lake resembles a large crescent, with its convexity turned northwards; but it were a great mistake to imagine that this form is actually the form of the shores, or that it is repeated upon every point. On the contrary, the general outline of that lake is the accidental result of the combination of many details, of many geological events which have followed each other at different periods, have modified the tract of land where the lake now exists, and have cut up its foundation in such a manner as to break the continuity of the solid rock, and allow it to be decomposed. Thus an extensive crescent-shaped hole with innumerable islands has been formed, in which the islands, in their various bearings, still indicate the direction of the intersecting masses, and appear at present as the fragmentary remains of a continuous tract of land, which is now replaced by a deep lake.

For many weeks I had been tracing the dykes which intersect the shores of Lake Superior in almost all directions, when I was one day most forcibly struck with the fact, that these dykes agree, in their bearings, with the bearings of the shores; and that even the greatest complications in the outlines of the shores could be accounted for, by the combinations of dykes intersecting each other in different directions. And indeed, now that I have the key for such an analysis, I find no difficulty in referring, even short lines of the coast, to the different systems of dykes which I know to exist there, and wherever my memoranda are sufficiently full, I find indications of dykes running in the direction of the coast. As soon as my attention had been called to these phenomena, I lost no opportunity of investigating the nature of the rock of these different systems of dykes, and I ascertained, to my great astonishment, that there are considerable differences in their mineralogical characters; some being amphibolic trap; others being injected with epidote; others having more the appearance of pitchstone; and, what is particularly interesting, the dykes which run in the same direction preserve the same mineralogical character, as well as the same bearing.

The systems of dykes which run directly north and south, and which form the inlets between Neepigon Bay and the main lake, and intersect the large island of St. Ignace, and separate St. Ignace stielf from the main land, all run north and south, and consist of very hard, tough, unalterable hornblende trap, of a crystalline aspect, and a grayish color; while the dykes, which run east and west, and mark out the northern and southern shores of those same islands, consist mostly of a greenish trap extensively injected with epidote, and breaking with the greatest ease into angular, irregular fragments. The northern shore east of the Pic has the same general bearing, due east and west; and here, also, we find the dykes more or less epidotic, and the metamorphic rocks talcose.

Again, the long shore running due east and west from Michipicotin westwards, is, also, along its whole extent, intersected by epidotic dykes running east and west.

The dykes of the north-eastern coast of the lake between the Pic and Michipicotin Island, which run north north-east to south south-west, consist of a pitchstone trap, like black glass, which, notwithstanding its external hardness, readily decomposes, and forms almost everywhere along these shores, coves, deep coves, narrow, straight inlets, small caves, and gives to the whole extent of that shore that peculiar aspect which distinguishes it so much from the other parts of the lake.

The more precipitous shores—almost vertical walls, and those peculiar modes of decomposition of the rocks which have left strange appearances in the masses, some of which have even been noticed by the Indian voyageurs, as Otter Head, for instance—the numberless exceedingly small islands of these shores, and the striking baldness of the overhanging rocks, are all of them most remarkable features. Though these examples are very striking, and may at once satisfy the mind that the most minute details in the peculiar features of the lake may be ascribed to geological agency, we nevertheless find still more striking evidence of this connection between the geological structure of the country and its form, along the north-western shore, west of St. Ignace, and between Isle Royale and Fort William. Three other systems of dykes here intersect the rocks, and give to the whole shore an entirely different aspect. At first sight, the bearings of the north-westerly shore appear already different from those of the northern shore proper, and the eastern shore, as their general course is north-east and south-west from the southern extremity of St. Ignace to Pigeon Bay, to which Isle Royale is parallel. But upon a close examination of these shores, it becomes obvious that this general feature is modified in various ways by the lines of the shore intersecting each other at acute angles, in three directions, and each of these different directions correspond exactly to as many systems of independent dykes. The eastern and western shores of Thunder Bay, or rather of the peninsula of Thunder Cape, run north-east, and parallel to them we have the cliffs of the shores south of Fort William, and west of Pic Island, which present the same bearings, as well as the shores of Black Bay also. The dykes which run in that direction are narrow belts of black trap. Nearly in the same direction, and very different in their mineralogical character, we find another set of dykes which run almost due north-east and south-west. The direction of these dykes is best indicated by a series of islands south of Sturgeon Bay, forming several parallel ridges, one of which consists of a series of small islands known under the name of Victoria and Spar Islands, and the other islands continuous with Sturgeon Island, in the prolongation of which we meet the most prominent dykes of Pic Island itself. The whole of Isle Royale lies in that direction, and the numerous promontories of its eastern extremity are particularly remarkable for their agreement, both in direction and geological structure, with the Victoria group of islands. The system is particularly rich in copper ores, and presents the most beautiful development of spathic veins. As I have not myself examined Point Keewenaw, I cannot say how far the prominent ridges there agree with those of Isle Royale and the Victoria Islands; but the agreement in the direction of the promontory itself is most striking; and the fact that this is the main centre of copper injections suggests the probability that Point Keewenaw also belongs, in its principal features, to this system; and I should not be in the least surprised if La Pointe and Whitefish Point derive their main features from dykes of the same system, though their solid foundation is concealed by accumulations of sand. The third system in this north-eastern shore runs east north-east near east, and is particularly marked along the southern shore of Thunder Cape peninsula, along which the dykes are nearly east and west, as just mentioned, deviating sufficiently to the north, however, to be clearly distinct from the dykes which form the shores from the Pic to St. Ignace, or from Michipicotin to Otter Head. And the nature of the rock of these dykes differs widely from the last, there being no epidotic injections accompanying them, and the trap being, on the contrary, of a light grayish color, resembling more the system which runs due north and south than any other.

So we have here six distinct systems of dykes, which contribute mainly to the formation of the northern shore of Lake Superior.

1. System of Michipicotin, running east and west. (See the annexed chart of the Outlines of Lake Superior.)
2. System of the Pic running north 30° west.
3. System of Neepigon, running due north and south.
4. System of Black Bay, running north 30° east.
5. System of Thunder Cape, running east 30° north.
6. System of Isle Royale, running east 45° north.

The large group of islands on the southern and eastern side of Black Bay, and south-west of St. Ignace, consists of innumerable islets, separated from each other by the close intersection of the three systems of dykes, which appear more prominent and strongly marked in their features further west, in Isle Royale and Victoria Islands, and about Thunder Bay.

But besides these six clearly defined systems, there seem to be two more, or at least one other distinct system running due north-west and south-east, cutting at right angles through Spar Island, and reappearing, as I understand from verbal communications of Mr. Foster, further south upon Point Keewenaw. This system is perhaps the cause of the bearing of the shores between Keewenaw Bay and Dead River; also of the outlet of Lake Superior between Point Iroquois and Gros Cap along the river St. Mary, unless this eastern system of intersection be distinct from the more western one.

But however this may be, so much is plain;—that at least six distinct systems of dykes, with peculiar characteristic trap, forming parallel ridges in the same system, but varying, for different angles, between the different systems, intersect the northern shores of Lake Superior, and have probably cut up the whole tract of rock, over the space which is now filled by the lake, in such a way as to destroy its continuity; to produce depressions, and to have gradually created an excavation which now forms the lake, and thus to have given to it its present outline. This process of intersection, these successive injections of different materials, have evidently modified, at various epochs, the relative level of the lake and land, and probably also occasioned the modification which we notice in the deposition of the shore drift, and the successive amphitheatric terraces which border, at various heights, its shores.

A more minute analysis of the mineralogical character of these dykes would no doubt afford satisfactory evidence of their original independence, and perhaps lead, in connection with a fuller investigation of their intersections, to the means of ascertaining their relative age. But I became fully aware of the geological importance and independence of these different systems of dykes only during my return, after leaving the neighborhood of Thunder Cape, the ground where this part of the subject might be best studied, and therefore I can now only call the attention of geologists to these facts, in the hope that they may, at some future time, be more fully investigated.

The whole range of rocks which constitutes the northern shore of Lake Superior is so extensively metamorphic, and so thoroughly injected in all directions by veins intersecting each other, that it is no easy task to analyze their relations; and for a full illustration of this subject, minute maps of well-selected localities are required, such as travelling geologists on an occasional visit can scarcely prepare. But I should be perfectly satisfied to see these hints more completely wrought by others, satisfied, as I am, to have shown, at least, how a minute investigation of the geological phenomena of a restricted locality may lead to a better understanding of the origin of the geographical features of a country.

But let me repeat that it were a great mistake to ascribe the present form of Lake Superior to any single geological event. Its position in the main is no doubt determined by a dislocation between the primitive range north and the sedimentary deposit south.

But the working out of the details of its present form is owing to a series of injections of trap dykes of different characters, traversing the older rocks, in various directions, which, from their mineralogical differences, have no doubt been produced at different successive periods.

The diversity of rocks which occur on Lake Superior is very great, and there are varieties observed there which seem to be peculiar to that district, presenting innumerable transitions from one to another, of which the Alps even do not present more extensive examples.

Of these we have new red sandstone passing into porphyries, into quartzites, granites, and gneiss, the metamorphism being more or less perfect, so that the stratification is sometimes still preserved, or passes gradually into absolutely massive rocks. Again, the dykes intersect other rocks almost without altering them, or the alterations in the immediate contact are so intense as to leave no precise lines of demarcation between the dyke and the injected rock. But here again, the phenomena are so complicated, that unless the illustration be accompanied by a very detailed map it were useless to enter into more minute descriptions.

29

The collections I have made of these rocks are sufficiently extensive to afford materials for such an illustration, and I may, perhaps, on another occasion, publish a more detailed account of the geological features of the northern shores, unless the expected publication of the geological survey of Canada by Mr. Logan, renders this essay superfluous.

I would here acknowledge the benefit I have derived in my investigations from the published reports of this survey, and also from the verbal communications of Mr. McLeod of Sault St. Marie. The rocks which occur on the northern shores are so characteristic that they cannot be mistaken, and even should the materials which I have collected not be published more in full, they will at all events afford to those who study the geological distribution of erratic boulders, valuable means of comparison, which will show that most of the erratics which occur in the northern parts of the United States are derived from the primitive range extending north of the lakes reaching along Canada and the United States to the Atlantic Ocean.

Among these rocks there is a variety of deep red felspar porphyry speckled with epidote, which, from its brilliant color, particularly attracts attention, and which occurs all along the northern shore from the Pic to Thunder Bay. This variety I have not observed farther east, and it may perhaps be taken as a guide to ascertain the range of erratics derived from the northern shore of Lake Superior.

XII.

GEOLOGICAL RELATIONS OF THE VARIOUS COPPER DEPOSITS OF LAKE SUPERIOR.

The general distribution of the different copper ores in the region of Lake Superior, presents some facts which seem to me to have a direct bearing upon the theory of their origin. It is a very remarkable circumstance that the largest masses of native copper should occur upon Point Keewenaw, and that the non-metallic ores should be diffused at various distances from the central region where the largest masses of native metallic copper occur. The various sulphurets and carbonates are found on the northern shores and about Lake Huron, in far greater proportion, and over a wider extent, than anywhere nearer the metallic centre. The black oxide itself is found beyond the limits of the large metallic masses, and nearer to them than the other ores. I cannot help thinking that this particular distribution has direct reference to the manner in which these various copper ores were diffused in the country where they occur. They seem to me clearly to indicate that the native copper is all plutonic; that its larger masses were thrown up in a melted state; and that from the main fissure through which they have found their way, they spread in smaller injections at considerable distances; but upon the larger masses in the central focus, the surrounding rocks could have little influence. New chemical combinations could hardly be formed between so compact masses, presenting, in comparison with their bulk, a small surface for contact with other mineral substances capable of being chemically combined with the copper. But where, at a distance, the mass was diffused in smaller proportions into innumerable minute fissures, and thus presented a comparatively large surface of contact with the surrounding rocks, there the most diversified combinations could be formed, and thus the various ores appear in this characteristic distribution. The relations which these ores bear to the rocks in which they are contained, sustain fully this view, and even the circumstance that the black oxide is found in the vicinity of the main masses, when the sulphurets and carbonates occur at greater distances from them, would show that this ore is the result of the oxidation of some portion of the large metallic masses exposed more directly to the influence of oxygen in the process of cooling. Indeed, the phenomena respecting the distribution of the copper about Lake Superior, in all their natural relations, answer so fully to this view, that the whole process might easily be reproduced artificially on a small scale; and it appears strange to me that so many doubts can still be expressed respecting the origin of the copper about Lake Superior, and that this great feature of the distribution of its various ores should have been so totally overlooked.

Printed in the United States
By Bookmasters